PAPERMAKING SYSTEMS AND THEIR CONTROL

PAPERMAKING SYSTEMS AND THEIR CONTROL

TRANSACTIONS OF THE SYMPOSIUM

HELD AT OXFORD, SEPTEMBER 1969

IN TWO VOLUMES: **2**

EDITED BY FRANCIS BOLAM

TECHNICAL SECTION

THE BRITISH PAPER AND

BOARD MAKER'S ASSOCIATION (INC.)

PLOUGH PLACE, FETTER LANE, LONDON, EC4

1970

For contents see Volume i, *page viii*

MULTI-VARIABLE PAPERMACHINE CONTROL

C. C. JOLLIFF, Southern Kraft Division, International Paper Company, Mobile, U.S.A.

Synopsis Supervisory control of average basis weight and moisture in the machine-direction must be achieved by different control procedures, depending upon the importance of other objectives such as production rate and dimensional stability. Maintenance of close control of refining must be modified, depending upon the number and arrangement of refiners. The control of basis weight, moisture and refining all interact with each other to increase the complexity of the control models required, since the adjustment of a single input variable may influence all three areas of control. Some of the experiences of International Paper's Southern Kraft Division control group in working with these control areas are described.

Introduction

IN A well-operated mill, where careful records of the test results obtained during routine sampling for quality control are kept so that the data may be properly ordered in time, analysis of the records for assignable causes of variation will almost certainly reveal that a large component of the variation can be attributed to differences between shifts. A little investigation of the operating practices of the machine crews will soon reveal the reason for this rather unexpected situation. When one shift crew relieves another, it is customary for the new crew to make changes in the settings of various controls on the papermachine. These changes are made, not because the paper made during the last eight hours has been of bad—or even poor—quality, but because the new crew believes that it can more consistently make paper of uniform quality with the machine adjusted to settings of the controls that they have found serviceable in the immediate past. It is thus demonstrated daily on papermachines not computer-controlled that there are many possible ways of adjusting the flows, speed controllers, temperatures and other variables that will result in acceptable paper.

There is a multiplicity of tests that are routinely run on paper in order to determine its suitability for some purpose. Most of these tests are empirical;

Under the chairmanship of Dr D. B. Brewster

the instruments used to perform the test are designed to approximate under carefully controlled conditions some particular stress or strain to which the paper may be subjected. In most situations, it is not possible to describe the properties that the paper must have to give satisfactory test results in terms of fundamental physical or chemical properties of the sheet. A wide range of paper weights, fibre content, chemical additives and so on may usually be found that will give empirical test results within a satisfactory range. Because the sales market for the manufactured paper depends only upon its ability successfully to meet the arbitrary limits specified for these empirical tests and on its price, it is natural to concentrate upon making certain that the test results exceed the minimum values set. Those fundamental properties or empirical tests not specified are permitted to seek whatever level will maintain the important tests at satisfactory values with minimum operator interference.

Analog controllers give an operator important assistance in maintaining the settings of various inputs to the papermachine close to their desired values. Digital computers can in many cases give even more precise control. If an upset in an input occurs, however, requiring that some other input be changed to compensate for the upset, the adjustments made may interact in complicated ways to change the final test results. In the absence of any other method of co-ordinating these compensating adjustments, the operator must decide what inputs to change and by how much. It is small wonder that operators tend to select levels of inputs that they believe to give them the maximum freedom of variable adjustment without causing a major upset in end product test results and stick to them regardless of how well the papermachine may have performed during the previous shift.

One of the justifications for adding a computer to the control hardware used on a papermachine is that much closer control of input levels may be achieved, with consequent savings in raw materials and improvement in product uniformity. Yet, as closer control over individual inputs is achieved, the interactions of changes in input levels become more readily apparent and their influence upon final product test results are not readily detectable. It is the purpose of this paper to enumerate some of these interactions.

Basis weight

THE first papermachine variable to which a digital computer control was applied in the Southern Kraft Division of International Paper Co. was average basis weight. The papermachine made a single quality of paper, tabulating index card stock. The paper was dried to a very low moisture content in a calender dryer. The basic scheme for basis weight control used a continuously scanning beta-gauge to determine the average fibre content of the web and a moisture gauge to determine the percentage moisture content. The average fibre weight was then used to control the opening and closing of the stock valve to maintain a uniform average fibre weight and the average moisture weight was used to control steam flow to the calender dryer section to maintain a uniform average moisture content. The result of holding average fibre content and average moisture content more nearly constant than had been possible was to make average basis weight more uniform. The uniformity in weight obtained was sufficiently impressive to justify application of the computer to basis weight control on the papermachine adjacent to the one first controlled.

The second papermachine manufactured a different product. The grades made on this second machine were unbleached wrapping papers and linerboard. Because such papers fetch a much lower price per ton on the market, it is important to make these papers in the largest volume possible. It is not feasible to bring the paper to a low water content, add back an excess of water, then remove the excess to obtain a permissible level; rather the machine must be operated at the maximum possible production rate. The limiting condition is the amount of steam available for drying the paper. Because of the way in which steam is delivered to the dryers, only a large number of drying rolls could be affected by adjustment of the steam flow and response of dryer temperature to changes in steam flow was very sluggish. It was accordingly decided to control basis weight solely by adjusting the quantity of oven-dry fibre per unit area. Since the moisture content of the sheet was not critical, so long as the maximum permitted was not exceeded, it was thought that this would result in improved control. This did not turn out to be correct. It was found that changes in oven-dry fibre weight resulted in wet or dry areas in the paper, because no adjustment in steam flow was made to compensate for the increased or decreased fibre content and the dryers were operating so close to maximum steam capacity that small changes in fibre content resulted in large changes in the percentage of water retained in the sheet. The control scheme accordingly had to be modified to include some adjustments for moisture before successful control could be achieved.

The two examples mentioned above show the importance of considering as as many aspects of the process as possible. On the first papermachine, uniformity of the product was essential and the value of the paper produced was sufficiently high to make it feasible to operate the machine at less than maximum capacity in order to achieve the required uniformity. The points at which control was applied were ones at which change in level could be effected rapidly, response to adjustments could be detected quickly and sufficient capacity was available to encompass a wide range of adjustments. On the second machine, a second important variable—production rate—had to be considered. Since this variable had to be maintained at maximum, if at all possible, a new control strategy had to be introduced. It appears fairly obvious that the most useful way to control basis weight on this second machine should be by operating it with maximum steam to the dryers, adjust dry fibre weight by controlling the stock valve and maintain uniform moisture content by varying the machine speed. Unfortunately, varying the machine speed introduces complexities into the model, since the lag from adjustment of the stock valve to detection of the change in fibre weight is then variable and the feedforward portion of the model becomes very complicated.

There are other difficulties to be overcome besides those requiring modification in strategy to maintain objectives other than uniform fibre weight. The achievement of uniform fibre weight in the finished sheet requires that the quantity of fibre laid down per unit area be maintained constant. If the proportion of fibre per unit volume in the stock line (consistency) is uniform, the weight of fibre delivered to the wire can be controlled by adjusting the volume of stock delivered to the head box. If the volume of fibre suspension flowing from the head box on to the wire is constant, the fibre basis weight of the sheet produced will vary with the percentage of fibre in the suspension flowing out the slice. In fact, the consistency of the suspension in the stock line is not constant, but shows a periodic fluctuation induced by the consistency controller, a fluctuation characteristic of feedback control loops. This fluctuation leads to periodic opening and closing of the stock valve. As the stock valve opens and closes, the volume of liquid supplied to the head box increases and decreases, changing the pressure that forces the suspension through the slice on to the wire. The opening of the stock valve to admit more fibre to the head box is thus accompanied by an increase in the volume of suspension in the box at the same time that the percentage of fibre per unit volume is increasing. The resulting increase in pressure in the box increases the flow per unit time through the slice of a suspension carrying a higher percentage of solids. The fibre per unit area of the sheet is thus increased both by a change in volume and a change in flow, so the change in weight may be greater than that required. Thus, much more careful control of both consistency and effective head box head is required. It is equally obvious that periodic fluctuations in wire speed because of the action of speed governors will have a similar effect upon fibre weight. All of these factors will, in turn, affect the amount of water retained in the finished sheet and consequently affect total basis weight to a greater degree.

Refining

THE fibre furnish to the papermachine must be mechanically worked to varying degrees to achieve the necessary bonding of fibres together, giving the finished paper the required stiffness, resistance to tear and puncture and other properties. This was the second important part of the papermaking process to which we attempted to apply computer control. In Southern Kraft Division, all papermachines to which we have applied computer control have used either conical or disc refiners for this work. The amount of work done in refining the fibre is ordinarily measured in the laboratory by determining the amount of water that will drain from a slurry of standard fibre concentration in a fixed time or the time required for a standard quantity of water to drain from the slurry. There are numerous conceivable ways of measuring this continuously on the papermachine. Several manufacturers have available continuous freeness testers. The quantity of water draining through the wire per unit time may be measured. It may be possible to relate the water draining per unit time at a single position on the wire to the degree of refining. Alternatively, it may prove more convenient to measure the change in vacuum applied to the wire at a single point or the change in the total vacuum applied to the wire and use one of these variables as a measure of refining. As usual, the choice of the response variable will be dictated by consideration of the ease of measurement, rapidity of response to changes in refining, sensitivity of the response to changes in refining, as well as reproducibility and repeatability of the measurement.

Our first installations for refiner control considered the problem of refiner control as separate and distinct from any other computer control on the same machine, but we soon discovered that such a position was indefensible. Adjustments in stock flow to the wire to maintain uniform fibre weight are accompanied by changes in water flow to the wire. Regardless of what response variable is used to measure degree of refining—water drainage or vacuum change—the variable responds to changes in stock flow for fibre weight control. We therefore found it necessary to incorporate into the refiner model terms to compensate for fibre weight adjustments. Other factors that profoundly affect the variables used for refiner response measurements are the temperature and pH value of the fibre suspension flowing to the wire. An important indirect influence upon the refiner response variable may be the speed of the papermachine, since increasing machine speed will require an increase in the quantity of fibre suspension delivered to the wire per unit time, hence an increase in water drainage per unit time and an increase in vacuum required.

Because of the interaction between the refiner response variable and adjustments made for fibre weight control, we have found it necessary to require that execution of the program for fibre weight control should trigger execution of the refiner control program, rather than permit refiner control action to be taken on a timed sequence basis, independent of other controls, as our original strategy was planned.

The degree of refining of the fibre has an important relationship not only to the ease with which water may drain or be sucked from the forming sheet, but also to the ease with which the sheet may be dried by heating. It is therefore necessary to include in the moisture control model terms associated with the degree of refining done upon the fibre and transfer this information from the refiner control program to the moisture control program. Depending upon the grade of paper being made, it may in some cases be necessary to control the degree to which the pulp is worked by the amount of drying capacity that is available, if it is necessary to run the papermachine at near maximum production rate.

The heavyweight paper grades made by most machines in Southern Kraft Division use mixed fibre furnishes. In almost all cases, this means that a series of refiners must be used to perform the required amount of work upon the fibre. In some cases, separate trains of refiners must be used for pine and hardwood, with retention or mixing chests inserted between different groups of refiners. The time required for a change in work load upon different refiners or refiner groups to be detected by the response variable is then obviously different. It is important in these cases to keep the range of adjustments possible on those refiners nearest to the head box at a maximum, so that a high degree of freedom is maintained for compensating for process upsets. If the upsets are of short duration, no further action is required. If a long-term trend is encountered, requiring a major adjustment to those refiners having the fastest response time, it then becomes necessary to re-establish a then maximum adjustment range on these refiners by redistributing the work load to refiners further removed from the papermachine. This redistribution must be accomplished gradually and the stock in transport carefully tracked so that no upset in freeness of stock delivered to the head box is encountered.

Moisture content

Most of the comments that can be made on moisture content, from our experience, have already been included parenthetically in the above two control areas. To recapitulate briefly: control of moisture content in the finished sheet is an essential part of basis weight control. On paper grades when the primary consideration is the maintenance of uniform quality, moisture content may be adjusted by controlling the temperature of one or more dryer sections, compensating for adjustment in fibre weight by increasing or decreasing dryer temperature as required. If variations in moisture content tend to become unusually large, it may be necessary to transfer part of the work load on the last dryer section to previous sections in order to preserve maximum freedom of adjustment in the last section. On those paper grades when dimensional stability is of particular importance, careful attention to the drying profile may dictate different and more complex strategies. On those paper grades for which maintenance of production rate at or near the maximum is a primary objective, it may be necessary to control moisture by adjusting the degree of refining, so that drainage rate increase will take the load off the drying sections with which water may be dried out of the sheet by heating. It is therefore necessary to leave communication between the refiner control models and the moisture control models.

The major control areas of basis weight, moisture content and refining encompass the major percentage of control endeavours on the papermachine by Southern Kraft Division of International Paper Co. We are at present involved in attempting to develop a strategy for the control of coat weight and cross-direction basis weight and moisture content, but any comments on total control of these or other areas would be, at this time, speculative.

Prepared discussion contribution

Basis Weight Control—Sub-Optimum Approach

W. T. WHIGHT, Empire Paper Mills and R. E. JONES, G.E.C.-A.E.I. Automation Ltd.

AUTOMATIC basis weight regulation on No. 1 fine paper machine at Empire Paper Mills has been achieved by using a substance gauge at the dry end, connected to a central control digital computer, this computer performing direct digital control (DDC) upon various areas of the papermachine.

The basis weight feedback control is a dead time algorithm, which acts upon the set point of the thick stock flow DDC loop.

A pseudo-random binary sequence (PRBS) trial was initially run in order to ascertain the process dynamics. Data collection of thick stock flow changes and corresponding basis weight fluctuations indicated that the process could be described accurately enough by a first order lag plus dead time. A flexible dead time algorithm was developed from this model by assigning a suitable single time constant and setting up a series of values of dead time and process gain in a 'look up' table.

This basis weight digital control program has an execution rate of approximately one quarter the transmission time taken for the first effects of changes in the thick stock control valve to be observed at the substance gauge.

Although the process time constant varies a little with the grade of paper in production, we felt an adequate controller could be constructed, based on a constant value. The dead time part of the model is tuned by using a 'look up' table of current thin stock (head box) flow and wire speed (the principal contributors to changes in process dead time) to determine the correct dead time parameters for the algorithm.

Similarly, the loop gain varies between grades and a further 'look up' table of substance and thick stock is used to extract the current gain parameter for the control loop.

We consider that the effect of this method of achieving basis weight regulation has been to produce about 90 per cent of the possible improvement that could be obtained by a full-scale system modelling exercise and development of an optimum algorithm. Closer system identification and control accuracy can be achieved only at a greater cost and with longer time for systems analysis.

Discussion

Mr H. B. Carter Would Mr Jolliff please clarify whether or not the Southern Kraft Division operator has some discretion on the head stock factor and carriage?

Mr C. C. Jolliff The operator enters the desired basis weight through a manual station. He enters the desired moisture content in the same way. In controlling refining, he ordinarily keeps control himself until the laboratory tests show paper properties to be within specifications. He then signals the computer to maintain refining at that level.

Dr I. B. Sanborn You mentioned that you have no moisture sensor before the size press. Is that correct?

Mr Jolliff We had a moisture sensor there, but not at the present time.

Dr Sanborn In essence, you predict the change that you have to make in your steam pressure to compensate for fibre changes?

Mr Jolliff That is correct.

Mr W. L. Daniel Has any consideration been given to the different refining effect obtained from refiners on light load from those on heavy load in the total load redistribution program?

Mr Jolliff No, as no difference in refining had been found.

Mr M. A. Keyes Initially, you did some work on the Georgetown installation that involved on-line measurement of wet end table flow patterns. Has this been incorporated in any control schemes?

Mr Jolliff All our controls are based on fact.

A Speaker You mentioned the Box & Jenkins method with the process parameters for basis control, but what methods were used to determine the model parameters used for the controller?

Mr Jolliff The parameters estimated by the Box & Jenkins method were minimising the variance and maximising the likelihood function.

Dr A. R. Farmer Please indicate the extent of the functions performed by the computers at the mills you mentioned. You refer to basis weight, moisture content and refiner control. In fact, what other functions does the computer perform?

Mr Jolliff The six computers installed in the Southern Kraft Division have a number of other tasks, but not all performed in any one mill. So far, we have used computer control on bleaching lines, lime kilns, multiple effect evaporators, recovery boilers and in the controls I have mentioned.

Mr R. E. Johnston In as much as the controllers suggested by the methods of Astrom and the methods of Box & Jenkins are almost the same controller—basically a dead time algorithm—it surprises me that no one has used any of the commercial units available for this purpose. Has anyone here had any experience for using them?

The Chairman This might tie in with a question that I wanted to ask Mr Jones—whether or not CONRAD included the facility for varying the dead time compensation automatically as the process dead time changes? Does the hardware offer the facility of automatically changing the parameter when the machine speed changes?

Mr Johnston No, it would have to be a manual change on the basis of a look-up table.

Dr Sanborn I should be pointed out that dead time in the papermachine does not correspond to an even number of sampling intervals. Hence, one must either truncate or round off when choosing the correct sample lag. With traversing gauges working on average basis weight, this can cause serious errors in dead time compensation. These errors seem to cause no serious problems of control on a practical basis. Thus, we have another illustration of a point that, though apparently important theoretically, is in fact not significant practically.

Would Mr Jolliff please comment on his remark that he needed to readjust the refiner controller each time he adjusted his stuff gate.

I presume because of this that you are unable to adjust the stock valve as frequently as once every 8 s.

Dr Farmer I should like to come back to the question I asked Mr Jolliff originally and his answer to it. Are facilities also provided for management information data processing or production scheduling? Please comment on the justification of the systems installed and the principal areas of pay-off.

Mr Jolliff We have not made a practice of printing out trend lags or other operator guides. Most of the installations have programs available for data collection, which are used as required. Our first installation in Georgetown was a research project of the New York office and did not have to be justified economically. We have had with that and subsequent installations attempted justification after the event because of very poor knowledge of the cost of running on and before the computer was installed. The only time measure of success I believe is the acceptance of computer control by management and ours continues to appropriate money for additional control computers.

Mr O. Alsholm Most of the DDC installations in U.S.A. and in Europe are equipped with almost complete analog back-up. There is no criticism in this statement, as it corresponds to the level of confidence we have achieved so far. On the other hand, our purpose is to make computer installations yield the most economical return possible. Analog back-up is expensive, thus the cheap 'limited' controllers used at Sittingbourne appeal to me, especially as I question the necessity of a complete bumpless transfer, if the computer isolates only a few times per year. Having available both Mr Keyes (who has been involved in most of the recent American DDC projects) and Mr Cyprus, it seems likely that we could get advanced comment on the matter of analog back-up in general, especially on the point of bumpless transfer to analog back-up. Not to be misunderstood, I would like to state that I do not mean the switch between computer control and manual back-up and vice versa.

The Chairman I am interested in CONRAD, which takes ten blocks to do basis weight control. How many words of storage are required for the block and what is the average number of blocks?

Mr R. E. Jones Ten blocks do the basis weight control. This includes, of course, the filtering operation I showed, which is using two of those and one to regulate the loop gain. On this particular installation, we had a storage allocation of 12 words per block of 12 bits each. We are working on a CONRAD package of somewhat greater power, perhaps taking 15 words of 18 bit length, but this will give greater facilities and a larger number of functions.

The Chairman I am sure that this áspect of programming to those involved in installations is recognised as a very important one, particularly this CONRAD approach and the one that Mr Alsholm has used. In Mr Jolliff's paper, there is reference to using machine speed as the manipulated variable for moisture control on a dryer-limited machine. Speed is a very important variable economically and I was rather concerned when he said that it could not be manipulated, because it requires a difficult model and the reason for this was that the head time changed when speed changed. I would like to question this point of view. For regulation purposes, the speed changes would not be great, so one would not expect the dead time to change very much. I consider that the major problem of manipulating speed is that the gain between the speed and the dependent variable moisture content is rather indeterminate; it can be either positive or negative, maybe even zero.

Mr Jolliff Our development group is at present finishing work on a program to maintain the papermachine at optimum conditions for the making of a grade of paper. This will include speed control, of course, but not specifically as a way of controlling moisture content.

Mr Keyes Two comments. One is about adjusting speed as an integral part of a moisture control system. Mr Robert Fritchie while at Albemarle Paper Co., Roanoke Rapids, N.C. implemented a system of this kind. It is particularly valuable for dryer-limited situations such as are encountered on linerboard machines.

My second point is a question for Mr Jolliff on the use of z-transforms for the design of a basis weight controller. As has been pointed out, use of the transforms in the case of dead time assumes that the dead time is an integral multiple of the sampling interval. Has anybody applied advanced or retarded z-transforms to basis weight control system design and, if so, what were the results?

Mr R. E. Jones If I understand the question, it is about the modified transform when the process dead time is not an integral number of sampling intervals. The way we have tackled this is to use the modified z-transform. One of the parameters in the string of blocks in my last figure is not unity, it is a parameter that we call M. The range of M is between nought and unity; its purpose is to handle the fractional part of the dead time.

Dr Sanborn To control speed, we do not adjust machine speed for the control of sheet moisture content, but we do consider speed to be the most important variable affecting productivity. If computers are to pay for themselves, productivity must be increased. Hence, speed must be controlled if money is to be made. We have a machine, for example, that in certain circumstances is speed-limited; under other circumstances, it is drying-limited. In this case, we are monitoring the speed and steam valve opening on this machine. If they are both not at some specified maximum value, speed is automatically increased by 10–20 ft/min, provided the drive is capable of such an increase.

AN ADVANCED INFORMATION SYSTEM AS A FUNCTIONAL PART OF A NEWSPRINT MACHINE

J. MARDON, J. E. BARRETT, W. C. CRIPPS*, R. E. MONAHAN, Technical Division, MacMillan Bloedel Ltd., Vancouver, B.C., M. A. CHATWIN, Alpulp Division, MacMillan Bloedel Ltd., Port Alberni, B.C., B. I. HOWE, Powell River Division, MacMillan Bloedel Ltd., Powell River, B.C., R. A. C. STAPLES, Bailey Meter Co. Ltd., Montreal, P.Q., Canada

Synopsis The paper describes an information system applied to a newsprint machine as a functional addition to the operator's capabilities. Examples of its use by the operators are given.

The system was designed, as part of an initial machine installation, after the intensive study of the mode of operation of two generally similar machines. The objective was to retain the skilled operator as part of the control system, but to provide him with more information more conveniently displayed to enable better and faster decisions to be made. The stored program system permits the operator to manage the machine on an information by exception basis.

The system is a Bailey 754 with alarm monitoring data display and future arithmetic capability. There are 150 analog inputs, 200 contact closures, 26 speed and draw measurements and 15 integrations. Data acquisition is continuous providing data to the parallel programmed and independently operating subsystems.

Introduction

A PROGRAM for achieving centralised control of a modern newsprint machine was undertaken in 1965 by MacMillan Bloedel Ltd. The program as developed is based on a concept of process control improvement following the principles of systems engineering.

This concept applied to papermachine instrumentation was described in the paper *A systems approach to papermachine instrumentation*[1] and has

*Graduate Student, University of Lancaster, England

Under the chairmanship of Dr D. B. Brewster

formed the basis for the efforts at MacMillan Bloedel Ltd. Briefly, the approach follows the step-by-step program stated below—

1. Establish the ultimate objective and a program to achieve this objective in a logical, flexible and time acceptable program.
2. Carry out a systems analysis, dividing the system into appropriate subsystems based on—
   (*a*) Parameters of production quality and production throughput
   (*b*) Interdepartmental relationships
   (*c*) Methods of supervision
   (*d*) Economic factors of the process
3. Define the control problems as they become apparent from step *2* above. This will require investigation and evaluation of the specific process subsystems in so far as—
   (*a*) What measurements are needed
   (*b*) How are they to be measured
   (*c*) What is being controlled
   (*d*) What are the relationships between variables
   (*e*) What effect does the individual control problem have on other parts of the whole process
4. Centralise supervision of first the process subsystems and, subsequently, the total system; automate this supervision when feasible and economical.
5. Develop and implement automatic management control as the need, the technology and economics dictate.

This approach may appear conservative and differs from the one many take today as a result of the availability of increased flexibility of process control computer hardware and software. In these instances, the process control computer is utilised for on-line control analysis, in addition to providing certain defined control actions. Whereas we recognise certain advantages to this latter approach such as familiarisation with the computer, use of the computer as a process learning tool and the forcing of discipline on operating and engineering procedures (when possible); we believe the step-by-step approach has greater long-term advantages, particularly for the situations in MacMillan Bloedel. This belief is strengthened when compared with published reports of the various applications within the industry. The advantages are—

1. The application of new techniques is kept within the realm of learning for existing personnel. Special skills are developed internally as anticipated, thus keeping the requirement for hiring hard-to-get skills to a minimum.
2. Full utilisation is made of present techniques and equipment when feasible. This is possible, since process control requirements are determined before a system is defined.

3. Process modifications are made as a result of process studies—before application of any advanced system. Consequently, any improvements resulting from the process modifications are known before application of the specified control system.
4. As process and control modifications are made, the program to meet our long-term objectives is reviewed and revised when appropriate. Thus, a 5-year plan permits realistic planning for capital investment and the staffing and training of personnel ultimately required for system application.
5. With the current dynamic state of the computer industry, we believe it prudent clearly to define objectives and requirements before committment is made to software and hardware. This is not because it is intended to await the availability of a greater performance-to-cost ratio, which, once the advantages of better control have been defined, could be an extremely costly procedure. The relative status of process operation, process investigation and advanced control system development is such as to permit leeway in choosing the time for involvement with more sophisticated systems.
6. If the number of complexity of processes making up the total process of an integrated pulp and paper mill be considered, it is seen that the computer capability and software complexity required to provide the desired control functions and systems analysis capability would be very large with comparably high capital investment. A number of computers appears a more practical solution. As an alternative to such an investment, we have chosen to develop a hybrid simulation facility, which will be used for study of all major areas of our integrated mills. In addition to the process and control studies, this facility will provide the means of gaining familiarity with basic software requirements, as well as a clearer understanding of real time demands. The simulation facility is proving our most valuable tool.
7. A step-by-step approach permits diversification of our personnel to a much greater degree. The systems employed to date have not required large team efforts from our central systems group and has thus permitted a carrying out of studies on several major process areas simultaneously. In addition to establishing process control requirements in areas that are interacting in the integrated mill, this approach has had the added advantage of developing the interest and the recognition of needs in the major areas of the mill (and all our mills) during the same time period. This is proving to be extremely valuable, as we all know a positive attitude toward process control improvement is imperative to the success of any system application.
8. Being freed from the added complexities of hardware and software and from the pressures of justifying an installation, it is possible for us during our systems and control analysis periods, to include the investigation and evaluation of special sensors for various critical measurements. Although we agree that the application of advanced control need not be held up by lack of certain measurements, it should be understood that these measurements might alter the control problems and most certainly would be desirable if available.

A study team surveyed the instrumentation and control system that were employed in the operation of two machines. The study team consisted of 7 full-time members, 4 part-time members and 7 corresponding members. The full-time members included 3 specialists; 1 experienced with on-machine instrumentation, 1 experienced with analog control applications in the paper industry and 1 knowledgeable in digital system technology.

The scope of the study included a systems analysis, as well as an analysis of control problems. Its objective was *operator-oriented* rather than *computer-oriented* and had the purpose of determining the instrument requirements for the centralised operation of the stock preparation and papermachine that would be installed in eighteen months.

The study consisted of the following activities—

1. 25 meetings with operating personnel on the No. 9 machine at Powell River and the No. 5 machine at Port Alberni for the purpose of defining instrumentation requirements for a modern newsprint machine.

2. On-machine studies to determine the instrument needs and included—
   1. Dry end measurements on the two machines studied.
   2. Head box total head measurements on both machines.
   3. Head box and wire pit consistencies and Mersey drainage tests for one machine.

The study also summarised the extent of instrumentation existing on the two machines, together with a layout of read-out stations in order to determine the requirements for and advantages of centralised information.

Instruments and procedures for accounting were also considered, as it is necessary to provide information for this function, as well as for operations.

The recommendations proposed a plan of action (Fig. 1), which started with what we knew and permitted expansion as time, manpower, experience and funds allowed. Each step provided a base for the next step, thus minimising the chance for loss of effort on the technical side and loss of confidence by operations as new techniques were learned by both parties.

The report included specific recommendations for the application of on-machine basis weight, moisture and caliper instruments; various analog control systems for stock proportioning and the papermachine; and the application of a digital information system to provide centralised data presentation on a 'management by exception basis' and which would provide the flexibility for expansion to include future performance monitoring and supervisory control. These recommendations took into account the current status of instrumentation technology within MacMillan Bloedel at that time, the staff and experience available for system application, the time and funds available for the new machine installation. Thus, the step-by-step approach (Fig. 1) was submitted as the recommended plan of action.

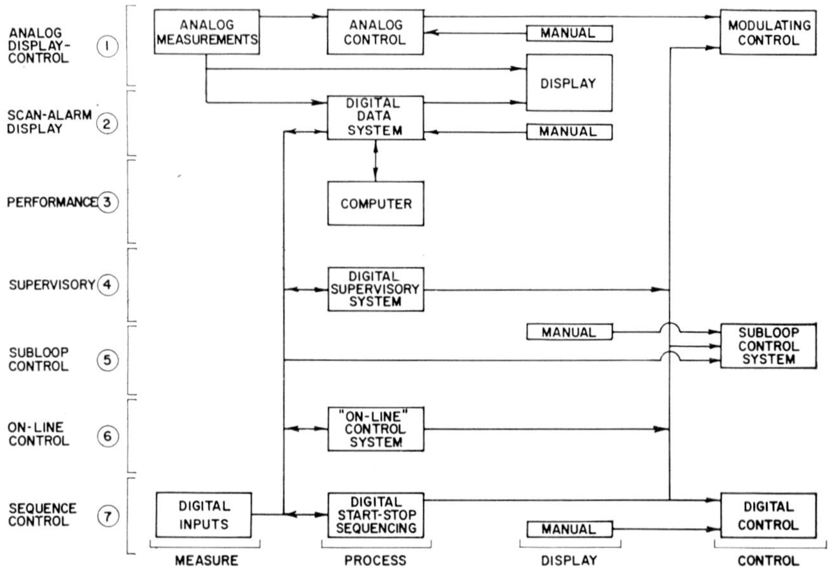

Fig. 1—Step-by-step approach to papermachine control submitted by the study team as the recommended plan of action

The digital information system recommended was to satisfy two major objectives—

1. To assist operating personnel in direct on-line action—
   (*a*) To assist the operator in on-line monitoring of the process.
   (*b*) To collect the display information to assist the operator in making decisions for on-line corrective actions.

2. To provide the means of gathering accurate, reliable data for on-line process analysis—
   (*a*) To collect and record data for accounting and quality control purposes.
   (*b*) To collect and provide records (printed or punched cards) to establish and implement long-range objectives. In order to implement long-range objectives, this system must be designed readily to accommodate future expansion.

These objectives defined the design criteria of the information system to meet the immediate as well as the long-range needs for achieving more efficient production, producing a uniform product and obtaining better utilisation of manpower.

Functional description of the information system applied to No. 10 papermachine

IN ACCORDANCE with the recommendations made by the study group, a multi-channel, parallel-stored program digital data acquisition system (a Bailey 754) was installed, which best met both the primary and secondary objectives. This system concept differs from the process control computer in that systems functions are performed by means of independent hardware subsystems rather than by timesharing a central processor.

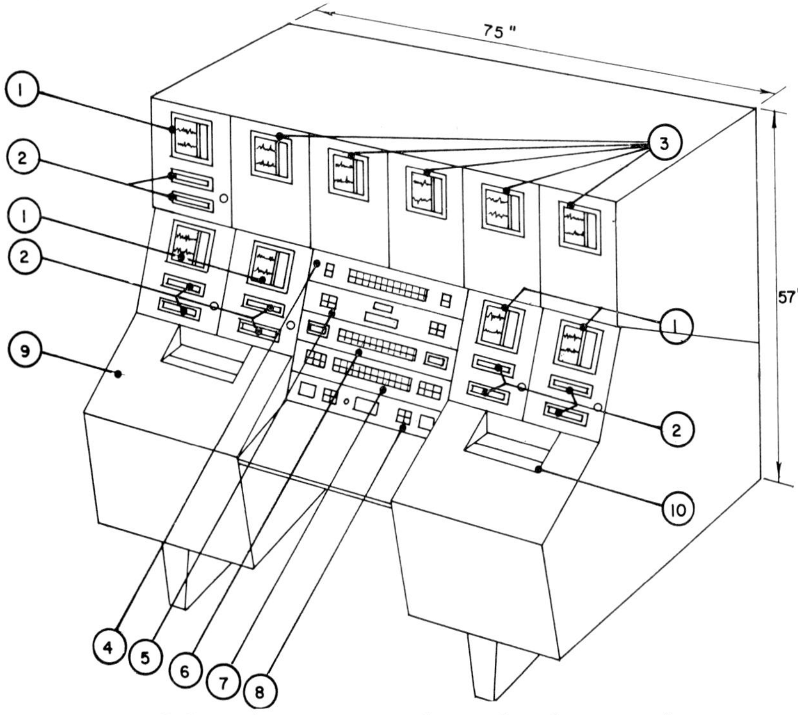

Fig. 2—Information system console: numbered arrows refer to—

1. Addressable analog recorders (red and blue pens)
2. Thumbwheel switches to select point address, zero suppression and range for each pen
3. Utility recorders (red and blue pens)
4. Back-lighted alarm annunciator buttons
5. Digital display lights select point number, input value of time
6. Back-lighted push-buttons to select display of common variables, thumbwheel switches to select variables not on push-buttons
7. Back-lighted push-buttons to select group review print outs
8. Thumbwheel switch and push-buttons for entering alarm limits and reviewing alarm status
9. Alarm printer
10. Utility printer

The particular advantages of this approach are—

1. Simplified programming—each subsystem has effectively continuous access to its own program. Thus, no executive routine is required and the programs are straightforward statements of sequential action or data.
2. The system reliability and, consequently, up-time is greater as failure of any one subsystem does not default the operation of other subsystems (with the exception, of course, of data acquisition).
3. System expansion capability and flexibility is extremely good—other subsystems such as the computer and control subsystems may be added to the present system with a minimum of system upset *if and when* required or desired.
4. This system required a minimum staff (one systems engineer) to implement. System programming did not require a trained programmer. (A two-year technical school graduate with no previous training or familiarity with the system modified, updated and documented the entire program in approximately four man-months.)

For our situation, the system described met the requirements for the plan of action dictated in Fig. 1.

Physically, the system operator's console (Fig. 2) is located in an air-conditioned control room (Fig. 3), with the stock preparation control console,

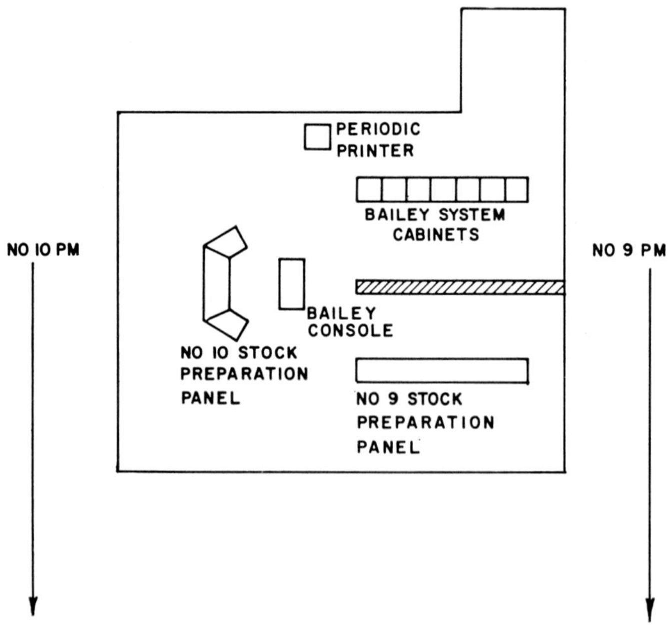

Fig. 3—Equipment located in the system information centre

approximately at the head box location for the No. 10 papermachine. This room also contains the stock proportioning panel for the No. 9 machine. The system cabinets are located in a room attached to the control centre.

The information system provides the following functions—

1. Data acquisition and storage for approximately 150 analog inputs. It also monitors approximately 20 integration inputs (pulse), 26 speed and draw measurements (binary coded decimal words) and up to 200 alarm contact inputs.
2. Alarm monitoring for all data inputs to the system. With the exception of adjustable alarm limits for all variables in data storage, this function has a hard-wired program. It reports new alarms, return to normals and alarm reviews on the alarm printer. Alarm reviews have time intervals select capabilities. A 20 window addressable annunciator located on the operators console is fed from the alarms monitor.
3. Periodic logging of selected variables as required. This subsystem has its own printer.
4. Value displaying in engineering units of any variable stored in data storage. Time may also be displayed by selection.
5. Analog trend recording for any variable as selected with up to 10 variables displayed at any one time. Each pen may be assigned to any variable.
6. Single point digital or analog trending of any selected variable.
7. Single-point digital or analog trending of historical data for any one of 100 programmed variables.
8. Group review, digital or analog trending of current data for any one of 20 programmed groups.
9. Group review, digital or analog trending of historical data as selected from one of 20 programmed groups. The historical interval may also be selected for up to 1 h.
10. Programmed data retrieval by tape punch.

The system is described more fully in Appendix 1.

Experience in using the system

IMPLEMENTATION of the system began in the autumn of 1967 with concentration on the first objective. This required working with the operators, developing procedures and demonstrating ways in which the information presented could be useful to the operators for on-line operations. One systems engineer was assigned full time to carry out this task.

Both the stock preparation people and the machine operators had undergone a training program before the systems arrived. For future installations, every effort will be made to secure time for 'hands on' training. Because all information pertaining to stock proportioning is available only through the system console, the stock proportioning personnel adapted to the use of the system very rapidly.

The machine tenders are not required to be in the control room and, since sufficient conventional instrumentation for operation is located around the papermachine, the machine tenders did not adapt as readily to the functional capabilities of the system. To achieve this goal, a systems engineer worked closely with the production superintendent in demonstrating the usefulness of the data presentation. The machine tender was then left to experiment for himself (which was found to be very important). Machine room personnel are now utilising the system to an effective degree.

To facilitate developing the procedures for the operators and as a basis for preliminary off-line analysis, a program of relatively simple but important process investigations was established. This programme is listed as Appendix 2 and consists of two types of project. The first type are those projects that could be undertaken by the systems engineer with only the information available from the system, production reports and other instrumentation. The other type of project required tests and other assistance from the mill technical department.

From October 1967 to June 1968, a record was kept of the situations in which the particular means of centralised information display assisted the operator in preventing improper operation or wastage. Appendix 3 lists a number of these situations and a summary is given below—

1. *Chest levels*—Whereas chest levels may be alarmed in the conventional manner, the fact that the alarm conditions are made apparent immediately to the stock preparation man and data presentation supports the alarm by appropriate information from which he can take proper corrective action has proven extremely valuable. Such action has prevented a number of incidences when chests might have overflowed or been pumped dry. In addition, with the variable alarm limits, the normal operating conditions can be altered to meet the current situation by approval of the appropriate supervisor.
2. *Stock consistencies*—Close monitoring of the consistency control loops has prevented a number of upsets to the machine. It should be pointed out here that 10 of the available 20 pens on the operator's console normally record stock consistencies and valve positions.

   For example, the mixed stock consistency transmitter appeared to shift its calibration. At this time, the dilution valve moved rapidly to a more open position. The problem was traced to the consistency regulator via the information system by noting that the basis weight did not change (on automatic control), but the mixed stock flow increased considerably. This situation had been previously noted on several occasions. On one of those earlier dates, the regulator dilution valve began creeping open, as shown in Fig. 4, at the same time as the mixed stock flow was being increased by the automatic basis weight control. Although the consistency indicated the same value, it must have been decreasing as indicated by the increasing mixed stock flow. The set point was raised by 2 per cent of chart and this brought the dilution valve and mixed stock flow down slightly, but they both remained higher than their previous values before the drift began. This indicates a shift in the mixed stock regulator set point by more than 2 per cent of chart.

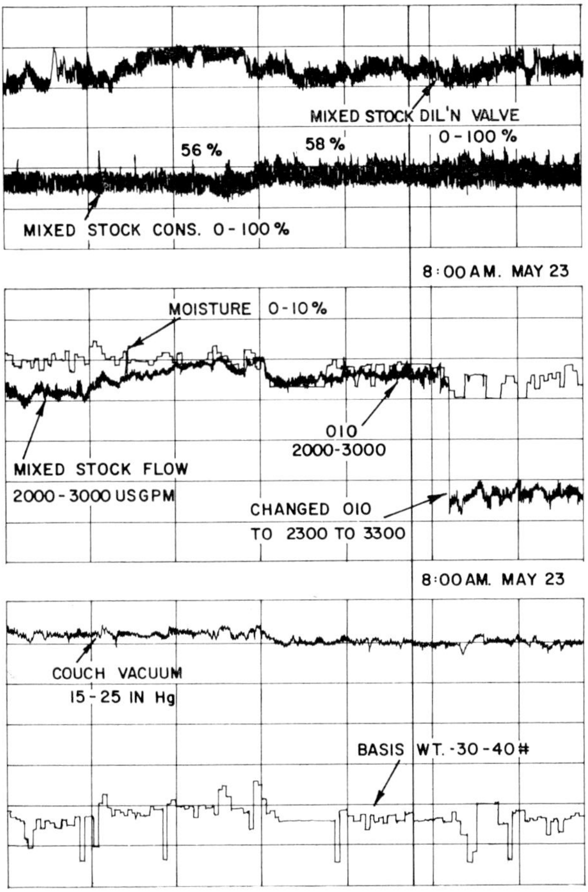

Fig. 4—Calibration drift in the mixed stock consistency regulator readily detected by means of the system analog recorders

At the second occurrence of the situation, a similar disturbance occurred, but this time it was much faster as shown in Fig. 5. Over a 1 h period, the set point was changed from 63 per cent to 70 per cent to 74 per cent to compensate for the drift in the instrument and to bring the dilution valve into a control range. Note that, during the time the set point was being raised, the mixed stock flow was increasing as well to maintain a constant basis weight. The instrument appeared to shift back as noted by the abrupt drop in the dilution valve position and the peak in basis weight. It finally stabilised at a set point of 5 per cent of chart higher than previously.

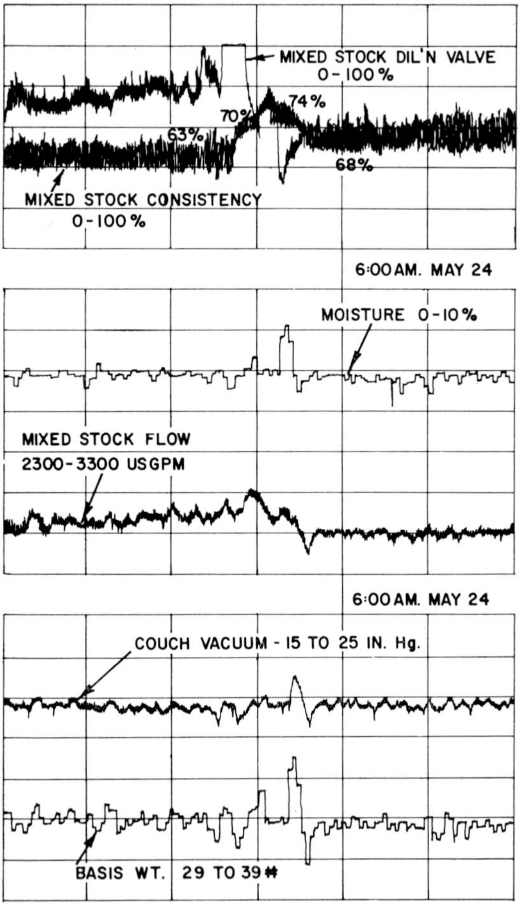

Fig. 5—System analog recorders readily detect a sudden calibration shift in the mixed stock consistency regulator and monitor the compensating set point changes made by the operator

3. *Stock flows* Alarm monitoring enabled immediate action to prevent off-specification paper on at least one occasion.
4. *Stock cleaning and de-aeration* During start-up, machine problems were analysed via the historical review function and it was found that the trouble

originated with a loose belt on the vacuum pump of the de-aerator. Other examples in this area were also resolved through use of process analysis by means of data presented from the historical review function.

5. *Head box* Monitoring of the head box variables saved machine down-time on several occasions.

Data from the information system was used to show that a small change in head box temperature will cause a significant change in head box total head. This relationship was studied using the digital trend facility on the utility printer. Fig. 6 is a record of this analysis. During this time, no adjustments were made at the head box, as attention was being concentrated on getting the sheet to the dryers and stacks. The jet-to-wire drag increased from 30 ft/min to 70 ft/min over this time period.

HEADBOX

TIME	HEAD BOX TOT HEAD	HI COUCH VAC	DRAW JET WIRE	FWD DR ROLL SPD	STRM FLOW VALV POSN	HEAD BOX LIQ LEVL	HEAD BOX TEMP	HOT WATR SHOW TEMP	B WT AS MADE O M	MIXD STOK FLOW
08:29	265.	20.09	0032	2289	39.9	14.4	069.	064.	31.8	2898
08:30	264.	20.29	0034	2288	39.9	14.8	069.	068.	31.8	2898
08:31	264.	20.70	0035	2287	39.9	14.7	069.	073.	31.8	2885
08:32	265.	20.90	0032	2288	39.9	14.7	069.	078.	31.8	2920
08:33	265.	20.91	0033	2288	39.9	14.7	070.	082.	31.8	2899
08:34	264.	21.02	0036	2288	39.9	14.9	070.	085.	31.8	2868
08:35	264.	21.10	0039	2289	39.9	14.9	070.	086.	31.8	2844
08:36	263.	20.81	0039	2287	39.8	14.9	071.	089.	31.8	2862
08:37	263.	20.82	0042	2288	39.8	14.9	071.	090.	31.8	2838
08:38	262.	20.81	0042	2288	39.9	14.8	071.	090.	31.8	2849
08:39	263.	20.72	0044	2288	39.9	14.8	072.	092.	31.8	2854
08:40	262.	20.68	0045	2286	40.0	14.8	072.	093.	31.8	2854
08:41	263.	20.93	0046	2287	39.8	14.9	072.	093.	31.8	2843
08:42	261.	20.81	0048	2288	39.9	14.8	073.	094.	31.8	2827
08:43	261.	20.84	0050	2287	39.9	14.9	073.	096.	31.8	2827
08:44	261.	21.10	0053	2286	39.9	15.0	073.	096.	31.8	2850
08:45	260.	21.22	0054	2287	39.8	14.9	073.	097.	31.8	2842
08:46	260.	21.27	0056	2287	39.9	14.9	074.	098.	31.8	2836
08:47	260.	21.25	0055	2286	39.9	14.9	074.	097.	31.8	2845
08:48	259.	21.21	0062	2287	39.9	14.7	074.	098.	31.8	2822
08:49	259.	21.57	0054	2285	39.4	14.9	075.	099.	31.8	2789
08:50	259.	21.22	0054	2286	39.0	14.9	075.	099.	31.8	2774
08:51	259.	21.09	0054	2286	39.0	14.9	075.	099.	31.8	2785
08:52	260.	21.04	0053	2286	39.0	15.0	075.	100.	31.8	2787
08:53	259.	20.97	0057	2287	38.9	14.9	076.	099.	31.8	2781
08:54	258.	20.93	0061	2287	39.0	15.1	076.	100.	31.8	2753
08:55	258.	21.04	0061	2285	39.0	15.2	076.	101.	31.8	2752
08:56	258.	20.99	0062	2285	39.0	15.2	076.	102.	31.8	2759
08:57	258.	20.79	0060	2286	39.0	15.1	077.	102.	31.8	2757
08:58	257.	20.70	0059	2287	39.0	15.2	077.	103.	31.8	2760
08:59	258.	20.83	0062	2286	39.0	15.2	077.	103.	31.8	2756
09:00	257.	20.64	0066	2287	39.0	15.1	077.	104.	31.8	2755
09:01	258.	20.55	0066	2285	39.0	15.1	077.	103.	31.8	2738
09:02	257.	20.44	0068	2285	39.0	15.1	078.	102.	31.8	2754
09:03	257.	20.18	0067	2287	39.0	15.2	078.	102.	31.8	2767
09:04	256.	20.01	0068	2285	39.0	15.1	078.	101.	31.8	2757
09:05	256.	20.12	0068	2285	39.0	15.2	078.	101.	31.8	2750

DECREASING TOTAL HEAD (HEAD BOX TOT HEAD column) — INCREASING TEMPERATURE (HEAD BOX TEMP column)

Fig. 6—Head box temperature as an influence on head box total head

6. *Fourdrinier machine* In one instance, an alarm on the flat box header vacuum led to discovery of a leak on the end of one flat box. Adjustment of the end seal position by the operator corrected the problem.

7. *Press breaker stack loadings* The digital indication of press and breaker stock loadings appears to provide a superior method for setting these values.

8. *Dryer section* The No. 10 dryer steam system has been the subject of considerable analysis since start-up. An interesting example of a disturbance to the steam system is shown in Fig. 7(*a*) (Note the increase in sheet moisture). At this time, the main header pressure dropped to 14·0 lb/in^2, thus causing an increase in indicated steam flow (up to 135 500 lb/h) and a subsequent loss of individual dryer section pressures shown in Fig. 7(*b*). The dryer steam balances group review is also included as Fig. 7(*c*).

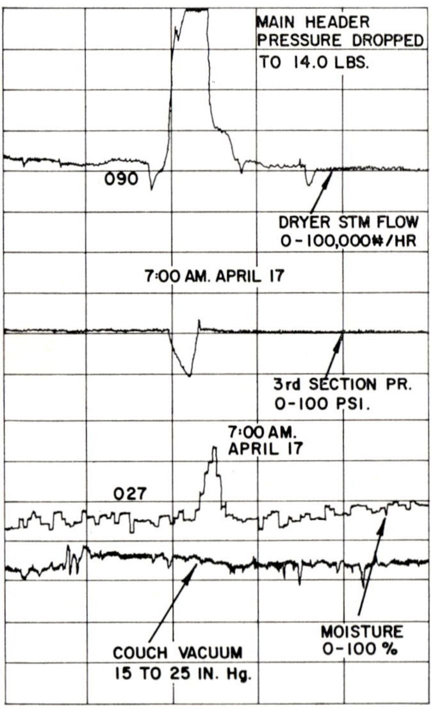

Fig. 7(a)—System analog recorders, used to monitor upsets in the dryer steam system, detect a sudden drop in the main header pressure resulting in a sudden rise in indicated steam flow and sheet moisture

STM SPLY

TIME	NO 1 DRYR PR	NO 1 DRYR DIFF PR	NO 2 DRYR PR	NO 2 DRYR DIFF PR	NO 3 DRYR PR	NO 3 DRYR DIFF PR	NO 4 DRYR PR	NO 4 DRYR DIFF PR	FELT DRYR STM PR	FELT DRYR DIFF PR
07:52	10.1	09.2	09.5	09.3	20.7	09.9	06.9	09.0	25.8	09.2
07:11	09.8	09.2	09.5	09.3	20.7	09.6	06.7	08.9	26.0	09.1
07:12	09.7	08.9	09.4	09.4	20.7	09.3	06.7	09.0	25.7	09.2
07:13	09.7	08.9	09.3	08.9	20.6	08.8	06.7	08.8	25.6	09.1
07:14	09.6	08.4	09.3	08.6	20.6	08.7	06.6	08.9	25.3	09.1
07:15	09.5	09.0	09.4	08.6	20.6	08.7	06.6	08.9	25.3	09.1
07:16	09.5	09.0	09.4	08.7	20.7	08.0	06.7	08.9	25.3	09.1
07:17	09.6	09.0	09.3	08.7	20.6	07.0	06.6	08.8	25.2	09.1
07:18	09.6	09.1	09.3	08.5	20.6	06.1	06.6	08.7	25.0	09.1
07:19	09.6	09.1	09.4	08.1	20.7	05.5	06.6	08.7	24.9	09.1
07:20	09.6	09.2	09.4	07.8	20.6	05.1	06.7	08.7	24.9	09.2
07:21	09.6	09.1	09.4	06.4	18.7	03.4	06.5	08.0	22.9	08.9
07:22	09.6	09.0	09.5	05.8	16.9	03.3	06.6	07.5	21.4	08.9
07:23	09.5	09.1	09.5	05.5	15.8	03.3	06.8	07.2	20.4	08.8
07:24	09.6	09.1	09.5	05.3	15.1	03.4	06.7	07.1	19.6	08.8
07:25	09.5	09.1	09.5	05.2	14.5	03.4	06.7	07.0	19.1	08.4
07:26	09.4	09.0	09.5	05.0	13.9	03.4	06.7	07.1	18.5	08.2
07:27	09.1	08.8	09.5	04.8	13.4	03.4	06.7	07.0	17.9	08.1
07:28	08.7	08.5	09.4	04.6	12.7	03.2	06.7	07.1	17.1	08.1
07:29	09.0	09.0	09.1	04.4	12.0	03.2	06.7	07.3	16.4	07.6
07:30	09.3	09.2	08.8	04.3	11.4	03.1	06.7	07.5	15.8	06.7

DRYER MAIN HEADER PRESSURE DROPPED

DRYER SECTION PRESSURES DROPPING

Fig. 7(b)—Steam supply group reviews records the decreasing pressure in each of the individual dryer sections caused by the drop in the main header pressure

DRYR STM BAL

TIME	DRYR STM FLOW	THER STM FLOW DRYR	HEAT VENT STM FLOW		DRYR COND FLOW	DRYR COND TEMP	MOIS AS MADE O M	NO 4 DRYR DIFF PR	NO 4 DRYR SPD	DRYR MAIN HEAD PR
08:36	0620	282	123	000	054	190	07.4	09.0	2205	59.5
07:26	0996.	188	141.	099.	076.	189.	07.7	07.0	2208	18.4
07:27	1019.	186	143.	101.	071.	188.	07.7	07.0	2206	17.8
07:28	1057.	184	145.	105.	065.	188.	08.1	07.1	2206	17.1
07:29	1082.	180	146.	108.	068.	187.	08.2	07.3	2206	16.3
07:30	1098.	180	144.	109.	074.	186.	08.2	07.5	2207	15.8
07:31	1109.	178	146.	110.	073.	185.	08.6	07.6	2207	15.3
07:32	1120.	178	150.	112.	064.	185.	08.6	07.5	2206	15.0
07:33	1128.	177	151.	112.	073.	184.	08.6	07.3	2206	14.7
07:34	1141.	176	151.	114.	068.	183.	08.9	07.2	2207	14.3
07:35	1152.	176	157.	115.	067.	182.	08.9	07.1	2208	14.0
07:36	1212.	185	159.	121.	072.	182.	08.9	07.5	2207	14.9
07:37	1353.	206	175.	135.	069.	181.	08.9	08.8	2207	18.9
07:38	1355.	217	173.	135.	072.	181.	09.3	10.1	2207	22.0
07:39	1302.	225	169.	130.	069.	181.	09.3	10.7	2206	24.2
07:40	1185.	236	166.	118.	071.	182.	09.3	10.9	2207	27.7
07:41	1154.	241	169.	115.	075.	183.	08.6	11.1	2208	30.0
07:42	1128.	259	185.	112.	067.	184.	08.6	11.7	2207	35.5
07:43	0787.	286	173.	078.	074.	186.	08.6	12.1	2206	51.7
07:44	0738.	274	159.	073.	062.	187.	07.8	11.5	2207	52.3

RESULTING UPSETS IN DRYER STEAM SYSTEM

MAIN HEADER PRESSURE DROPPING

INDICATED STEAM FLOW RISING

SHEET MOISTURE RISING

Fig. 7(c)—Dryer steam balance group review records the drop in main header pressure and the subsequent upsets in steam and condensate flows and temperatures through the dryer steam system

9. *Utilisation by machine room personnel* The information system has been a valuable tool for analysis of machine conditions during normal or disturbed operating conditions, primarily because of the availability of a maximum amount of information in a central location. There are many examples to be cited when the system was used as a study tool—
   (*a*) Analysis of the steam system with and without press vacuum.
   (*b*) Analysis of relationship between couch vacuum and mixed stock consistency.
   (*c*) Analysis of start-up conditions compared with previous operating condition.
   (*d*) Lubrication system—start-up checks.
   (*e*) Analysis of variations in basis weight.
   (*f*) For providing data for analysis by the technical department.
   (*g*) For accounting functions.
   (*h*) General usage includes a check on changes to the machine by machine operators and controller loop tuning in the stock proportioning areas.

To summarise our experience in using the system to date, recapitulation of the results compared with the system objectives should be in order.

The first objective of this information system was to assist the operator in on-line action. This objective has been attained, as can be seen from the examples cited in the preceeding section. There is still further work to be done here, including—

1. Additional operator training.
2. Further implementation of a function termed the critical alarm (another operator tool).
3. Addition of an arithmetic unit for on-line performance monitoring.

The second objective was to 'provide the means of gathering accurate, reliable data for off-line process analysis'. This objective has been realised in part, as indicated by reference to Appendix 2 and some of the previous examples. Further work is required in this area—

1. Calibration of all input data.
2. Addition of certain important inputs.
3. Addition of computing capability in the system.

From the philosophy (which way to go in digital systems) mentioned at the outset of this report, it is very apparent to us that the system purchased was the correct choice for this project. Had the technical aspects of the equipment been more complex, the manpower available for the project would have been unquestionably insufficient, both in regard to numbers and training. This would most certainly have increased the implementation problems and reduced operator acceptance.

Practical consideration for the application of digital systems

THERE are a number of factors that one should plan to deal with when considering the application of any digital system—

1. Hardware and software malfunction.
2. Training of operation and maintenance personnel.
3. Appropriate programs of routine maintenance.
4. Related instrument reliability.
5. Process operating experience on the system team.

The objective, of course, is to hold any upset as a result of these factors to a minimum. It was considered that the application of a simplified system to perform the functions required was desirable. Resolving the problems caused by these factors during implementation convinced us that the right choice had been made.

Comments* relating to hardware malfunctions and software errors are intended to point out 'occupational hazards', which will undoubtedly occur—in varying degrees—in the application of any sophisticated system.

At the time, the system was considered ready for on-line use, the stock preparation operators were knowledgeable about the system and its functions, although the papermachine operators had not really begun to use it and therefore did not understand its function or its capabilities. Stock preparation operators were using the system to nearly its full capability relative to their job. Yet, since the system is probably 80 per cent oriented to the papermachine, it was being used at something like 15 per cent of its total capability. The figures quoted are qualitative only.

This situation remained basically unchanged for approximately 3 months, at which time the papermachine operators began to pick up knowledge and interest in the information system. The breakthrough at this point can be completely attributed to the Production Superintendent, Powell River Division, who made a considerable effort to train each of the No. 10 operators individually on the functions of the information system and how these functions can be utilised to assist machine operation. After this training period, papermachine operators have been utilising the information system to approximately the same degree as have the stock preparation operators.

It is appropriate at this time to discuss the variety of problems that occurred during the particular system implementation. This topic is often discussed relative to any system installation and most of the problems are

* These comments should not be misconstrued as being a complaint against the particular supplier. On the contrary, the support received from the firm producing and installing this system under the circumstances is to be commended.

well known, but the experiences on this system represent the first case of what will be a continued factor in the start-up and implementation of advanced systems by our systems engineering group—

1. One of the main roadblocks to a much faster and smoother implementation phase was the hardware malfunctions at start-up and the continued malfunctions during approximately the first six months of operation. Although the number of equipment problems was not excessive, the operation was not as smooth as would have been desired. The result is a common feeling of mistrust among the operators of the information system reliability. With this feeling prevalent, it is most difficult to instil enthusiasm for using the equipment, regardless of how obvious the benefits are. It is essentially the age-old story of the human quality of remembering the bad points about something and tending to let all the good points slip away.

   The only way to offset this difficulty is to ensure that advanced systems of this type are shipped with a minimum of hardware problems. This can be achieved only by a thorough factory inspection, which requires sufficient lead time properly to specify and design the system.

   Once installed, equipment of this type must be subjected to regular thorough maintainance (preventive maintenance) in order that hardware malfunctions are kept to an absolute minimum. On the system described, it was demonstrated time and again that small failures (for example, erroneous type-out of a single group review) reinforced the feeling in the operators' minds that the information system was unreliable. Preventive maintenance is extremely important to the continued reliable operation of peripheral equipment.

2. Software errors obviously contribute to the feeling of non-confidence among the operating personnel just as greatly as hardware malfunctions do so. Here again, the only way to prevent problems of this kind is to provide sufficient time for the equipment and software to be thoroughly inspected on the factory floor; with its inspection being monitored by a user representative, if at all possible.

3. Advanced system installation on a process must include primary instrument problems as part of the total system implementation. Experience with our application indicates that this type of problem can also adversely affect the reputation of the system in the eyes of the operators. In any mill that is production-oriented, instrument malfunctions are not necessarily corrected immediately. Problems of spare parts, maintenance scheduling, job priorities and 'brush fire' work often delay maintenance for long periods of time, thus creating erroneous system data. One solution to this problem is to assign an instrument mechanic to the project full-time for the duration of the implementation phase.

   Calibration is another factor that must be mentioned. It is mandatory that the demands on a instrument mechanic's time do not preclude his ensuring that all system inputs are in calibration.

4. A serious problem in the implementation phase of the system described was the lack of process understanding (papermaking) on behalf of the systems engineering group.* Intimate process operating knowledge is needed during system implementation. It is in sufficient to have a basic familiarity with the process, as this has little to do with the hour-to-hour or day-to-day operating situations. An important key to implementing this type of equipment is to be able to show the operators, in their own terms of reference, how they can utilise the information in a way that is beneficial to their operation of the machine within their job responsibility.

   There is no way around this problem. As previously mentioned, the implementation phase of this project was effectively at a standstill, until the production superintendent stepped in and personally trained the individual operators in the uses of this equipment.

5. Associated with the above problem is the fact that the No. 10 information system was not required by the papermachine personnel to run the machine *as machines had been run in the past*. This being the case, there is a natural reluctance on the part of the operators to go out of their way to learn something that is not a definite requirement of their job responsibility.

   Conversely, the stock preparation personnel were forced to use the system as it provides the only records and/or indications of chest levels, stock flows and stock consistencies.

   It is obvious that equipment that can be considered as redundant will tend to be ignored. Future systems of this type should be an integral part of the process operation, thus assuring a starting point, at least, for complete system implementation.†

6. A more general problem that is always related to a staff project is the staff-line relationship in a business organisation. Even though both line and staff play an important role in company plans, the line officials retain full authority and responsibility for execution of these plans, once they become a firm part of company policy.

   The line and staff relationship must be considered in a project such as an advanced control system implementation. It is essentially a staff project and must be carried out without line authority or responsibility. A direct result of this situation is an inefficient project. It will be our recommendation in the future that consideration be given to active line participation and responsibility in the project. This would provide the authority within the project group and result in improved implementation for both operations and maintenance.

7. Another important problem was the lack of and timing for operator training. The only operator training for the system was carried out 6 months before operation of the system. This was an extensive course given to the majority

* This knowledge was available, but at too high a level in the company to be directly utilised.

† Consideration was given to this in the design stage, but it was too radical a step for acceptance.

of the papermachine and stock preparation operators. The course effectiveness was reduced because—

(*a*) Too much time elapsed from the time the course was given to the start-up of the machine. Candid comments from the operators indicate that, despite the course being good and an understanding of the system being achieved, most of this was lost in the following months before the system was installed.

(*b*) Lectures were given without equipment present. It is difficult to teach material of this type without having the actual equipment available to provide immediate experience.

The training problem is compounded by the fact that the No. 10 papermachine is the senior machine and is subject to more frequent changes of crews. This means that new operators without the familiarisation course as background frequently operate the machine.

It is suggested that continued 'hands-on' training be considered for this type of equipment. Training should be given to both the normal operators and the relief operators and should include any instrumentation or control that is part of the system's entire scope—that is, the scanning basis weight, moisture, caliper systems.

8. When planning an advanced system installation, parallel installation of other new equipment must be considered for the effect it will have on the ability of the operators to learn the operation and the capabilities of all systems. Without a program of learning, too much can be forced upon the operators at any one time.

It is also worthy of note that operators will tend not to experiment with an advanced system when the 'system expert' is around. It is advantageous to the success of advanced system application to permit the operators free time to operate the system after appropriate instruction and direction has been given.

Conclusions

THE primary objectives of the No. 10 information system were—

1. To assist operating personnel in direct on-line action.

2. To provide the means for gathering accurate, reliable data for off-line process analysis.

In the first objective, the operating personnel believe that, with proper and continued attention to the system, a substantial 'assist' to papermaking is possible. For example—

(*a*) The system is used extensively by stock proportioning personnel for monitoring stock flow and chest level.

(*b*) Group trending on the utility printer has assisted in improving start-up procedures and for speeding up and slowing down by presenting the operator with related variables such as basis weight, vacuums and stream flow valve position.
(*c*) Historical review of groups of data such as stream flows assist in determining points of upset during normal operation and during speed trials.
(*d*) Accurate data (and trending of this data) on speeds, draws and press section loadings is an assist in determining when changes were or should be made.
(*e*) Speed variation analyses are readily made on the system's trend recorders.
(*f*) Alarm monitoring provides an important key to resolving upsets (or potential upsets) before the danger point.

In order to obtain the full benefit of the system, it is mandatory that the system be maintained in first-class working order. This can be accomplished only by means of a thorough preventive maintenance program. Such a program must include all peripheral equipment associated with the system.

Routine checks and calibration of instruments providing data to the system are necessary to ensure validity of data presented (and to be used for calculations). Those instruments such as the beta-gauge that require certain mechanical adjustment (such as changing of cables) must be kept in proper operating condition if the confidence of the operator is to be gained and maintained.

One of the primary considerations at the time of system purchase was the manpower required to implement and use the system effectively. In spite of the simplicity designed into this system, which we had intended would render even one full-time man unnecessary (and which is technically the case), the normal resistance to change from past practices by operating personnel and the shortage of technical personnel at Powell River have shown that, for full utilisation, one man is needed to support the system and act as liaison between the system and operations. This man becomes even more important when the computing function is added to the system, although this should be compared with the 2–4 men required on other reported process control computer applications.

The ultimate intent of the process analysis is to determine specific control requirements for the No. 10 papermachine and to develop from these requirements generalised control system recommendations for newsprint machines.

It was—and is—imperative that the above objectives be met within the staff capabilities available within the company for such effort and that the use of the system for off-line work will not interfere with on-line implementation. The type of system employed satisfied these requirements in that a step-by-step approach was feasible—each step to be accomplished by degrees of system implementation—and that the internal system programming was relatively simple. As a result, special programming talent was not required to implement and maintain the desired system functions.

Steps planned to be taken in the complete program include the addition of an arithmetic section for on-line calculation of performance indices and, should the results of these indices indicate the desirability, some degree of supervisory control. The system has demonstrated its capability to provide the input output (1/0) function for such expansion capability satisfactorily.

There are several factors to be considered should the aforementioned steps be approved and carried out—

1. Although special programming talent is not deemed necessary, a systems engineer must be employed by the division to work with the systems engineering department in the implementation of the computer and, subsequently, of the control. His responsibility would carry on, after the systems engineering department has completed its activities, providing the necessary supervision and liaison with appropriate mill departments to ensure future successful system operation. This man could be a young process engineer with suitable abilities and trained on the job for the position.

2. Plans must be made and action taken properly to calibrate all input measuring devices—both as components and as a system—before completion of the computer expansion. Subsequently, periodic checks will be required to ensure that accurate and reliable data is being fed to the computer. Success of the system depends on this requirement being fulfilled. Some leeway is possible in data accuracy when used only for display of instantaneous data, but the best accuracy possible is absolutely essential when data is to be processed and meaningful indices established.

Given the continued co-operation of all departments involved, there is considerable knowledge yet to be gained by the expanded implementation of No. 10 information system. With the knowledge gained, there is every reason to expect additional economic benefits—primarily, through improved uniformity of operation and as foundation information for design of future papermachines and control systems.

Reference

1. Mardon, J., Barrett, J. E. and Meheffey, W. H., 'A systems approach to paper-machine instrumentation': *Appita*, 1965, **18** (6), xvii–xl

Appendix 1

The 754 system functions are as follows—

1. Data acquisition The data acquisition subsystem existing in the No. 10 system has the capability to monitor 150 analog inputs at 15 points/s. It also monitors up to 20 integration inputs (pulse inputs), 26 BCD digital inputs (speed and draw) and up to 200 contact type inputs.

2. *Alarm monitoring* This subsystem monitors all analog and digital inputs at a rate of 3 000 points/s and compares their stored value against a programmed alarm limit. At the same rate, it also monitors the status of the external contact closure alarms. The status of both types of alarms is stored to allow a review at any time or at a set interval of 1 min, 5 min, 30 min or 60 min of the existing alarm conditions. New alarms, return to normal messages and alarm reviews are presented on the alarm printer. This alarm subsystem also has the capability for comparing all analog or digital values against a second set of alarm limits, designated questionable alarms. This final function of the alarm monitoring subsystem is to provide area annunciation on a 20 window annunciator. Each alarm condition can be programmed to alarm in any area.

3. *Periodic logging* The periodic log subsystem provides a print-out, automatically or at regular intervals, of the values of approximatly 125 system inputs. These values are typed on a preprinted log sheet consisting of two pages bonded together to give a 'roller towel' typewriter sheet.

4. *Value displaying* The value of any system input, high or low alarm limit and high or low questionable alarm limit can be displayed on a digital read-out on the system console. This display is 4 digits with decimal point and sign. The 20 most frequently used variables can be displayed directly by depressing a push-button engraved with their alphanumeric identifier.

   Time can be displayed on these same digit lights by depressing a momentary push-button.

5. *Analog trend recording* Five two-pen recorders are mounted on the console to provide for trend recording of any 10 selected system inputs. Any pen can record any system input at a desired zero suppression and chart range.

6. *Utility functions* A utility subsystem is included in the No. 10 information system. This subsystem consists of a further five two-pen recorders and a separate printer. The following functions are possible—
   (*a*) Single point digital print-out of any system input, high or low limit or questionable high or low limit.
   (*b*) Single point digital trend print-out of current data for any analog input. Trend interval can be 1 min, 5 min, 15 min, 30 min or 60 min.
   (*c*) Single point analog trend of current data on one pen of the analog recorders. This pen is identified by having the printer first type out the variable and one set of current data.
   (*d*) Single point digital trend of historical data for 100 selected digital or analog inputs collected over 1 h at 1 min intervals.
   (*e*) Single point analog trend of historical data for up to 100 analog or digital inputs collected over 1 h at 1 min intervals. This pen is identified by having the printer first type out the variable and one set of current data.

(*f*) Group review* digital trend of current data for anyone of a possible 20 logging groups on operator demand. The trend interval can be set at 1 min, 5 min, 15 min, 30 min or 60 min.

(*g*) Group review analog trend of current data for any one of a possible 20 logging groups on operator demand. The variable presented on the recorders are identified by having the printer first print out the group review with one set of current data.

(*h*) Group review digital trend of historical data at 1 min intervals for the previous 60 min on operator demand.

(*i*) Group review analog trend of historical data at 1 min intervals for the previous 60 min. The analog values are identified by having the printer first print out the group review with one set of current data.

7. *Tape punch* The No. 10 system has a paper tape punch available as an output device. This paper tape can be initiated automatically or on demand to record the values of up to 100 system inputs.

8. *Critical alarm review subsystem* This subsystem was set up so that an alarm condition on any variable could initiate, via programming, a group review analog trend of historical data for a selected group, automatically followed by an analog trend of current data for the same group. The operator has the ability to stop this review by—

(*a*) Stop utility push-button if the critical variable is no longer in alarm.

(*b*) Critical alarm disable keylock switch if the variable is still in alarm.

Some modifications to the system have been made that enhance its usefulness—

(*i*) The five utility recorders were given an alternate function. Those recorders may now be directly connected, through a push-button on the information system console, to the following permanent inputs—

Recorder 1 Red—Mixed stock consistency
Blue—Mixed stock dilution valve position
Recorder 2 Red—Groundwood consistency
Blue—Groundwood dilution valve position
Recorder 3 Red—Kraft consistency
Blue—Kraft dilution valve position
Recorder 4 Red—Broke consistency
Blue—Broke dilution valve position
Recorder 5 Red—High couch vacuum
Blue—(Not yet assigned)

If the push-button designated *enable recorders* is backlighted, these five recorders will be connected to the above variables. If this push-button is not backlighted, the recorders are then available for use with the utility subsystem as designed.

The reason this modification was made was to utilise these recorders. Previously, they were sitting idle 99 per cent of the time for the following reasons—

(*a*) If the recorders are being driven from the utility subsystem, the utility printer is tied up. Hence, display of current data in the analog trend format is impracticable, as the utility printer cannot be used.

* The system has been organised such that groups of 10 variables can be displayed together in the various formats

(*b*) The scaling of the analog recorders is unsatisfactory, hence they are very difficult to interpret. The scale is either 1:1 or 1:4 and this can be determined only by knowing beforehand what you expect to see on the recorder. Besides, because the one scale must accept both very large and very small signals, the resolution is very poor.

(*c*) The noise from the chart motors causes intermittent resetting of the system time. This is also a deterrent to their use.

(*d*) The chart speed switching from high to low was not fast enough and sometimes 15 min was required before the chart would move.

(*ii*) The compiler subsystem was modified to permit data collection at either 1, 2 or 3 min intervals, thus providing a 1, 2 or 3 h historical review. The sampling interval is changed via a thumbwheel switch in cabinet 8.

This feature provides a system memory that can be altered to suit a specific purpose—for example, long-term studies (up to 3 h) can be recorded through a historical review at the end of the study period.

(*iii*) The historical review capability in the utility subsystem was modified to allow the print-out to skip from 1 to 59 items of the oldest data. This feature allows any part of the historical record to be displayed without having to wait for all 60 readings to be typed out.

This feature is implemented by setting the number of readings to be skipped on the two righthand digits of the thumbwheel switch on the right side of the console. Once the group review is selected, the *historical review by-pass* push-button is depressed and the correct number of readings will be displayed.

(*iv*) The alarm subsystem was modified to provide an alarm dead band on all programmable alarm limits. The system as installed had two sets of alarms, one for standard process monitoring and one designated *questionable* to provide alarming for such things as instrument failure. Since the questionable alarms were not implemented and indications were that they would not be, it was decided to utilise this memory and hardware to provide an alarm dead band. Thus, the system now has an alarm limit and a return to normal limit and both must be programmed for each point alarmed. Return to normal limits are programmed as *questionable* alarm limits. The purpose in applying the dead band concept in digital system alarming is to eliminate annoying conditions when a process variable is oscillating around the alarm point and causing repeated print-outs of alarm and return to normal conditions.

(*v*) The critical alarm subsystem was modified to correct difficulties inherent in the original design. The main modification provided the facility for initiating a critical alarm group review only on a *new* alarm condition. This feature allows operator acknowledgement of the group review at any time after it has been initiated, regardless of whether the alarm condition still exists.

Appendix 2

Process implementation projects

THESE projects can be divided into two main types. The first type are those projects that can be undertaken by the systems engineer with only the information available from the system, production reports and other instrumentation. The second type of project involves work that must be done by the Powell River technical department.

The project descriptions to be given here are brief and merely serve to outline the objective of the work.

Type 1 projects

1. *Lost time analysis* The project objective will be to show how the data system can be used as an effective tool for lost time analysis and reporting.

An example of a lost time analysis for a one month period will be produced. The project will take into consideration the lost time reporting system already in use at Powell River and will determine where the data system can be of advantage.

2. *Head box variables correlation to basis weight* The objective is to determine any relationships between variations in the head box variables and basis weight (machine-direction).

   The basis weight gauge will be used in conjunction with the continuous monitoring of the head box variables.

3. *Heat and material balances over the dryer systems* The objective is to utilise the information available from the data system plus any other necessary information such as hood dry bulbs and fan amperage in order to calculate a heat and material balance over the dryer system.

4. *Relationship between dryer steam consumption, moisture and felt age* At fixed moisture levels, cross-plots between steam consumption and felt age will be made to determine a relationship between the two. All dryer felts will be taken into consideration.

5. *Relationship between average basis weight and mixed stock consistency* The objective will be to determine the relationship between average basis weight and mixed stock consistency for the No. 10 papermachine.

   The data will be collected at periods of constant mixed stock flow and machine speed.

6. *Relationship between suction box vacuum and amount of broke* This relationship will be defined for the No. 10 machine at times when the other factors influencing drainage are constant—that is, constant basis weight and groundwood freeness.

7. *Temperature of system as related to couch and flat box vacuum* Temperature will be plotted against these two vacuums to determine this relationship for No. 10 papermachine. Care must be taken to ensure that other influences such as groundwood freeness are relatively constant.

8. *Relationship between couch vacuum, flat box vacuum and moisture at the reel* The objective is to define the relationships between couch and flat box vacuums and reel moisture for the No. 10 papermachine.

   Data will be plotted for periods of steady dryer conditions.

9. *Felt tension versus steam consumption* With other related variables constant, a relationship between felt tension and steam consumption will be looked for.

10. *Felt age versus press vacuum for three presses* The objective is to determine what effect the felt age has on press vacuum. Data will be plotted for periods of two successive felts on the machine.

11. *Head box consistency as a function of dryer steam* At a fixed moisture and machine speed, the relationship between No. 10 head box consistency and the dryer steam consumption will be determined. This can be done directly using the manual test results.

12. *Basis weight valve position versus consistency* At different levels of machine speed, the curve showing basis weight valve position as a function of consistency will be determined.

13. *Speed change responses* Data will be gathered to determine the response of the process to a change in machine speed. Particular attention will be paid to basic weight and moisture.

Type 2 projects

1. *Calibration of consistency regulators* The objective here is to obtain data that can be used to define the absolute calibration of the regulators on the No. 10 stock preparation system, particularly kraft and groundwood stocks.

2. *Relationship between formation and total head* At a fixed drag on the wire, formation will be cross-plotted against machine total head (speed).

   Formation samples will be taken and run off in co-ordination with the mill technical department.

3. *Relationship between formation and drag* At fixed speeds, formation will be cross-plotted with drag to define the operating point at which the formation is the optimum.

   Formation samples will be run off in co-ordination with mill technical department. At the same time, the relationship between drag and CD & MD tearing strength will be evaluated.

4. *Press moisture profiles* A correlation between press moisture profiles and dry end basis weight and moisture profiles will be made in conjunction with the mill technical department.

   At the same time, the effect of differential loading on the presses will be evaluated.

5. *Effect of loading on the breaker stack* Caliper and smoothness variations at the dry end will be correlated with the breaker stack loading. Samples will will be taken at reel turn-ups by co-ordination with the mill technical department.

6. *Basis weight, moisture and smoothness with and without steam to the calender stack* The effect of steam in the calender stacks on these three variables will be determined. Samples will be taken at a reel turn-up for each of the two conditions.

7. *Effect of whitewater on formation*

8. *Suction box vacuum as a function of freeness* To correlate changes in groundwood freeness to suction box vacuum at stable machine conditions. Testing will be done by the mill technical department.

Appendix 3

Area	*Incident*
Chest levels	*1.* Machine chest level was alarmed at a high level condition on start-up and subsequent overflowing was prevented. *2.* A similar situation occurred on start-up with the kraft stock chest. *3.* The pump feeding the No. 10 kraft stock chest kicked out, thus causing the kraft stock chest level to fall. A system alarm alerted the operator and the pump was started again, but not before the level dropped to 3·2 ft. Without this early warning, the kraft stock flow to the machine chest would have been disturbed and the trial upset. *4.* In another instance, a high alarm condition was noted on the high density groundwood storage. The operator was able to take action to prevent overflow and stock loss. *5.* The power to the pumps feeding the No. 10 kraft stock chest and the dye head box was lost. A low level alarm on the kraft stock chest was recorded at 13.08 h, but the operator was not available. At 13.26 h, the dye head box emptied and a dye flow low alarm was initiated. This was picked up by the operator and steps were taken to remedy the situation. Although the operator was working on the dye flow, a kraft stock flow alarm was initiated at 13.41 h as the kraft stock chest had also emptied. At this point, all flows to the machine chest were shut off. The alarm monitoring system provided the warning to the operator and, had the first alarm condition (13.08 h) been acknowledged, the disturbance would probably have been less. Without the alarm system, the problem would have been far more severe.

Area	*Incident*
	6. The stock preparation operator requested that the low level alarm on the kraft stock chest be raised to permit more time on occasions when the trouble occurs. He reported that a recent power cut-off on the kraft stock supply pumps had caused No. 9 machine to run out of stock while No. 10 machine was just keeping up. The limit was raised after discussion with stock preparation supervision. This example represents how the alarm monitoring system can be used and how the operating personnel have utilised it.
Stock consistencies	*1.* One start-up, a failure in the groundwood stock consistency controller was picked up immediately by the alarm subsystem. Action was taken to prevent a machine disturbance.
	2. The groundwood dilution water for groundwood consistency control was shut off when the groundwood mill was shut down. A high alarm on the groundwood dilution valve position was recorded at 12.32 h. At 12.42 h, the mixed stock consistency regulator was out of control as noted by a high alarm on the mixed stock dilution valve position (due to machine chest becoming heavy). At 12.48 h, the machine went down due to the increased consistency. Although operator action was not fast enough to prevent the machine from going down, the warning was available and alternative action could have been taken to prevent the downtime. For example, stock flows to the machine chest could have been shut off temporarily while the dilution water was returned.
	3. A similar problem occurred at a later date, when maintenance crews shut down the groundwood water pump not realising that it was being used for consistency control on No. 10 machine (the groundwood mill was down). The groundwood dilution valve position high alarm warned of the situation. The operator was able to get the dilution water back before the machine was put down. It was noted on this occasion that the mixed stock dilution valve position started to increase rapidly 5 min after the upset and went from 55 per cent open to 84 per cent open in the following 18 min. This indicates a 5 min dead time in the machine chest.
Stock flows	*1.* During start-up, the dye line plugged and flow was

Area	*Incident*
	cut off. The system alarm for low dye flow enabled immediate corrective action and prevented off-specification paper.
	2. The low limit on dye flow was removed by persons unknown. This was picked up by the operators and replaced at their request. This example points out that, once a system is installed, its additional features —above and beyond conventional instrumentation— come to be depended upon in the same fashion as other instruments.
Stock cleaning and de-aerator	*1.* A low alarm on the de-aerator vacuum warned the machine tender of trouble. Subsequent investigation pointed out a slipping belt on the vacuum pump. The vacuum did not drop more than 1 in Hg over the trouble period. Had the operator not been warned at this early stage, it is possible that the vacuum could have dropped further and upset the machine.
	2. An alarm condition in one of the supply pressures was picked up. Historical review of the centrifugal cleaner group indicated that a sump level was varying. Subsequent investigation showed a troublesome control valve, which was then repaired.
	3. An alarm on the tertiary cleaners supply pressure was initiated. Investigation via the centrifugal cleaners historical review pointed out a low secondary rejects sump level. This level could not be increased with the level control because of the control valve sticking. The instrument department was notified and and the condition rectified.
	4. An alarm on the secondary centrifugal cleaner supply pressure was recorded at 9.19 h. At the same time, a sharp drop in basis weight and couch vacuum occurred as shown in the analog record of Fig. 8. Investigation of the centrifugal cleaner group review as shown at the top of Fig. 8 showed that the primary rejects sump level had jumped from 35 to 96, whereas the secondary cleaners supply pressure apparently remained constant. It is assumed that, at the jump in sump level, a surge in light stock to the secondary cleaners was caused and this was the reason for the low couch vacuum and basis weight. The transient in the secondary supply pressure was likely very fast, hence a sustained offset in the secondary supply was not recorded over the 1 min interval sampled. The

Area	*Incident*
	low couch vacuum and low basis weight are also shown on the head box historical group review in Fig. 8. The machine tender, who had been analysing the problem on the charts in Fig. 8, investigated the cause of the sump level change and found that a cleaning man had turned the set point knob on the locally mounted level controller while cleaning it.

TIME	C-C PRIM SPLY PR	C-C SEC SPLY PR	DECU LATR VAC	C-C TERT SPLY PR	C-C TERT DSCH PR	C-C PRIM SUMP LEVL	C-C SEC SUMP LEVL	STRM FLOW VALV POSN
09:15	30.3	35.6	27.4	46.7	0.71	35.5	37.3	39.7
09:16	30.3	35.6	27.4	46.7	0.70	35.1	37.3	39.6
09:17	30.3	35.6	27.4	46.7	0.71	35.1	37.3	39.6
09:18	30.2	35.5	27.4	46.8	0.71	35.2	37.3	39.7
09:19	30.2	35.7	27.4	46.9	0.71	35.3	36.7	39.6
09:20	30.2	35.6	27.3	46.8	0.71	86.8	38.8	39.6
09:21	30.3	35.6	27.3	46.9	0.71	94.4	37.1	39.7

SUDDEN RISE IN SUMP LEVEL

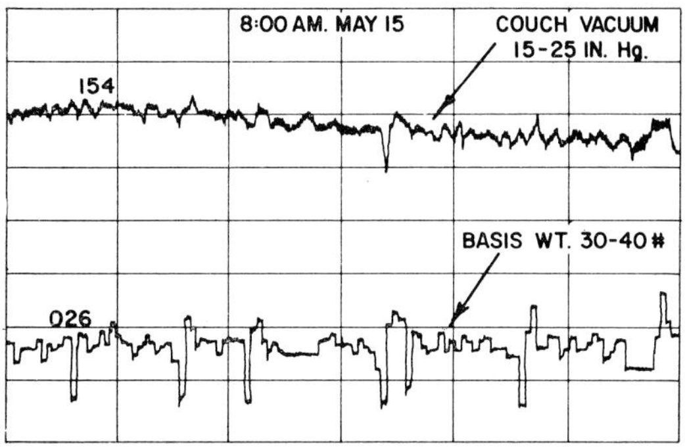

TIME	HEAD BOX TOT HEAD	HI COUCH VAC	DRAW JET WIRE	FWD DR ROLL SPD	STRM FLOW VALV POSN	HEAD BOX LIQ LEVL	HEAD BOX TEMP	HOT WATR SHOW TEMP	B WT AS MADE O M	MIXD STOK FLOW
09:14	269.	20.87	0014	2290	39.7	15.1	089.	105.	31.8	2548
09:15	270.	20.77	0016	2291	39.7	15.0	089.	106.	31.8	2540
09:16	269.	20.87	0012	2291	39.6	15.0	089.	105.	31.7	2542
09:17	270.	20.90	0014	2291	39.6	15.1	089.	105.	31.7	2537
09:18	270.	20.80	0016	2289	39.7	15.1	089.	105.	31.7	2550
09:19	270.	20.83	0015	2291	39.6	15.1	089.	105.	31.6	2565
09:20	269.	20.84	0014	2290	39.6	15.1	090.	106.	31.6	2561
09:21	270.	20.45	0013	2290	39.7	15.1	089.	105.	31.6	2567
09:25	269.	20.90	0014	2291	39.6	15.1	089.	105.	30.6	2597
09:26	270.	21.03	0015	2291	39.7	14.9	089.	105.	30.6	2604
09:27	269.	21.07	0013	2298	39.6	15.1	089.	105.	30.6	2601
09:28	270.	21.13	0000	2291	39.7	15.1	089.	105.	32.1	2572

DROP IN COUCH VACUUM

DROP IN BASIS WEIGHT

Fig. 8—The centrifugal cleaner group review (top) records a sudden rise in sump level: the resulting surge of light stock to the head box causes a drop in couch vacuum and basis weight as indicated in the system analog record (centre) and the head box historical review (bottom)

Area	*Incident*
Head box	*1.* On several occasions, the head box level went into high alarm, owing to an inadequate air supply. On these occasions, the sheet on the wire became very wild, but the machine did not break down. The problem was corrected by supplementing the compressor air with enough mill air to provide a stable level control at the hornbostel. Had the problem not been noticed at an early stage, it is conceivable that prolonged operation in this state could have caused some down-time.
	2. On start-up, the information system was useful in recording deviations in the head box total head. At this time, when the head was generally constant because of a fixed machine speed, programmable limits were put on total head. This alarming procedure provided immediate corrective action and eliminated down-time.
	3. Several pens of the selectable trend recorders were put on the head box variables and the whitewater flow to the groundwood mill to determine if fluctuations in this whitewater flow are causing temperature upsets to the head box during normal operation. The trend recorders were set up recording the key variables over a period of several days in order to provide data for a correlation analysis.
	4. The jet-to-wire draw measurement is a useful variable to have displayed and recorded. Early calibration problems in the differential pressure sensor providing the measurement caused a considerable feeling of mistrust with the calculated reading, but this has since been overcome and the draw value is now reasonably accurate. This figure can be utilised by mill management and supervisory personnel to monitor the machine operation. For example, on one occasion, management personnel requested the 35 ft/min drag existing on the machine be reduced as part of an investigation of low tear values on the machine.
	5. On another date, the machine tender used the calculated draw value to set up the machine. This was the first recorded example of operator acceptance of and confidence in this variable.
Fourdrinier machine	*1.* On start-up, the flat box header vacuum kicked out, forcing the machine down. Subsequent analysis by

Area	*Incident*
	means of the utility subsystem historical review indicated that the flat box header vacuum had failed first.
Press and breaker stack loadings	*1.* The digital indication of press and breaker stack loadings appears to provide a superior method for setting up these values. It is possible to set up the nip pressure values exactly, using the information system, whereas it is very difficult to interpret the pressure gauges on the press and breaker stack consoles. Boss machine tenders have been using this information system facility for both setting up and recording loading values.
	2. On one occasion, a machine room supervisor had been able to solve the problem of a soft back edge on the reel by using the information system to analyse all those factors that can affect it. It was found that an uneven breaker stack loading had been the cause.
Dryer section	*1.* The No. 10 dryer steam system has been the subject of considerable analysis since start-up. It has been pointed out that the information system has been used for monitoring changes in dryer pressures and corresponding changes in steam flow. From this information, there is some evidence that the steam system is undersized and could be the cause of some paper breaks.
	2. In a technical department survey into the possible causes for poor drying over a specific operating period, the data recorded on the periodic log sheets was used to study the steam system operation.
Examples of use by supervisory personnel	*1.* On machine start-up, the basis weight valve had to be set 100 per cent open in order to obtain a 32 lb sheet at the start-up speed. At the previous shut-down, the basis weight valve had been only 80 per cent open at the same mixed stock consistency. After approximately 4 h operation, the mixed stock consistency appeared to rise and the basis weight valve had to be cut back to 80 per cent open. Stock preparation and machine room personnel analysed the problem via the information system. Based on later events, it appears that the mixed stock consistency regulator had drifted over the shut-down.
	2. On another date, the fourth dryer section speed was put on a trend recorder, owing to a recent problem of it slowing down. Later that day, the boss machine tender noticed the speed beginning to fall off again

Area	*Incident*
	and was able to correct the problem before the sheet broke down. Before the speed was brought back up again, the draw after the breaker stack was extremely slack.
	3. The pump feeding the wire pit kicked out. The result was a falling level and rising consistency in the wire pit, which led to an increased basis weight. Although an alarm was not recorded by the system, it would have been had there been a correct high limit on the basis weight value. The machine was producing a 30 lb sheet on this occasion and the 33·5 lb alarm limit for the 32 lb sheet had not been changed. The head box historical review in Fig. 9 shows the increasing couch vacuum and an increased basis weight at 16.06 h.

TIME	HEAD BOX TOT HEAD	HI COUCH VAC	DRAW JET WIRE	FWD DR ROLL SPD	STRM FLOW VALV POSN	HEAD BOX LIQ LEVL	HEAD BOX TEMP	HOT WATR SHOW TEMP	B WT AS MADE O M	MIXD STOK FLOW
16:11	257	21.53	0022	2234	42.4	15.2	086	104	29.6	2452
14:27	256.	20.83	0018	2235	42.3	15.1	085.	111.	29.0	2512
14:30	256.	20.81	0017	2235	42.4	15.1	085.	112.	29.0	2513
14:33	256.	20.76	0017	2235	42.4	15.2	085.	111.	29.0	2505
14:36	256.	20.78	0017	2234	42.3	15.3	085.	111.	29.0	2511
14:39	256.	20.68	0016	2235	42.4	15.1	086.	111.	29.0	2512
14:42	257.	20.68	0014	2235	42.5	15.2	086.	111.	29.0	2509
14:45	256.	20.58	0015	2234	42.4	15.2	086.	111.	29.0	2503
14:48	256.	20.72	0015	2234	42.5	15.1	086.	110.	29.0	2501
14:51	257.	20.40	0014	2235	42.5	15.3	086.	110.	29.0	2515
15:12	257.	20.47	0013	2233	42.5	15.2	086.	110.	28.9	2522
15:15	257.	20.73	0014	2234	42.5	15.0	086.	110.	29.0	2497
15:18	257.	20.50	0013	2234	42.4	15.1	086.	110.	28.9	2510
15:21	257.	20.53	0014	2235	42.5	15.2	086.	110.	28.6	2522
15:24	257.	20.62	0013	2234	42.5	15.2	086.	110.	29.0	2530
15:27	257.	20.79	0000	2235	42.4	15.2	086.	110.	29.0	2512
15:30	257.	20.63	0000	2234	42.4	15.2	086.	110.	29.2	2490
15:33	257.	20.71	0000	2236	42.5	15.2	086.	110.	29.0	2495
15:36	257.	20.61	0015	2235	42.5	15.1	086.	110.	29.0	2532
15:39	257.	20.80	0013	2236	42.5	15.2	086.	110.	29.1	2521
15:42	256.	20.65	0013	2235	42.5	15.0	086.	110.	29.0	2510
15:45	257.	20.79	0013	2235	42.5	15.2	086.	110.	29.0	2516
15:48	256.	20.69	0014	2235	42.5	15.1	086.	110.	29.1	2513
15:51	257.	20.83	0000	2233	42.5	15.1	086.	110.	29.0	2510
15:54	257.	20.80	0000	2235	42.5	15.2	086.	111.	29.1	2501
15:57	257.	20.89	0000	2234	42.5	15.1	086.	111.	29.1	2517
16:00	258.	21.31	0013	2235	42.4	15.2	086.	110.	29.1	2512
16:03	256.	21.63	0014	2235	42.5	15.1	086.	110.	29.1	2510
16:06	257.	21.61	0014	2234	42.4	15.2	086.	110.	29.9	2484
16:09	257.	21.76	0014	2235	42.5	15.1	086.	110.	29.9	2480

RISE IN COUCH VACUUM

DROP IN MIXED STOCK FLOW

RISE IN BASIS WEIGHT

Fig. 9—The head box historical review records a sudden rise in basis weight caused by a level drop and consistency rise in the wire pit following the kick-out of the Broughton pump; the subsequent rise in couch vacuum and drop in mixed stock flow are also recorded

Area	*Incident*
	4. The basis weight average increased rapidly for a short period of time. Coinciding with this increase were falling de-aerator levels and a de-aerator vacuum, which went from 25·6 in Hg to 25·4 in Hg. Also over this period, the mixed stock consistency was out of control with the dilution valve 100 per cent open.
	Yet, all input consistencies to the machine chest were well controlled and did not show any disturbance. No conclusion was drawn from the analysis other than questioning a possibility of poor mixing in the machine chest, which would show up in the sheet with the mixed stock regulator out of control.
	5. A low basis weight alarm was recorded at about 8.38 h. The head box historical group review is shown in Fig. 10. The machine tender picked up the alarm and subsequent checking showed that the knotter had kicked out, thus causing the wire pit consistency to drop.

HEADBOX

TIME	HEAD BOX TOT HEAD	HI COUCH VAC	DRAW JET WIRE	FWD DR ROLL SPD	STRM FLOW VALV POSN	HEAD BOX LIQ LEVL	HEAD BOX TEMP	HOT WATR SHOW TEMP	B WT AS MADE O M	MIXD STOK FLOW
08:58	271.	21.23	-0000	2286	39.0	14.3	098	-321	31.5	2685
08:32	272.	20.85	-0000	2285	39.1	14.-	098.	110.	31.5	2668
08:33	272	21.02	-0001	2286	39.1	14.2	098.	110.	31.5	2670
08:34	272.	20.75	0002	2288	39.1	14.2	098.	141.	31.5	2655
08:35	273.	20.27	0007	2286	39.1	14.2	098.	400.	31.9	2646
08:36	273.	19.97	0007	2287	39.1	14.1	098.	110.	31.9	2642
08:37	273.	19.91	0009	2286	39.1	14.3	098.	111.	31.9	2638
08:38	273.	19.86	0007	2286	39.1	14.2	098.	102.	29.2	2667
08:39	272.	19.90	0006	2285	39.2	14.2	098.	109.	29.2	2663
08:40	273.	19.90	0005	2286	39.1	14.3	098.	110.	29.2	2647
08:41	273.	19.82	0008	2286	39.1	14.3	098.	111.	28.7	2649
08:42	273.	19.86	-0006	2287	39.2	14.2	098.	108.	28.7	2680
08:43	273.	20.05	-0008	2287	39.1	14.1	098.	108.	28.7	2654
08:44	273.	19.98	-0008	2286	39.0	14.3	098.	110.	28.5	2632
08:45	272.	20.03	-0004	2287	39.1	14.2	098.	110.	28.5	2661
08:46	272.	19.93	-0005	2287	39.1	14.3	098.	109.	28.5	2652
08:47	273.	19.89	-0008	2286	39.1	14.3	098.	108.	28.5	2656
08:48	273.	20.02	-0008	2285	39.1	14.3	098.	107.	28.8	2765
08:49	274.	20.56	-0005	2286	39.1	14.4	098.	109.	28.8	2678
08:50	274.	20.34	-0007	2286	39.1	14.3	098.	109.	28.8	2659
08:51	273.	20.42	-0011	2287	39.1	14.3	098.	109.	29.5	2668
08:52	271.	20.75	-0000	2286	38.9	14.2	098.	109.	29.5	2662
08:53	272.	20.92	-0000	2286	39.0	14.2	098.	109.	29.5	2682
08:54	271.	20.95	-0000	2286	38.9	14.3	098.	109.	30.1	2671
08:56	271.	21.08	-0000	2287	39.0	14.3	098.	107.	30.1	2679
08:57	271.	21.11	0000	2286	38.9	14.3	098.	109.	31.5	2682

DROP IN COUCH VACUUM

DROP IN BASIS WEIGHT

Fjg. 10—The head box historical review records a sudden drop in basis weight caused by a drop in wire pit consistency following the kick-out of the Jonsson knotter

Area	*Incident*
	6. During parts of the breaker stack moisture survey the information system was put on the 3 min compiler sample interval in order to compile experimental data for subsequent recall.
	7. The 3 h recall in the utility subsystem was used to analyse a speed up to 2 705 ft/min. Historical group reviews were used to gather data on all machine conditions over the speed-up period.
	8. During the implementation of the automatic basis weight control, the information system was utilised to monitor machine conditions—for example, the system was used to show the backlash in the basis weight valve.
	9. Since start-up, the integrated values on the eighth page of the periodic log sheet have been used by the statistics department for accounting purposes. These are primarily steam and condensate values.
	10. Lost time has been used for accounting, but reliability problems with the reel amperage contact have not allowed complete acceptance of this value. The problem is one of location for the reel amperage contact. Operating changes such as a new doctor or changed draw from the stacks to the reel can cause changed loads and an erroneous contact, which leads to an incorrect accumulation of lost time.
	11. The machine operators began using the on-machine instrumentation for slice adjustments based on the oven-dry basis weight profile.
	12. It was pointed out during a meeting that the head box and draw group reviews are useful in analysing machine speed-ups so that they can be done in a minimum time with a minimum disturbance.
	13. The information system can also be utilised by machine room supervisory personnel to check on changes made by machine operators—for example, on one occasion, historical information was used to question a machine tender on changes he had made to the head box.

Discussion

The Chairman First of all, I would like to raise a question. You have given a number of examples of data being used to substantiate the existence of some problem. One of these was the relationship between air pad temperature and total head. This relation (Boyle's law) is well known and I believe that one of your papers several years ago included this. I wonder why you did not install analog control of total head in anticipation that temperature and other disturbances would be present. By not doing this, did you not contravene your philosophy of proceeding one level at a time?

Mr J. Mardon The main problems in papermaking today are almost entirely psychological. Consequently, it is not enough to know yourself what has to be done, you have to convince everybody to go along with you. In fact, we are now starting systems of the kind you are describing. One has to remember that the fellows who run these big machines, especially on the west coast of N. America, have something to be proud of. They have large production records. The machines today are running at speeds of 2 800–3 000 ft/min and, when we were doing this work, they were running at 2 400–2 500 ft/min. There is a tremendous inertia and desire for people to stay with what they know about. What might seem obvious to us is not obvious to them and it has taken three—four years to install our first successful system. All that you need to have in these large companies is one successful system and you very rapidly spread it through the rest of the company. I warn you young men whom I see in front of me, however, that it is also true that all you need is one unsuccessful system and you are out on your ear.

Mr T. J. Boyle I came into the paper industry five years ago and it has taken me almost that length of time to realise fully how little the papermaker relies on recorded data of any kind. Your system seems to suggest that you have succeeded in gaining acceptance of rather elaborate presentations of tabulated data. What steps have you taken successfully to add the use of digital data to the much sound information already available to operators?

Mr Mardon These things take time. It is rather like going through the Suez Canal, now shut down. During my first passage on the way back from India, I was conscious that the captain was proceeding at exactly the right speed. Had he gone any faster, he would have sucked the banks of the canal in and all the water would have run out. This is the kind of pace that one has to try to maintain—as fast as you can, but not so fast that everything collapses around you.

In this particular case, two facets of our general philosophy are involved. First of all, the capably senior papermakers in this large company were employed in the study, on the basis of which the system was designed. Mr Chatwin (in my view, one of the best papermakers in the world) put his personal attention to the very point you are speaking about. We have had four fairly skilled papermakers involved therefore in persuading everybody that the digital record was satisfactory and it is mainly due to Mr Chatwin's very hard work and dedication that we were able to make this point. One of our difficulties was failure of peripherals and maintenance. We had to work over the system again. These malfunctions, because of inadequate maintenance organisation nearly wrought failure. Unlike other systems, ours had these troubles.

Dr H. Sandblom I recall a paper published several years ago by some people from I.C.I. Ltd., in which it was reported how easily the operators had switched over to use digital information. Everybody was, in fact, surprised to see how easily they had converted from the analog to the digital system. A further study showed that, even though the operators had had analog recordings to work from, they had apparently converted these to digital information.

Mr H. B. Carter My question may well be asked of Dr Sandblom. Our papermakers, I have found, like those in other mills, are completely oblivious to many written records even in circular chart form. We also have found that our papermakers have asked that we stop plotting moisture and basis weight profiles and give them the figures. Mr Mardon made a statement that he expects his system to enable him to upgrade the papermakers—does he mean that he cannot foresee any possibility of this system being controlled automatically, but that we will always have the machineman there? I interpret this as meaning that papermaking is still an art, in which case we are wasting our time in talking about automatic control.

Mr Mardon I advise asking Mr Staples about the architecture of this system, which he designed. We are all very proud of it. For machines under discussion, the machine tender should be maintained as part of the control loop. One of the dangers of putting the machines on to completely automatic control is that people who are running the process forget how to handle them in any other way. This particular system was carefully designed as explained and can in fact be regarded as a test bed. Our present work involves the installation of an arithmetic unit and checking out different things to see how much we may gain from various expansions and changes. For example, we have heard much about basis weight control by computer, but one of the necessary questions is that, if you have a fully functioning analog system, far better basis weight control can be obtained than by manual control. Is it worth the extra effort to put the basis weight control through the computer and get a very small additional improvement? It is very difficult to find out the answer to this, but, in the kind of equipment that we set up (this flying test bed), these and other things will all be investigated. We are taking things very slowly and in a very conservative manner.

THE INSTALLATION OF A COMPUTER ON A FINE PAPER MACHINE

H. D. CYPRUS and D. ATTWOOD, Bowaters U.K. Paper Co. Ltd., Engineering & Scientific Services Division, Northfleet, Kent

Synopsis The paper examines the various stages in a project to implement control of a papermachine and its associated stock preparation plant using a digital computer. The project is looked at from the initial economic feasibility through to the commissioning and implementation stage.

A general description of the computer system and the control strategy employed is included. Particular mention is made of those aspects of the project not immediately associated with the control of the process, yet making important contributions to the whole system.

Reference is made to the more significant problems faced during the project, giving when possible indication of lessons learned for future applications.

Introduction

IT IS generally accepted that the availability of on-line measurements and conventional control devices has had a considerable effect on the profitability of the papermaking process. It is also widely believed that the digital computer has a further significant contribution to make by its application to the control of those interacting process variables that remain largely uninfluenced by conventional devices.

The full potential of improved control is difficult to estimate in advance, especially as many benefits are of an intangible nature. The more obvious areas of economic interest are—

1. Saving in fibre costs.
2. Improvements in operational efficiencies.
3. Reductions in broke at breaks and grade changes.

The intangible returns are largely associated with the computer used as a learning tool and as a means of centralising control and information.

Under the chairmanship of Dr D. B. Brewster

Economic justification

DURING the early part of 1965, Bowaters carried out a series of feasibility studies aimed at justifying the introduction of a digital computer for control purposes on to one of the U.K. papermachines. The studies covered a number of machines, producing a range of papers from fluting to quality coater base. The results of the studies were, in the main, very favourable and this was particularly true for No. 16 machine at Sittingbourne, manufacturing a range of base sheets for blade coating. This machine was therefore chosen for Bowaters' first computer control project. No. 17 machine, which is situated alongside No. 16, was soon to switch to a similar range of papers and it was therefore recommended that allowance be made early for its eventual inclusion in the entire control system.

Table 1 summarises the comparative benefits estimated for four different machines. There is considerable variation, but each would probably justify a computer control system of one size or another. Calculations for the purposes of the feasibility studies were based on the *shift sigma technique*, which is discussed below.

Typical production practice is to control product quality to some target value and to place upper and lower sample acceptance limits at two standard deviations (2σ) above and below this mean. The normal distribution curve has the property that 5 per cent of its area lies outside the 2σ limits, thus 5 per cent of production normally falls outside the specification limit. If process control is improved such that the standard deviation of the quality distribution is reduced to half its previous value, then the mean quality level can be altered

TABLE 1—SUMMARY OF COMPARATIVE BENEFITS

Area of development	*Mechanicals* Financial gain	*Fluting* Financial gain	*Newsprint* Financial gain	*Fine paper* Financial gain
	£	£	£	£
1. Substance control—reduction in materials demand due to lowering of average substance	No claims	13 618	No claims	28 000
2. Moisture control—increase in average moisture content	5 560	5 720	10 500	No claims
3. Furnish control—reduction in chemical pulp content	No claims	No claims	10 500	10 000
4. Machine efficiency—increase in output due to reduced time/broke losses	29 540	20 660	32 940	88 500
5. Machine speed — increase in general average resulting from raising of lower limits	4 600	41 600	7 200	61 500
	£39 720	£81 598	£61 140	£188 000

N.B. It has been assumed that the additional saleable tonnage will be sold according to the present mix for each machine

by an amount equal to the original standard deviation without more than 2·5 per cent of production falling outside the specification limits (Fig. 1).

The argument can be extended to aspects of process operation that do not involve specification limits. In this case, limiting factors are provided by the physical limitations of the plant. If process operation can be stabilised so that its standard deviation is halved, then the mean operating level can, as before, be moved by an amount equal to the original standard deviation without exceeding any of the plant limitations.

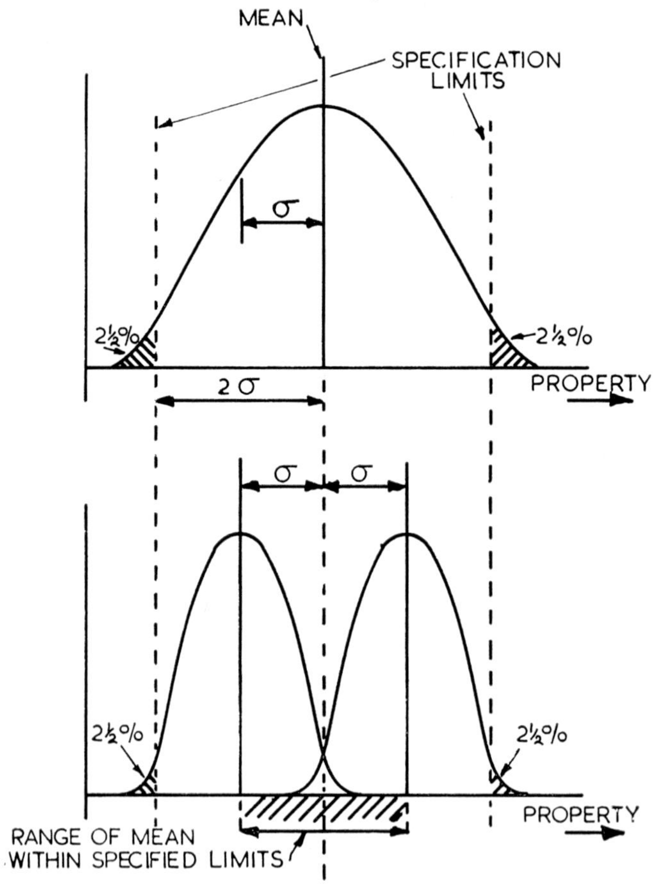

Fig. 1—Normal distribution curves

There are usually different economic penalties for different mean operating and quality levels. Thus, in the calculations for the feasibility studies, the economic implications of a shift in the mean were combined with estimates of current standard deviations to yield estimates of potential savings in the various areas of machine operation. Fibre savings were anticipated from improved control of basis weight, moisture content and loading proportion. Other economic values were determined on the basis of improved papermaking efficiency and a reduction in lost production through web breaks and grade changes. No value was placed on intangible benefits.

Alternative techniques for analysing the economic viability of computerisation are available.[1] In general, the more successful involve the comparison of process operation with achieved best performance or with theoretical limits. One approach that Bowaters have used attempts (unlike the shift sigma method) to look beyond the economic benefits of improved regulation of the various process stages by estimating the likely increase in process performance optimisation rate arising through the use of a computer.

This performance optimisation—which is, of course, going on throughout the life of a process as the natural outcome of increasing operating experience and knowledge—should show itself as an improvement in process profitability. Any major change to the plant or operating strategy will cause a discontinuity in the rate of optimisation and no such change should be considered successful unless the optimisation rate is increased.

It is necessary to analyse the performance of a process in economic terms and in ways that will reveal shortcomings and indicate areas of potential improvements.

The full description of the performance of a process involves a variety of interrelated factors such as quantity produced, purity, customer satisfaction with product quality, raw material costs and supervision costs. The most satisfactory single measure of process performance is one that measures the instantaneous profitability. This will be termed the *performance quantity* (PQ) and it can be expressed in terms of all the values and costs involved in the process—

$$\begin{aligned} PQ = {} & \text{Value (Products} + \text{Customer satisfaction} - \text{Returned products, etc.)} \\ & - \text{Costs (Materials} + \text{Labour} + \text{Maintenance, etc.)} \end{aligned}$$

The performance quantity as defined provides the best measure of process performance. In some cases, other criteria may seem more appropriate—for example, cost per unit production or production rate. Such sub-targets may be misleading, since they tend to ignore those factors that at particular times in the life of a process may become of extreme significance.

Performance trends may be studied by plotting the PQ for a plant for a number of past years. To be meaningful, the calculations must be carried out with fixed raw material and labour prices, etc. The prices and values in the current year are normally chosen for convenience. The procedure can be justified only when the plant and operating strategy have been unaltered during the whole period.

The growth of the PQ for most complex industrial processes without external disturbance shows a logarithmic rise. This is due to a sharp rise in performance as the plant is run up, followed by a steady rise, but with decreasing rate, as the plant operators gradually optimise the performance. Any major change to the plant or operating strategy will cause a discontinuity with another logarithmic curve rising from the time of change.

The logarithmic rise will have the equation—

$$PQ_t = PQ_m(1-e^{-t\ T})$$

where PQ_t is the instantaneous PQ value at year t,
PQ_m is the maximum theoretical PQ value or ceiling and
T is the time constant of the curve in years.

Curves of the component parts of the PQ may be plotted to aid interpretation of past performance variations. These components of the PQ can be broken down and presented as a diagram showing the flow of cash through the process, culminating in the final profit. Such a diagram is helpful when locating areas of the process that make a major contribution to the PQ.

Data collected by the accountants can usually be presented directly as a cash flow diagram. The items that contribute directly towards the production of the product—such as raw material and steam costs—can be separated from fixed costs such as rents, insurance and management overheads. Other items such as labour, maintenance and warehousing must be investigated at the source and a decision made on whether or not extra production will cause an increase in costs. This separation into fixed and throughput-dependent costs enables the profit associated with extra production to be established. If the extra product can still be sold for the original selling price and if the fixed costs are high, the profit per unit for the extra production will usually be much greater than the mean profit per unit.

A typical application of the performance quantity technique is illustrated by Fig. 2, which was generated in the course of a recent computer feasibility study. The past performance of the process involved was studied for the period 1962–67 and a PQ growth curve calculated that showed the expected logarithmic rise (Fig. 3). During the early part of 1968, a major process modification was implemented, thereby introducing a significant discontinuity to the curve. It was possible, of course, to determine the theoretical maximum for this post-1968 curve and, by assuming a growth rate similar to that of the

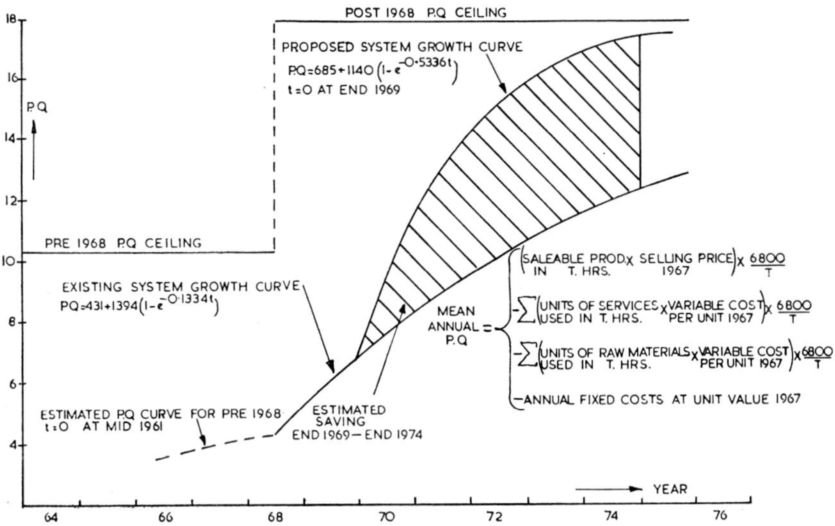

Fig. 2—Growth curves showing potential saving

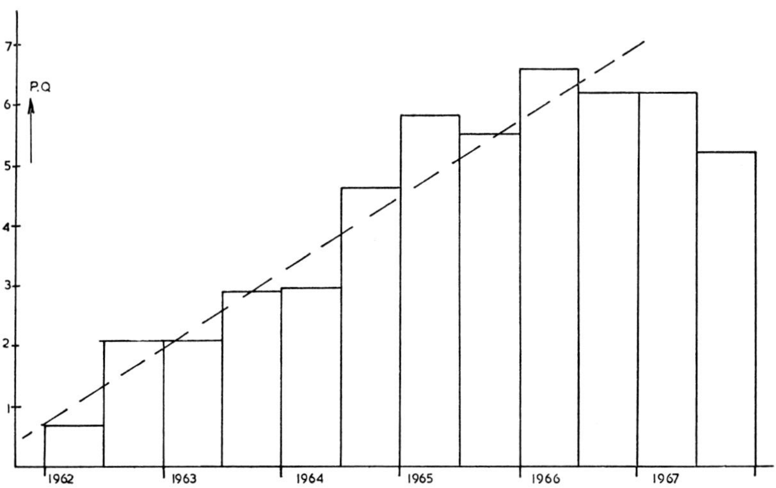

Fig. 3—Performance quantity trends 1962–67

pre-1968 period, the current performance curve extrapolated to 1974 was determined.

The study assumed that a computer system would be installed and available by the end of 1969. It was considered that, as a result of this computerisation, the process would be optimised at a faster rate. Firstly, process stability would enable process relationships to be established more quickly and would also make the assessment of performance at a particular operating state considerably easier. Secondly, an organised approach to optimisation, preferably involving frequent assessment of the PQ, would speed the rate of optimisation. It seemed probable that each of these factors would increase the rate of optimisation by at least a factor of two, hence both together should effect a fourfold improvement. The computer system would not directly affect the theoretical maximum PQ.

A growth curve for the process with the proposed computer system was therefore determined for the period 1970–74. The area between this curve and the post-1968 curve indicative of current performance was determined and this represents a sum of money equivalent to the increase in return over the years to be obtained by the computerisation.

Selection of the computer system

FOLLOWING the decision to proceed with the computerisation, various manufacturers were approached and invited to meet Bowater management and technical personnel. At that time, experience of computer control within the organisation was negligible and therefore the company was looking not only for a hardware/software package, but for a manufacturer interested in participating in a joint project that, although designed to be economically viable, was also to have a research and development aspect. It was our opinion that several computer manufacturers could provide the quality of hardware and software packages required, but we thought that it was in our best interests to purchase from a supplier who had the type of organisation and individual personalities best suited to match our own. The intention has always been that this first installation should lead into a long-term program, involving the application of computer control to some extent or other to each of the papermachines in the U.K. group.

The choice of actual system was therefore a secondary consideration at that time. Nevertheless, the more obvious details of software and hardware design were looked at and, as our understanding has increased, their significance has become more obvious. It is the less obvious points, however, that are now of particular interest to us and are those upon which we place greatest emphasis in our plans for the future.

In selecting hardware, along with details of core cycle speeds, instructions range, etc., we would consider carefully the following—

1. Degree of modularity.
2. Design for easy extension and modification.
3. Likely reliability and failure rate.
4. Quality of maintenance back-up and supply of spares.
5. Type of output stations and their compatability with standby equipment.
6. Type and quality of peripherals, especially operators' consoles.

The selection of software is considered more difficult, especially in the light of the apparently high rate of development and change. In general, however, it needs to be flexible and designed to be used by process engineers, not skilled programmers. Therefore, we would look for the following features—

1. Modular executive program.
2. System of autonomous packages.
3. Engineer-oriented language.
4. Simplest acceptable approach to a problem.

After some consideration, Bowaters selected a CON/PAC 4060 computer coupled with Kent interface equipment. The system was purchased with the standard monitor executive, together with various functional and interrupt driven programs, including one for direct digital control (DDC), which were written by the manufacturer's programmers. Bowaters chose at that stage to employ no full-time programmer of their own, but to set up a joint systems team with the manufacturer. The team was totally responsible for the project and consisted of two people from each company. This team has since been increased to include additional Bowater personnel during the commissioning and implementation stages.

Computer specification

Central processor Parallel arithmetic unit with 16 384 (extendable to 65 536) 24-bit word core store, having a 1·6 μs cycle time.

Locations up to 16 384 are directly addressable.

Above 16 384 words relative addressing permits preassembled programs to be located anywhere in the core store and be obeyed without address alterations.

Odd parity is written on transfer into memory and checked during read-out.

Seven index registers were available.

Program load was wired in.

Automatic priority interrupt system Continuous sequential scanning of 48 priority levels (extendable to 64).

Both inhibitable and non-inhibitable types available; the latter for use generally as pulse counters.

Operator communication system Digitronics paper tape reader—photo-electric sensing, 100 characters/s, seven channels plus parity.

Westrex paper tape punch—100 characters/s, seven channels plus parity.

2 IBM model 735 typewriters—15·5 characters/s.

Peripheral buffer capable of handling up to 8 items.

Programmer's console.

Operator's panel.

Process communication system Analog input scanner controller, low level amplifier with common mode rejection, successive approximation analog-to-digital converter capable of handling up to 1 024 inputs with up to 8 group advance.

Multiple contact output controller, capable of handling up to 32 groups (extendable to 64) of decimal outputs (4 per group), binary outputs (16 per group) or analog outputs (1 per group).

Digital input controller capable of handling up to 64 groups of 23 bits plus error bit.

Automatic stall alarm

Automatic sequencing of computer shutdown and start-up

Total project strategy

THERE are three categories of control function that will benefit a paper-machine. Firstly, those aspects of present practice that will be improved by more precise control and that can easily be automated. Secondly, those aspects that hold out promise of large returns based on development and the use of optimisation and integrated control techniques. Finally, those aspects that will yield only to the most powerful computer techniques just now being fully exploited in the field of process control.

With these ideas in mind, it was decided that a phased approach should be adopted towards the project. Five phases were planned and orders were placed during August 1966 for phases 0 and 1, together with the necessary facilities for easy extension at a later date. The phases were—

Phase 0—The planning and preparation phase

Phase 1—Implementation of control functions of immediate benefit using established techniques.

Phase 2—Development of new papermachine controls and application of steady state optimisation techniques.
Phase 3—The inclusion of associated finishing processes.
Phase 4—Application of advanced techniques.

Phase 0 was expected to last about 12 months (the time for delivery of the hardware). The first quarter of this period was given over by the systems team to the detailed design of the control system in terms of computer inputs and outputs, which was essential for the building of the hardware configuration. For the rest of the period, the team was involved with all the planning and preparatory analysis that needed to be completed before delivery of the computer. They controlled the project using a PERT analysis updated on a monthly basis. Through this, the various activities associated with programming, hardware manufacture, instrument selection and installation, cabling and numerous building and erection tasks were co-ordinated.

In addition to the basic task of designing a control configuration, all other aspects of the total computer system required to be considered and suitably phased into the project. As decisions here reflected the entire philosophy of the project; these aspects represent an important part of the total system and are therefore worthy of mention at this point.

The standby system

The existence of a reliable standby-to-computer system is considered essential and full consideration was given to all possible system types. These range from the fully redundant system, employing a second computer, to an entirely manual system in which, in the event of a computer failure, the operators are left with merely *raise/lower* facilities. It was agreed that the standby system should match in complexity the computer/plant system, although it should also be economically justifiable in itself. The design was tied therefore to economic calculations based on estimates of computer *mean time between failures* and *mean time to repair*.

The chosen design[2] is an external digital system (that is, non-supervisory) based on three levels of standby control. Important control loops are provided with full two-term or three-term, bumpless controllers, less important loops with more simple two-term, manual set point controllers and the remainder with purely manual controls. In order to minimise production losses, the system includes automatic transfer facilities, with switching to standby control being effected by various hardware and software links. Manual transfer is retained for emergency use. There is an 'all or nothing' philosophy, whereby no single loop can go to standby alone. Each loop, however, can be separately manually adjusted through the computer. Standby control on No. 16 machine is through stations situated around the plant rather than as a whole in the computer control room. It therefore involves the full machine crew.

There are six bumpless controllers acting on the following loops—

Stuffgate flow
Stuffgate consistency
Flow box head
Second dryer section pressure
Third dryer section pressure
Fourth dryer section pressure

Each controller, which is situated in a convenient local control station, has its set point automatically updated by the computer until it is brought into use when either a computer/local switch is adjusted to allow overriding local control or automatically in the event of computer system failure. Control constants are adjusted at the controller.

There are ten simple controllers acting on the following loops—

Rough chest consistency.
Broke chest consistency.
Fine chest consistency
Stock flow to proportioner
Broke flow to proportioner.
First fine chest level.
Second line chest level.
Flow box level.
First dryer section pressure.
Fifth dryer section pressure.

The controllers are situated in the interface cubicle in the computer room, but manual set point adjustments are made at local control stations. The controller output shares a path to the plant with the normal computer output.

The standby controllers represented in themselves a new type of control for the machine operators and it has therefore been necessary to implement a training programme that has included a period of continuous running with the equipment in use.

Operator information

Particular importance was placed on the need to compensate the machine operator for removing him from the machine floor. Therefore, considerable care was taken in the choice of design and facilities of the computer rooms.

Two rooms are provided—the computer room and the control room (Fig. 4). The former contains all the computer and interface hardware, tape preparation and editing equipment and other general facilities for technicians and programmers. The other room is intended essentially for the purposes of controlling the computer/plant system and is used primarily by the machine operator.

There are five television cameras sited along the papermachine and a further one in the quality control laboratory. The operator can therefore survey the process from the control room by switching among these cameras, using three monitors situated in the room. There is also an intercommunication system,

Fig. 4—General view of computer rooms

allowing up to a seven-way conversation, which covers the full papermaking process. The intercommunication control is situated on the main control desk, together with the operator's panel and a single typewriter (Fig. 5). This typewriter provides a permanent record of all actions taken through the operator's panel, together with alarm conditions and any logs that may have been put out either automatically or on demand. There is a range of special short length logs available to the operator specially designed to enable him quickly to ascertain the state of his plant. The use of abbreviations and mnemonics here is minimised to reduce the risk of confusion. Of particular interest are the alarm log (which lists all loops currently in a state of alarm) and a series of sub-logs, each giving data about perhaps only six or seven variables.

Although functionally adequate, the manufacturer's standard operator's panel was redesigned for both ergonomic and aesthetic reasons. For this task, the systems team was joined by an industrial designer.

The panel (Fig. 6) is divided into two sections. The left side facing, which has a blue background, is given over essentially to control functions and the other side, which is white, to information functions. Lamps, switches and buttons are variously coloured; data displays make use of correspondingly coloured floods. A special feature is the use of a dual flood for data input using the *check/execute/cancel* switch, whereby the value is checked using an orange background that turns green once the change is accepted or cancelled.

Fig. 5—Operator information and control station

There are three digits available for addressing and each address can refer to either a fully functional control loop or a simple software logic switch. In the former case, the following ten functions are available—

- Measured value (MV)
- Desired value (DV)
- Maximum change in DV allowed ($\triangle$DV)
- Three control parameters (P,I,D)
- Two absolute alarm levels (HA, LA)
- A control alarm band (CA)
- Manual positive valve positioning (MPP)

Each loop can be switched on or off individually, whereas depending on the status, either the DV or the valve can be stepped by various preset amounts. Usually, only the *on/off* feature applies to logic addresses, although MV or some other function may be used to relay data. A set of reference sheets is available to act as a directory to panel addresses (Table 2).

Fig. 6—Operator's panel facia

Other features of the panel include a digital clock, alarm lamp and buzzer, keyswitch protection on most functions, a *type displays* button and selection of up to 30 logs. Switching to computer control can be effected only through the panel; manual switching to standby control can also be achieved in the event of an emergency. There is a *not accept* lamp that is illuminated in the event that some panel action offends a wide range of safety checks.

Management information

A comprehensive range of operating records and efficiency reports was developed and designed to eliminate the need for other records for this particular machine. Typical of these are—

The papermaker's log A list of instantaneous measured values of all variables of interest to the papermaker, with those out of specification typed in red (Fig. 7).

TABLE 2—TYPICAL PANEL REFERENCE SHEET

Code	Title	Normal DV range	Scale	Units	Alarm group	Log code	Slave loops	Master loops	Notes
S.0 0.	Rough chest level	0–136	—1	Inches	1	0			
S.0 1.	Rough chest consistency	3–6	−2	Per cent	2	1		S.23	
S.0 2.	Broke chest consistency	3–6	−2	Per cent	2	2		S.28	
S.0 3.	Stock flow to proportioner	0–1 000	0	gal/min	2	3		S.32	
S.0 4.	Broke flow to proportioner	0–400	0	gal/min	2	4		S.29	
S.0 5.	Alum flow to proportioner	0–100	−1	gal/h	2	5		S.30	
S.0 6.	Size flow to mixing pump	0–200	−1	gal/h	3	6		S.31	
S.0 7.	Second fine chest level	0–196	−1	Inches	1	7			Overflow at 194.5
S.0 8.	Fine chest consistency	3–6	−2	Per cent	1	8		S.34	
S.2 0.	Hydrapulper water	0–15 000	+1	Gallons		21			
S.2 1.	Hydrapulper clay (cor.)	0–1 000	0	Gallons		22		S.22	1 box = 100 gal
S.2 2.	Hydrapulper clay	0–1 000	0	Gallons		23	S.21		
S.2 3.	Rough chest cons. comp.		−2	Per cent		24	S.01		
S.2 4.	First Shartle power	0–400	−1	kW	2	25		S.25.S.39	
S.2 5.	First Shartle power/flow ratio		−2	kW/gal/min		26	S.24		
S.2 6.	Second Shartle power	0–400	−1	kW	2	27		S.27.S.40	
S.2 7.	Second Shartle power/flow ratio		−2	kW/gal/min		28	S.26		
S.2 8.	Broke cons. compensation		−2	Per cent		29	S.02		
S.2 9.	Broke ratio	0–50	−1	Per cent		30	S.04	S.33	

PAPERMKR LOG 18/07/69 1205

MSF 5613E-1 GPM	STK 7395E-2 %	BRK 2005E-2 %	ALM -5000E-2 GPM	CLY 5534E-3 GPM	SZE 7875E-2 GPM
RF1 5000E-3 UNIT	RF2 5000E-3 UNIT	RF3 4825E-3 UNIT	RF4 1251E-2 UNIT	BKL 7212E-2 IN	BKC 6659E-3 %
SKC 2077E-3 %	HXC 4254E-3 %	FLC 1000E-3 %	FFX 3500E-4 UNIT	JET 7390E-1 FPM	TMP 1863E-2 DEGC
RH 5020E-3 UNIT	DRM 1230E+0 GPM	CHK -7500E-3 [illegible]	D01 3110E-3 %	D02 1520E-3 %	D03 -1150E-3 %
D05 -5100E-4 %	D06 -2600E-4 %	D07 1690E-3 %	D08 -1260E-3 %	D09 1300E-4 %	D10 2100E-4 %
SPD 8371E-1 FPM	SFM 2622E+1 PPM	PS1 7931E-3 PSI	PS2 6947E-3 PSI	PS3 7697E-3 PSI	PS4 1598E-2 PSI
PS5 1599E-2 PSI	FCC %	SBW 1951E-4 GSM	BSP 7180E-2 GSM	MSP 4662E-3 %	BWL 7622E-2 GSM
MDL 4122E-3 %	DKL 1585E-1 IN				

Fig. 7—Typical papermaker's log

The engineer's log A full list of current status, set point, measured value and control equation constants stored for all control loops (Fig. 8).

ENGINEER LOG 18/07/69 1241

CODE	STATE	MV	DV	MCDV	HIAL	LOAL	COAL	UNIT	PB	ITC	DTC
S00	AUTO	8664E-2	8640E-2	2004E-2	1350E-1	4898E-2	8040E-3	IN	1250E-2	1000E-1	1250E-4
S01	AUTO	3428E-3	3410E-3	1285E-3	5288E-3	1868E-3	1262E-3	%	2000E-1	2000E-2	1250E-4
S02	AUTO	6777E-3	6708E-3	1019E-3	1001E-2	1008E-3	3009E-3	%	3000E-1	2000E-2	1250E-4
S03	AUTO	3253E-1	3225E-1	1009E-1	8000E-1	1000E-1	6000E-2	GPM	1000E-1	5000E-3	1250E-4
S04	AUTO	1128E-1	1134E-1	4038E-2	3000E-1	5000E-2	4025E-2	GPM	1500E-1	1000E-2	1250E-4
S05	MAN	-5006E-2	1001E-1	2019E-2	2000E-1	-1001E-1	8000E-3	GPM	2500E-1	1000E-2	1250E-4
S06	MAN	7884E-2	1461E-1	3028E-2	2499E-1	-2002E-1	1803E-2	GPM	1500E-1	1000E-2	1250E-4
S07	AUTO	1941E-1	1942E-1	3024E-2	2000E-1	6998E-2	8004E-2	IN	1500E-2	1200E-1	1250E-4
S08	AUTO	3014E-3	3009E-3	1019E-3	5999E-3	2009E-3	1008E-3	%	2000E-1	2000E-2	1250E-4
F00	AUTO	1634E-2	1631E-2	1003E-1	8000E-2	5000E-3	2009E-2	GPM	6000E-2	2000E-3	1250E-4
F01	AUTO	5623E-1	5625E-1	2020E-1	9000E-1	3500E-1	4000E-2	GPM	2000E-1	2500E-2	1250E-4
F02	AUTO	5031E-3	5009E-3	1019E-3	8009E-3	3008E-3	5070E-4	UNIT	5000E-2	1500E-1	1250E-4
F03	MAN	1046E-2	1114E-2	5050E-4	1400E-2	8495E-3	2005E-3	IN	5000E-2	5000E-2	1250E-4
F04	AUTO	3131E-2	3108E-2	5184E-3	5000E-2	1999E-2	4091E-3	IN	2500E-2	4000E-2	1250E-4
D00	AUTO	7931E-3	8029E-3	1002E-2	3500E-2	9956E-4	3995E-3	PSI	5000E-3	1000E-2	1250E-4
D01	AUTO	6862E-3	7009E-3	1002E-2	3500E-2	9956E-4	3995E-3	PSI	1000E-2	1000E-2	1250E-4
D02	AUTO	7712E-3	7743E-3	1002E-2	3500E-2	9956E-4	3995E-3	PSI	5000E-3	1000E-2	1250E-4
D03	AUTO	1596E-2	1601E-2	1002E-2	3000E-2	5994E-3	3995E-3	PSI	5000E-3	7500E-3	1250E-4
D04	AUTO	1598E-2	1601E-2	1002E-2	3000E-2	1001E-2	3995E-3	PSI	2000E-3	1000E-2	1250E-4
D05	MAN							UNIT			
D06	MAN							UNIT			
S20	MAN		1300E+1	1000E+1				UNIT		2000E-5	1000E-5
S21	MAN		3000E-1	1000E-1				UNIT		3333E-4	1000E-5
S22	MAN		1000E+0	2000E-1				UNIT	1000E-3	1000E-3	
S23	MAN	5000E-4	3009E-3	2009E-3				%			
S24	MAN	2218E-1	2400E-1	5010E-2	3000E-1		2000E-2	UNIT	3000E-1	5000E-3	1000E-3
S25		0E+0	5000E-3	2025E-3				UNIT			
S26	MAN	2129E-1	2300E-1	1003E-1	3000E-1		2000E-2	UNIT	2500E-1	5000E-3	1000E-3
S27		0E+0	5000E-3	2000E-3				UNIT			
S28	MAN	5000E-4	5009E-3	2028E-3				%			
S29	AUTO	2600E-2	2605E-2	1000E-2				%			
S30		-1935E-4	1001E-3	1002E-3				%			
S31		2338E-4	1010E-4	2020E-4				%			
S32	AUTO	1180E-1	1181E-1	2028E-2	1800E-1	3000E-2	1008E-2	IN	7500E-3	1200E-1	1250E-4
S33	MAN	4554E-2	9000E-2	2016E-2	1500E-1	3000E-2	1008E-2	IN	7500E-3	1200E-1	1250E-4
S34	MAN	5000E-4	3999E-3	1018E-3				%			
S35	AUTO	2225E-1	2250E-1	1003E-1	3000E-1		2000E-2	UNIT	2500E-1	5000E-3	1000E-3
S36	MAN	3000E-4	4825E-3	2025E-3				UNIT			
S37	AUTO	1459E-1	1450E-1	1003E-1	3000E-1		2000E-2	UNIT	1000E-2	5000E-3	1001E-3

Fig. 8—Typical engineer's log

The shift or grade summary A summary of the average basis weight, machine speed and deckle, calculated production, a broke analysis and the average values and standard deviations for numerous process variables over the last shift or making.

Profile logs Typewriter plots of the cross-machine values of basis weight, moisture content and slice position (Fig. 9).

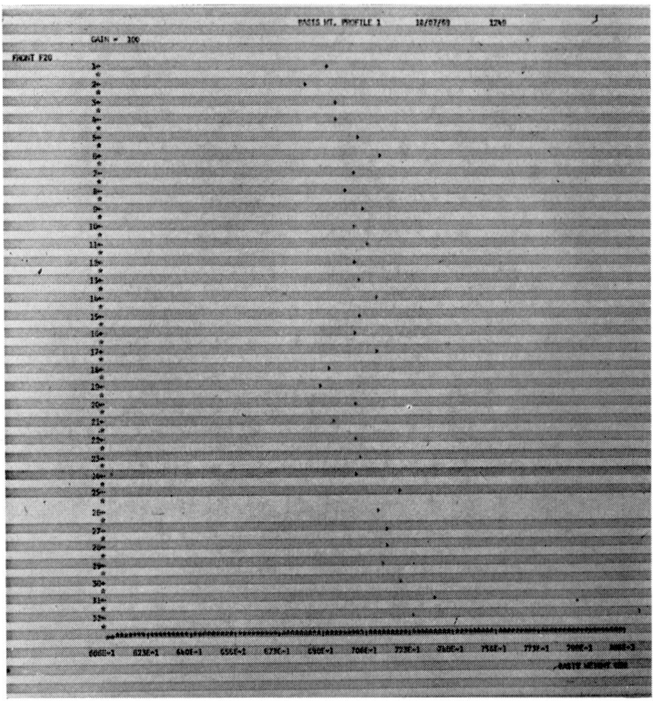

Fig. 9—Basis weight profile log

Trend logs Typewriter plots of up to four process variables at any time, sampled at one of four sampling intervals and containing the most recent 32 sampled values (Fig. 10).

Operator training

A programme for operator training was drawn up that, because of the lengthy time scale involved, had to be carefully planned in order to maintain interest.

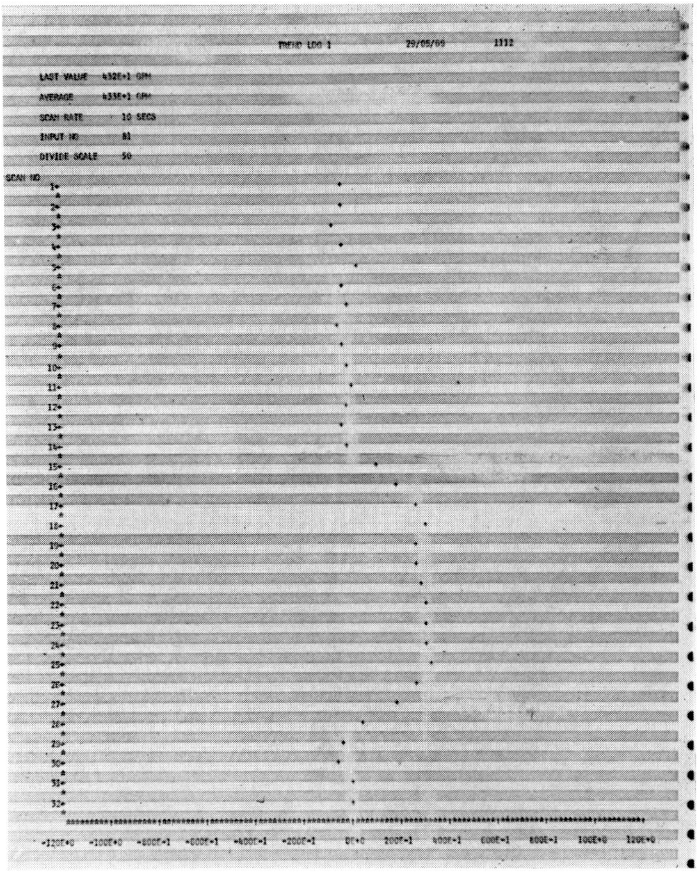

Fig. 10—Typical trend log

An operator's manual written by the systems team was presented in the first instance to the operators and supervisors, section by section, at a series of lectures. In general, these were given by the mill training officer. The sequence of 15 lectures was—

1. Introduction to a typical control loop.
2. Control room facilities.
3. The information and data system.
4. Simple on/off and manual controls.
5. Special and emergency routines.
6. Direct digital control.

7. Control of the vertical pulper and the proportioning system.
8. Control of consistency and the refiners.
9. Control of the flow box and speed and draw.
10. Control of basis weight and moisture content.
11. Manual standby facilities.
12. Simple standby control.
13. Bumpless standby control.
14. Start-up and shutdown.
15. Fault procedures.

Each lecture was given six times to allow full coverage from each shift and typical attendance was 12 men per lecture.

The programme of lectures was commenced in March 1968 and continued through to June. Following this, the manual was modified in light of comment and response. A copy of the final version was issued to each senior operative and supervisor during October-November, when a programme of panel instruction and practice was also carried through by the systems team. The manual is not designed for minute-to-minute use, for which reference sheets are of more value, but as background information, During December, detailed on-line instruction in the use of the standby system was given. Since January 1969, the machine operator has been responsible to some extent or other for the computer/plant system. Planned instruction continued through to June with full shift support being given by the systems team and extra cover by production personnel when necessary. It is anticipated that retraining and updating of the operators' understanding will be necessary for some considerable time, especially while changes or extensions to the computer system are being made.

Special routines

A range of special routines was developed aimed at providing a wide range of system checks. The group termed *Emergency routines* was designed to carry out some sequenced routine in the event of a dangerous or unusual condition arising on the plant such as certain stock valve closures, wire shuts and paper breaks. Others are more concerned with possible defects in the computer hardware or software. These latter routines were designed to be extremely searching and are in-line with the fail safe philosophy adapted for the project.

Instrumentation and cabling

The systems team joined with the instrument engineer in the selection and siting of instrumentation. In general, a policy of standardisation has been adopted. The task of instrument and cable scheduling was taken on by the systems team, together with the responsibility for ensuring that correct calibrations are maintained. Advice was obtained from the computer manufacturer on best cabling practice, as mistakes here are likely to be especially difficult to clear up.

All inputs and outputs to the plant are linked through a marshalling cubicle situated in the computer room. Each line is capable of being broken at this cubicle for the purposes of testing and maintaining plant safety. All inputs are capable of being relayed to a multi-point recorder also situated in the cubicle.

The control system

THE control system is based on the use of a standard DDC program used in conjunction with special programs written for control functions requiring a more complex approach than simple DDC allows.

The No. 16 machine control system comprises 106 control loops, only 19 of which are simple DDC. In addition, about 50 other measured variables are fed into the computer for purposes of logging and later phases and there is the ability to switch in excess of 30 pumps and actuators.

The DDC program calculates control outputs in a form suitable for output to the interface equipment based upon a two-term or three-term incremental control algorithm of the form—

$$\Delta VP = KA_1(\theta_n - \theta_{n-1}) + A_2\theta_n\tau + \frac{A_3(\theta_n - 2\theta_{n-1} + \theta_{n-2})}{\tau}$$

where ΔVP is the valve position increment,
suffix n refers to the nth sampling interval,
K is a function of the gain,
A_1, A_2, A_3 are constants and
τ is the sampling interval.

All changes to the control constants may be effected as keyswitch-protected options through the operator's console. More fundamental changes, such as the selection of control terms, are made on-line through the computer console via the block of storage associated with each loop. The program accommodates cascade control loops, prevents integral saturation and employs both input and output filters. Associated software enables each measured value to be checked for high and low alarm levels, control alarm levels and has various manual action facilities.

The complex control loops come under the following nine headings.

Pulper control

The addition of water and clay through pulse meters is controlled so as to present uniform stock to the first consistency controller and to the refiners (Fig. 11).

The operator is responsible for the addition of a constant total weight of fibre at each charge and the computer supervises preset water and clay additions initiated by a *batch commence* button. Once the additions are completed, the computer shuts the water and clay valves and illuminates a *batch complete* lamp on the pulper panel. For purposes of wash-down, the operator has local manual control over the water valve. This overriding control is available, however, only within each batch and the full preset quantity must be taken. There is a density meter situated in the clay ring-main and this is used to facilitate computer corrections for density variations by adjusting the DV of the clay addition loop.

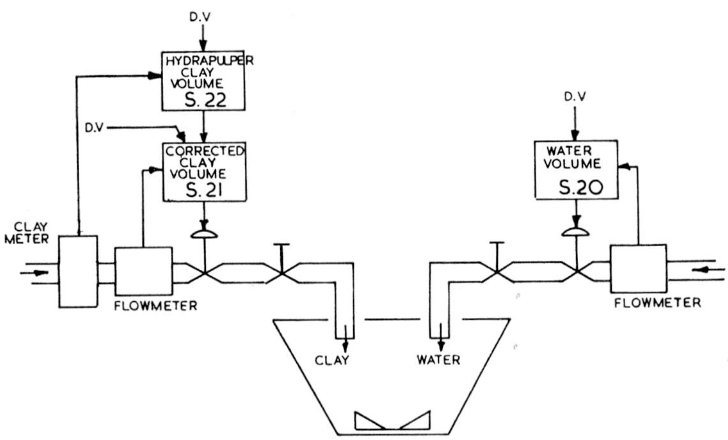

Fig. 11—Hydrapulper complex loop

Compensated consistency control

The simple consistency DDC loops, which are located on the rough stock, broke, fine stock and at the stuffgate are compensated for fluctuations in stock flow and temperature (Fig. 12).

The computer allows for easy adjustment of constants for calibration of on-line measurements and of the compensation curves, each of which is defined by an origin, flow or temperature increment and eighth incremental corrective factors over the range of flow or temperature concerned. The compensation is applied to the set point of the consistency loop.

Refiner control

The refiners can be controlled (Fig. 13) so as to maintain either the ratio *refiner load/stock flow* or the temperature differential across the refiners.

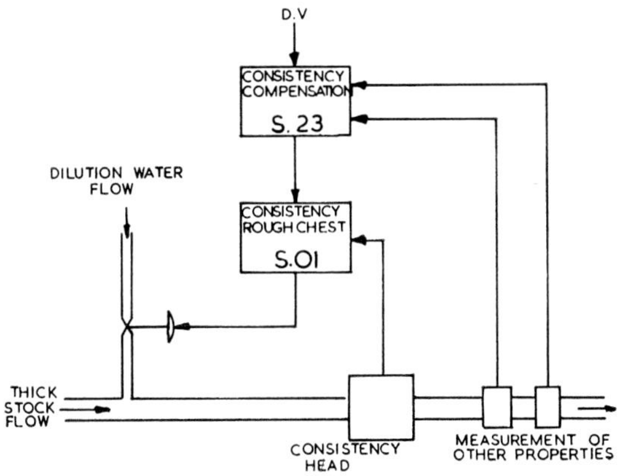

Fig. 12—Compensated consistency complex control loop

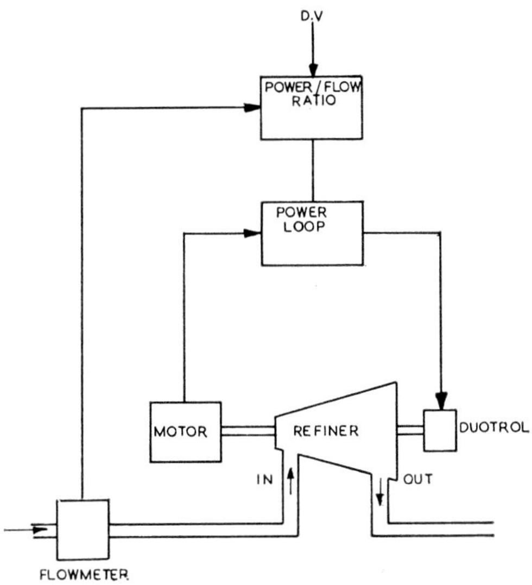

Fig. 13—Refiner complex loop

The control action takes the form of *raise/lower* signals to automatic plug positioners. Although these are equipped with suitable limit devices, the computer system critically surveys each output and checks for safety. Automatic back-out by computer is an important facility.

Proportional control

The flows of stock, broke and alum into the fine chest and size at the mixing pumps are controlled to obtain proportions specified by the operator (Fig. 14).

The chest level cascades to the stock flow and the broke and alum components are, in turn, ratio-controlled to this. Size is ratio-controlled to the stuffgate flow. The simple DDC loops on the various flows can of course be maintained independently of the complex loop.

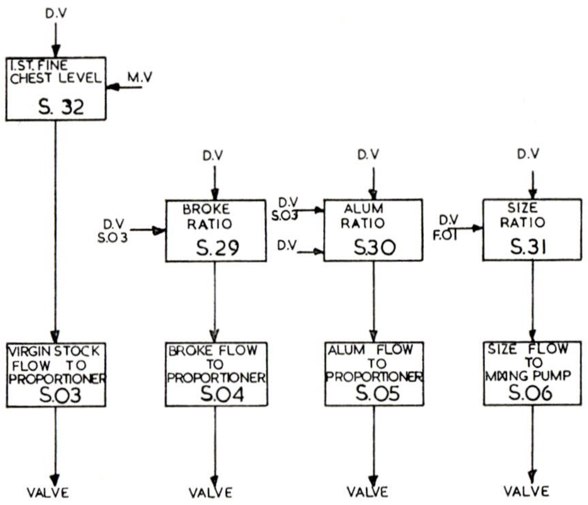

Fig. 14—Proportioning system complex loop

Flow box control

The closed, pressure flow box is controlled to enable the operative to adjust the stock consistency and efflux ratio independently The flow box is equipped with an adjustable slice (Fig. 15).

Various control configuration options are available and can be easily implemented by adjusting the complex loop-addressing format. It is possible, using simple DDC to supply set points to the head, level and slice controls alone.

Cross-machine slice gap adjustment

The cross-machine slice profile is adjustable through 31 individual adjuster motors, each capable of receiving *raise* and *lower* signals as required. There is a built-in safety check being continuously carried out, which prevents an excessive adjustment between two adjacent units. A display of the actual profile is available as a typewriter plot and plans are being prepared for closed loop operation in a later phase.

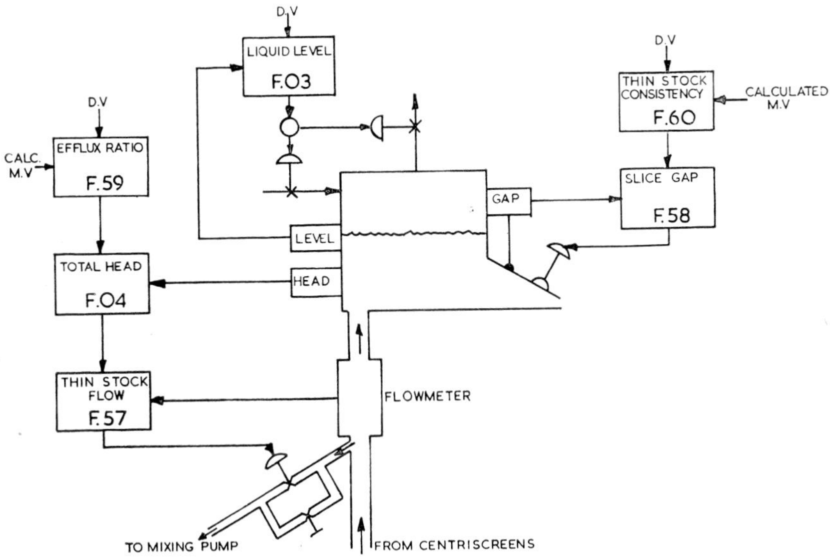

Fig. 15—Flow box complex loop

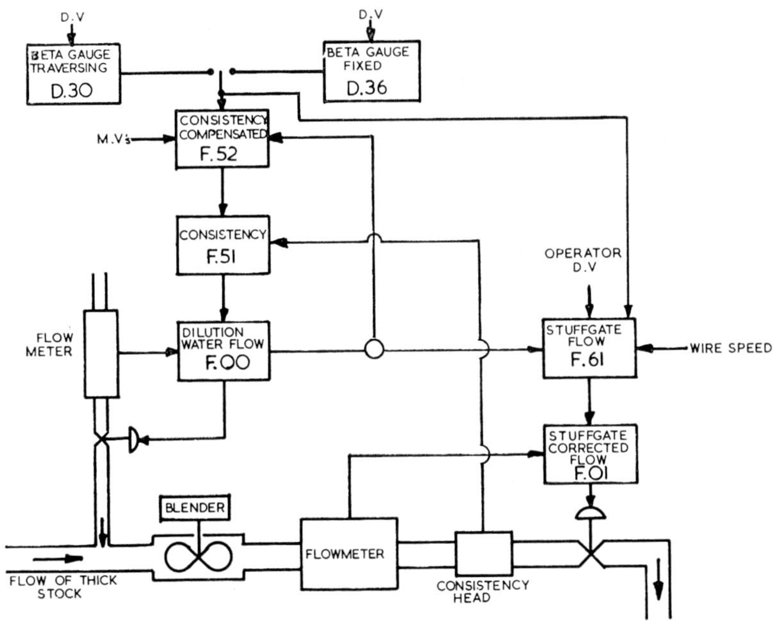

Fig. 16—Basis weight control complex loop

Basis weight control

Basis weight is measured by sensor heads at both the reel and before the size press. Although it is normal for both gauges to be traversing, control is available for the fixed head condition. In this case, the substance measurement is presented to the control system as a filtered, mean value and, for the purposes of traversing head control, as a traverse average (Fig. 16).

The trimming control action is through adjustments to the set point of the stuffgate consistency control loop, which cascades to a simple dilution water flow loop. The flow of thick stock at the stuffgate is initially set up for the current grade and is automatically tied to machine speed. If a change in basis weight DV is made, then the stuffgate flow set point is adjusted accordingly. Similarly, it is adjusted if the consistency control moves to a preset control range limit. In this event, an equal percentage adjustment is made to the consistency control set points.

An automatic calculation is done to check beta-gauge calibration. This is obtained from the weight of the steel core, weight of the full reel, sheet width and a precise measure of the length of paper on the reel. There are a series of checks carried out to ensure that the calibration is reliable. In addition, there is a laboratory data input channel that enables the latest laboratory value and the internally corrected value of, basis weight to be compared. If these differ markedly, a warning message is printed out.

Moisture control

The sheet moisture content is measured at both the reel and before the size press and is presented to the control system as a filtered, mean value. Moisture control is effected by means of adjusting the set points of the machine dryer sections (Fig. 17).

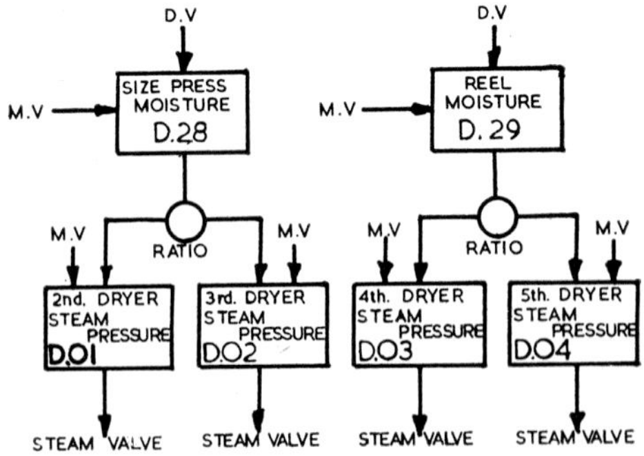

Fig. 17—Moisture control complex loop

The main dryers are in three sections and size press moisture control is through the second and third of these. The two sections are also ratio-controlled, thereby introducing an additional degree of flexibility. The two sections of the after-dryers control reel moisture in a similar manner.

Speed and draw control

The machine speed is controlled by a simple DDC loop operating on the main drive motors, whereas individual sectional draws, monitored by pulse interrupts, are controlled to preset values by means of adjustments to cone and pulley units. Sectional speeds are available as measured values (Fig. 18).

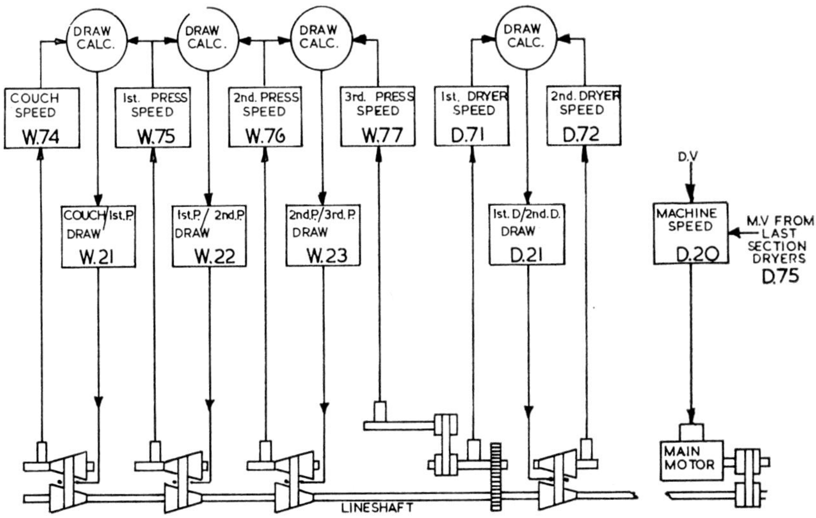

Fig. 18—Speed and draw complex loop

The speed of the first main dryer section, which acts as the master speed section, remains in constant ratio to that of the main lineshaft. On either side of the master, there is sectional draw control, with adjustments being made to that section furthest from the master. Under normal circumstances, the response of this system is slow and it acts as a trimming control. The local manual adjustments have been retained, but when the automatic system is working they only have a short-term effect. If an emergency draw condition is detected, for which the trimming action would be too slow, there is a facility for locking out the normal draw control and introducing a form of feedforward control that resets the entire speed and draw system.

Phase one

ONE problem during the early planning stage was the prediction of the duration of phase 1. It was possible to obtain estimates for hardware testing and proving, program testing and for the initial system commissioning, but the time necessary for the tuning of the system seemed more difficult to predict. In fact, phase 1 was planned to last 12 months from the date of delivery of the computer hardware, expected during August 1967.

The first major delay to the project arose, however, through late delivery of the computer. Arriving during the first week in January 1968, it was some 18 weeks late. This therefore largely invalidated the need for the PERT analysis, which otherwise had been extremely valuable, once it was generally accepted by the various groups associated with the project. The software was ready well in advance of delivery, but had to wait on the hardware before it could be tested. This need not have been the case had a suitable simulator been available at the supplier's site.

All further delays in the project have largely been due to hardware faults, which were particularly common during the period immediately after initial hardware commissioning. At the time of writing, availability since completion of hardware testing has been only about 70 per cent, but has fortuntately maintained an acceptable level over the last six months since the major sources of difficulty were overcome. Software has caused little or no delay. Systems work has gone according to plan, except for a period of delay caused by a total changeover of personnel on the supplier's side. This naturally involved the project in some loss of continuity and momentum. One other delay arose through the need for an additional reassembly of programs as a result of limited working space.

In general, hardware faults have first shown themselves intermittently and have therefore been difficult to isolate and repair. They seldom took long to put right once they were located, but the delays they brought about were damaging to both progress and morale.

Analog inputs and scanning difficulties have been responsible for many of the delays and have resulted in a modification to the low level amplifier and a number of grounding alterations. The delays were greater as a result of the unfortunate failure of the analog-to-digital converter, normally accepted to be a reliable piece of equipment.

In all, some 6½ weeks have been lost through arithmetic unit faults, but they have not been apparent for some considerable time and it is hoped that they were largely the teething problems of a new system. It seems very likely that their frequency was unduly exaggerated by the relatively high ambient temperatures common in the control room before the present cooling system was introduced.

Parity errors have accounted for a large part of the total delay. In general, these have been in data areas and could quite safely have been ignored. The computer system in any case checks for such errors and we have chosen that this check should stop the computer system, thereby enabling some action to be taken. It appears now that they are becoming increasingly less of a problem; unfortunately, the reason for this is not completely clear.

Comment on hardware faults would not be complete without reference to the high failure rates of peripherals, especially typewriters, which alone have accounted for several lost weeks of systems work.

Computer hardware, interface and peripheral equipment installation and and testing commenced immediately on delivery. During this period, the systems team commenced testing of the standby system in readiness for the phasing out of conventional controllers. Plant inputs and outputs were made available to the computer through the marshalling cubicle and all lines were checked for continuity. All this work terminated in early April 1968 with a week of testing, using a series of standard routines designed to check out fully the complete hardware system.

On completion of this first stage of phase 1, the programmers commenced their task of testing and initiating the software, starting with the basic organisational programs and following up with the various system programs. This work facilitated the systematic testing of all plant inputs, both digital and analog and involved the early proving of all basic alarm, console and information logging programs. Once these operated in a satisfactory manner, all inputs were linked permanently to the computer, measured values were accurately set up and the computer's basic data system was brought into full-time use. This availability was fitted into the operator training programme, which was proceeding in parallel.

Thereafter, specific functional programs (of which 18 were written) continued to be brought into use one at a time and, once proved by the programmers, they were commissioned by the systems team. When appropriate, checks were made on the output lines at the marshalling cubicle before they were linked through to the plant. One of the first programs handled was that one associated with the standby switching, which was essential for the well being of the production unit during later stages. This was followed by the program controlling the numerous special routines. The DDC package was then made available and each loop was tested, brought on-line and tuned individually. At this stage, the standby system (which had been fully commissioned) was totally phased in and became the first line of back-up to computer control. Training of the operators in the use of the standby system was immediately completed.

Numerous other activities were of course going on during these early commissioning stages. Important among these were instrument and cable scheduling, instrument proving and calibration, program modifications. Preliminary testing of the complex loops was also being carried out and serious effort was directed to these during December. The original phase 0 estimate for their implementation was 20 weeks.

Comment

THE commissioning of the complex control loops is nearing completion 18 months after delivery and coincides with the writing of this paper in June 1969. In light of the significant delays experienced in the early stages of phase 1, the progress of systems work has been quite good. The project team is particularly pleased by the degree to which the computer system has been accepted by mill personnel, particularly the machine operators.

Although practically every part of the system as designed in phase 0 has been seen to work and a large part is freely available for use, there are areas of disappointment. Speed and draw control testing has been difficult to fit into the production schedule, slice profile instrumentation has failed to keep working and compensation data has proved difficult to obtain. Altogether, the expected performance of the computer system is not immediately forthcoming and work is continuing on the tuning of the individual control areas. It is hoped that this will be effective by the end of the year and that the benefit to the process will be self-evident by that time.

The whole project has of course been a learning exercise for Bowaters and, now that higher system availability can be expected, this process must continue, although it may appear to slow down progress. The present stage of the project has been reached with only about 10 man-years of Bowater effort, plus a similar amount from the manufacturer; the Bowater team has recently been enlarged, particular effort is being put into the task of tuning and day-to-day responsibility for running the papermachine through the computer rests with the operatives. It seems likely that the tuning process may continue for some considerable time, especially as the long-term benefits of the computer, as a learning tool and means of freeing the operative from tedious control duties, become more significant.

A number of lessons have come out of the Sittingbourne exercise that will not be forgotten in future projects. The value of energetic project management cannot be overemphasised. Computer installations are all-embracing in that they inevitably involve all departments within the production unit concerned. Therefore, particular care must be put to organisation and information distribution if the whole project is not to founder through petty jealousies or local incompetence. Worthy of mention again, at this point, is PERT or critical path analysis, which is invaluable to a computer project.

A planned programme of operator training together with a reliable and relatively comprehensive standby system have proven to be useful in maintaining interest during early days and a source of confidence throughout.

The value of operating records and reports can be greatly overstressed. They are very heavy on computer storage and, except in the concept of the fully automated mill, should be reduced to a useful minimum. The paper industry should be striving for the fully automated mill.

The total dependence of a computer system on instrumentation has been confirmed. It is unwise therefore to commence a computer project, without this fact being generally accepted and the necessary action taken to ensure that good quality hardware and technical support will be available at all times.

It is intended that a full evaluation of the benefits of the computer system should be carried out at the right time. Bowaters have little doubt, however, that a suitably designed computer system will be an asset to each of our papermaking streams and this project has done much to clarify our thinking for the immediate plans for later phases at Sittingbourne and for at least one other machine in the U.K. group.

References

1. Stout, T. M., 'Economics of computers in process control': *Automation*, 1966, **13**, (10), 82–90; (12), 91–97
2. Cyprus, H. D., Morley, R. L. and Hitchin, D. F., 'Standby to a computer control system': *Paper Tech.*, 1969, **10** (5), T93–T101

Prepared discussion contributions

Some Results from the Computer Installation at Empire Paper Mills

G. D. MADELEY, Reed Group Ltd., Empire Paper Mills, Greenhithe, Kent

---

MY PURPOSE is to give a brief description of what we have achieved with our computer installation and to draw some comparisons in philosophy with the paper by Cyprus & Attwood.

The papermachine makes fine writing and printing papers at speeds up to 1 000 ft/min. Most of the papers are made with a dandy roll and a fair percentage are watermarked. The computer is a GEC Series 90, Model 2, with a 16 K, 12 bit word core store. The addition time is 3·5μs. There is no back-up store, but we have recently decided to double the core size. We use DDC on practically all loops.

The primary function of the installation is to control the papermachine from head box to reel-up, but future developments will take us back into the stock preparation, on into the finishing area.

The computer was installed in October 1967. By January 1968, thick stock flow was under computer control. By October 1968, substance was under continuous computer control and has been ever since. We studied the process using pseudo-random binary sequence trials and found that the system could be represented by a time lag and a first order exponential rise. Before the computer installation, the substance had a coefficient of variation of 3·06 per cent. Over the period April-June this year, this variation had been reduced to 1·89 per cent. This variation includes both between makings and within makings. The average for within makings is 1·35 per cent and on some orders is down to 0·5 per cent. The main factors contributing to the rest of the variation is partly between making variations and partly that the machineman will alter the computer set point on the basis of his weighed samples. Until we can use the reel weighing mechanism to update the substance gauge calibration, we are not in a strong position to stop this practice.

One of the major incentives to come out of the feasibility study was moisture

Under the chairmanship of Dr D. B. Brewster

control. This was tried in October 1968 and the moisture variations was reduced to $\pm\frac{1}{2}$ point. On many occasions during our experiments, the machine speed was lower than was necessary, not because of papermaking problems or lack of drying capacity. As a result of this, we tried a technique of controlling the machine speed to the maximum drying capacity. We were able to achieve appreciable increases in speed—in one case, 100 ft/min. Unfortunately, we have been unable to continue with moisture control, because of a series of problems that developed with the moisture meter. These we think are now cured and we are back on moisture control.

We also use the computer to control grade changes. Essentially, we follow the practice that the machineman used to do. The machineman feeds the computer with the data of new substance and speed required; the computer calculates the new required settings based on the existing running conditions. When he is ready, the machineman presses the *enter* button and the machine ramps to the new set points. We are limited in the slope of the ramp by the rate at which the machine changes speed. On our machine, the number of grade changes to which this is applicable is limited, as the majority involve dandy cleaning, dandy change, backwater change or wash-up. As a result, we see no need in practice for any greater sophistication.

The control room is mounted so that it overlooks the dry line. All the controls and information that the machineman may require is in this room. From here, he operates via the computer or, if necessary, via the standby controls. Although all the information is in the control room, the machineman in fact spends most of his time on the machine floor. He is completely confident that the computer will keep the machine under control in his absence.

After some initial instrument failures, we have gone to considerable trouble to check the inputs. Every input that is used for control is checked by the computer to make sure that it is reasonable—that the fuse has not blown or the reading is not too different from the previous one. If the computer finds that the input is outside set limits, then the appropriate loop is put to manual and an alarm sounded. The other loops are unaffected. The alarm continues until some corrective action is taken. By using this approach, the computer has been in continuous control of the machine, despite occasional failures elsewhere.

One of the factors that is influencing the current work is the supervisory structure in the mill. This has been traditionally a horizontal structure and so the common supervisor over the beaterman, machineman, dryerman and reelerman is the chief papermaker. We have always had excellent co-operation from the machineman, but it is not so good from the others. This is almost entirely because they tend to feel isolated. The facilities that they have to communicate with the computer have not turned out to be sufficient. As a result, we are having to increase the facilities available at the outstations beyond what was initially thought necessary. In retrospect, a more vertical structure would have made the computer operation much easier to put across. It has since been considered, but the problems of implementing it appear to outweigh the advantages.

Looking back over the project, one of the most prominent features is the role of instrumentation. We have had a number of problems with our instruments, which have caused serious delays in the progress of the project. The computer arrived while some of the instruments were being installed. Inevitably, the delays caused by the instruments gave a bad name to the computer, particularly to those people who watched the project critically and from a distance. Considerable time is necessary to ensure that the instruments are reliable before any attempt is made to use them with the computer. Without a computer, reliability is an abstract idea; with a computer, reliability is measured in sleepless nights and long weekends.

In conclusion, our computer project has had considerable successes and some great disappointments. Already we have beaten our target for substance control and with obvious potential for further advance. Moisture control has been a problem of instrumentation, but what we have seen is promising. Thick and thin stock and flow box level are under DDC. We have facilities for control of efflux ratio, but there is no incentive at present. Clay addition can be put under DDC and, when the new ash gauge is calibrated, sheet ash content will be controlled. We are more than ever convinced of the benefits of our computer.

On-line Development of Process Control Computer Programs

R. H. UNTHANK, Wiggins Teape Research & Development Ltd.

ALMOST without exception, process control computer systems have been to some extent development projects. Most have been applied to commercially operated production units, for which the improved control of the process provides adequate financial justification for the computer installation. In such cases, conflict arises between the use of the computer for the basic control of a continuously running process and the development of new programs. One solution is to provide a comprehensive conventional standby control system so that the computer can be taken off line at any time; alternatively, to limit the system to supervisory set point control. This approach is very expensive and there is a tendency for the computer to become purely a research tool, imposed upon a rather unwilling production unit. Another common approach is to use a relatively expensive computer system with backing store and full time-sharing facilities. This second method is justifiable when the process ocntrol computer is required to carry out a large amount of off-line work such as scheduling and invoicing, but is unnecessarily expensive if the computer's principal task is process control.

On the Wiggins Teape Argus 400 system at Dartford, a different approach was adopted. The standby system is little more than the absolute minimum required to run the machine during the extremely infrequent emergency repairs to the computer hardware. This arrangement has had the advantage that the process operators have rapidly come to rely upon the computer system on a full-time basis, but the system is essentially a development project in which new control and logging programs are frequently tested. The conflict for computer time between routine process control and development work has been largely overcome by means of a miniature 'background' operating system. This monitor system enables spare processor time, core store and peripherals to be made available for running development programs without interruption of the on-line control function. Its current uses include program editing, translation and test runs.

The Argus 400 has 24 K of core store, two teleprinters and a high-speed tape reader and tape punch. There are no hardware multiprogramming protection facilities or backing store so the hardware costs are very low for a system of its size. Most of the programs were specially written for the system in Assembly language (machine level code). The on-line system has continuous use of all process inputs and outputs and one teleprinter. The other teleprinter, the tape punch and reader are used only intermittently (for operator demanded process logs and at grade changes). Normally, under 50 per cent of the available processor time is used for on-line control, which is almost entirely DDC. The basic monitor progıam is part of the interrupt timing organiser. In use, it seizes control of the teleprinter, tape punch and reader from the on-line system. Monitor-controlled programs then have exlusive use of these peripherals, a 4 K block of core store and all free processor time. They also have access to all process inputs.

The monitor system is controlled by a single key-operated switch, which causes it to read and execute a directive punched on paper tape. There are three principal directives—

(*a*) Load the following program from paper tape to store.
(*b*) Run the program currently in store.
(*c*) Terminate use of the monitor and return control of the peripherals to the on-line system.

Absolute safety of the on-line system is achieved by using a special self-checking binary code for all program loading and producing tapes of only thoroughly tested programs. These include special logging programs, tape editing and copying routines, the Assembly language translator (tape to tape) and a specially modified program loader that incorporates an interpretative guard tracer for safe trial of untested programs. This will halt a test if an order affecting the on-line system is encountered.

The monitor system has virtually supplied the development team with a second Argus computer. Its implementation took place in the two months immediately following the commissioning of the basic control system. It has since been in constant use in the development of a tape editing program, data logging routines and production logging and grade change systems. There are currently four programmers working on system development. If each required the computer for two hours a day, the on-line time would be reduced by 30 per cent in the absence of the monitor system. Such a requirement would conflict directly with the need to build up operators' knowledge of and a confidence in the control system. Other benefits include the rapid practical training of new programmers and easy access to the computer for established people. Such a system is surely an essential part of any dedicated DDC computer control system in a production unit for which any further systems development is envisaged.

Discussion

Mr R. G. Nagro May I congratulate you on your paper and your computer system. You stated some estimates of the expected system payout. After having had some working experience, would you like to modify those or support them or is it still too early to tell?

Mr H. D. Cyprus I take it you are referring to the estimates shown in Fig. 1. I would not change them, except possibly to raise them somewhat. We have certainly not yet achieved those returns. This is partly reflected in the troubles we had last year; it also affects our lack of expertise, which is still being developed. I cannot see why we should not achieve those levels.

Harking back to my comments on performance quantity, I am not convinced we did not do ourselves, the management and the project a disservice by looking at it in this light—by isolating the return on improved regulation of basis weight, proportions of the constituents, etc., yet I cannot think of an alternative.

Mr J. Mardon My first point on Mr Cyprus's paper is my great difficulty in accepting the idea that one first chooses a computer, then undertakes the planning preparation phase. This seems to me a negation of what I might term the classical approach. First of all, the planning phase; then one should choose the pieces of equipment suitable for the specification when it is drawn up. I would appreciate Mr Cyprus's comments on that.

The second point is to say to Mr Nagro that the idea of having on-line access to the computer without in any way impairing its functionality for control has been taken into account in the design of the Bailey 855.

Mr Cyprus There seems to be a measure of misunderstanding. The planning and preparation phase that I referred to under the heading phase 0 was in fact detailed preparations and detailed planning in line with the decision to go ahead with this particular system. Of course, the company, working with more than one manufacturer, did a preliminary preparation and planning phase. This went on for some time before the decision to buy this particular system and it decided for us the general detail, for example, whether or not to buy a computer, whether an analog or digital, whether small, large or medium sized, etc. Phase 0 was detailed planning and detailed preparation, which must be done after deciding what equipment to use. Otherwise, I agree with your comments.

Mr W. D. Hoath Mr Cyprus mentioned that they had their own learning to go through in computer installation. What were the problems experienced? Were they hardware, software and how well did they keep within their budget costs? On the standby equipment, he is having to cope with several manufacturers. Was this a problem in itself?

Mr Cyprus With respect, I will not go into the detail on the first part of the question, because the answer lies fully in the paper. The troubles we had did not affect our costing, but I think the reason for this is maybe the way we organised the budget initially.

We had no great embarrassment because of different manufacturers. We know our instrument suppliers well and there were no delays through them. Although our interface and standby equipment was manufactured by one firm, it came as part of the supply from another firm in a satisfactorily phased manner. To be fair, the hardware was delivered late and therefore this may well have hidden problems that we might otherwise have had.

Mr M. I. MacLaurin Some of you may have heard me speak on the subject of computer project assessment earlier this year in New York.* Since that time, we have gained further experience in this area and, were I to speak again on the subject, I would be less assured of the practicability of the method I described, despite its theoretical merit.

The basic problem is not so much in comparing how well the papermachine performs now compared with the pre-computer period, but more in identifying how much of this improvement may properly be attributed to the effect of the computer.

Mr Cyprus I agree totally with you. It seems to me pointless to compare the 1965 performance (when we did our initial study) with running today. As I have suggested, this has no meaning. We must define a parameter, the performance quantity (which takes into account all those factors influencing performance) and monitor it continuously. Normally, it should show a logarithmic rise and a step to a higher level if some factor contributes significantly to the well being of the process. We have developed this growth curve for No. 16 machine for the last six years or so and will maintain it, hoping to see an effect attributable to the computer.

An alternative might be to take the computer off for a month, then put it back on, hoping to show the improvement. I consider this to be an impracticable proposition.

* *Tappi*, 1969, **52** (8), 1 480—1 483

AUTOMATION TRENDS IN PAPERMAKING SYSTEMS

M. A. KEYES and R. H. TEMPLE, Control Systems Division, Beloit Corporation, U.S.A.

Synopsis This paper discusses the current trends in the evolution of hardware for computer control and the implications of these trends for the papermaking industry. This discussion covers, in addition to computer main frames, the development of remote analog signal multiplexing, conversion and digital data transmission, the evolution of graphic displays as operator panel replacements and the elimination of long cable runs from instruments by the use of laser data links.

As an overview of the future, an indepth presentation is given of an integrated papermill computer control hierarchy (of the sort to be expected within five years). In this configuration, the actual process control is performed by a number of free-standing (that is, no special computer room), small (20 in $\times$ 20 in $\times$ 60 in) computers dedicated to the direct digital control of the process sub-units (for example, grinders, bleaching plant, head box, dryer section and coaters). These first level computers are subservient to a larger foreground/background computer, performing such functions as control parameter optimisation, management information reduction and presentation, as well as on-line optimum production scheduling.

This portion of the paper also includes a discussion of the control techniques and strategies that will be in common use at this date and the additional process variables that will be on computer control (gloss, air permeability, colour and caliper). A part of this discussion covers the future state of the art, which will be approaching the control of the distributed parameter aspects of the process (such as cross-machine moisture and basis weight control).

Introduction

ANALYSIS of current trends clearly indicates that the next few years will produce a maturing of control application philosophy in the pulp and paper industry. A major result of this will be the widespread acceptance and installation of computer-based control systems, because they offer an attractive return on invested capital, not because of the historical glamour motives.

In addition to consideration of the usual goals of higher production rates,

Under the chairmanship of Dr D. B. Brewster

improved product quality, lower reject rate, greater flexibility to meet changing requirements and so on, two increasingly important areas of need and, consequently, justification will come to the fore. The first of these is the result of the pressing current need for integrated management control and co-ordination of operations in conjunction with the upgrading and consolidation of personnel requirements.

The second and newer area originates from the increased emphasis placed on working conditions for operating personnel. Limits are being placed by regulatory bodies on allowable long-term noise and atmospheric contaminant levels for plant operating areas. In many cases, it will prove easier to remove the personnel from the process environment through automation and centralised control rather than to render the entire plant environment acceptable for long-term operator residence.

In the past, too many process control computers suffered from reverse justification (that is, install this powerful tool, then look for something for it to do). This approach rejected all sound project planning principles such as careful definition of goals, planning and enforcement of schedules and continuing critical evaluation. The result in a great many cases has been either explicit failure or an internal erosion of confidence and lack of measureable achievement.

Increasingly, control computers are losing the glamour and special treatment of the prima donna and are being recognised as a powerful production tool that must be controlled and utilised by production personnel to achieve the desired levels of operating success. In the end, the 'success' of a project can, after all, be measured only in terms of its net effect on plant operations.

In this environment of increasing acceptance on a rational basis, two outstanding characteristics will be particularly significant. One of these is the fruition of the use of hierarchical control in our industry. A comparative multitude of small computers will be utilised to control individual process areas and units. The use of advanced digital communication techniques will allow co-ordination of these subservient systems with a more powerful central unit.

The second important facet of future advanced control operations of this type will be the increasing relative importance of software, services and application engineering. This is due to the simultaneous realisation by paper users of the potential high hidden costs of software and applications (some pulp and paper users have utilised teams of 5 or 10, even 50 men and have yet met with only moderate project success) and the decreasing costs of hardware. This effect was illustrated in a recent Forbes Survey[(1)] covering data processing applications. The results of this survey are depicted in Fig. 1. As experienced industry people will concede, there is a good deal more effort in implementing, say, a continuous digester control system than there is to writing a payroll program and making it work smoothly. The result of this realisation will be an increased demand for packaged systems and turnkey applications.

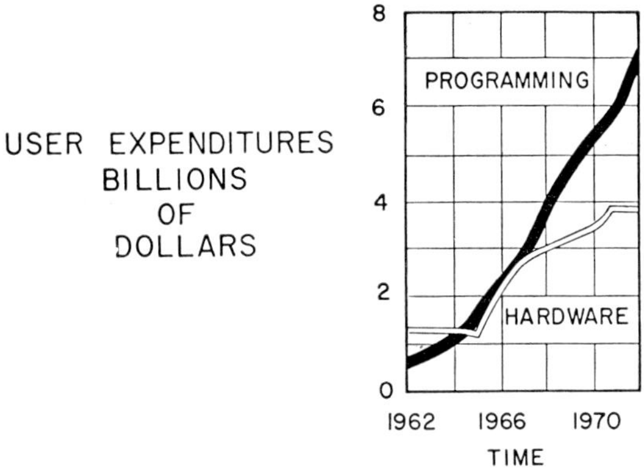

Fig. 1—Increasing relative software costs[1]

Hardware development

TO IMPLEMENT modern control technology, the digital computer is by far the most flexible tool available. Many times, it provides the only feasible approach to implementation of an advanced control system. Thus, the much-heralded advent of *medium-scale integration* (MSI) and *large-scale integration* (LSI) digital electronics are having and will have a tremendous impact on control systems. Since, through this circuit technology, electronic components are becoming much less expensive, smaller and more powerful, three basic new types of computer for process control applications are evolving. The first type will be physically smaller than present-day machines, will cost about the same, but will have from four to five times the capability. The second type will have the same capacity as current machines, but will be much smaller and will cost much less. An indication of this direction is seen in some of the present-day aerospace computers, one of which weighs 10 lb, measures 4 in × 4 in × 9 in, with 7 500 words of memory and a 38 instruction repertoire.[2]

The third type has been termed the micro-computer or the 'computer-on-a-chip'.[3] This digital subsystem will functionally replace the current electronic controller[4] with a physically smaller, lower cost, more capable unit. The typical micro-controller will be capable of controlling 4–8 DDC loops and, in a hermetically sealed box of approximately 4 in × 4 in × 10 in, will contain all the necessary logic circuitry, memory, analog/digital (A/D), digital/analog (D/A) and computer interface electronics. This micro-controller will implement a form of the general digital compensation transfer function and will operate asynchronously with respect to any other 'supervisory' digital computers, receiving data from these units only when it is desired to change a set point or a control algorithm parameter. In addition to a projected one for one cost advantage over analog systems, there are other important incentives in using this type of approach such as reduced instrument wiring costs, reliable drift-free operation and simplified master control computer software.

Implication for computer control systems

A SIGNIFICANT implication of these developments for the paper industry is that a given degree of control complexity and scope will be much less expensive in basic hardware. This has both good and bad aspects. It implies that the capabilities of the computer hardware will be much less of a barrier in the future and larger tasks will be feasible. Yet the systems aspect of the work will be no simpler and, in particular, the software and hardware interfacing aspects will continue to be complex. Many new application software packages will be available for purchase or lease from both computer vendors and systems houses. This gain will, however, be offset to a degree by the proliferation of small computer vendors and the specialised or unique characteristics of each installation and computer system.[5–7]

Specialised hardware is being developed to handle certain relatively well-defined functions that will simplify the interfacing problems to a certain extent as discussed above for the micro-controller. In other cases, the control computations will still be performed in a central computer. The digital/analog interface and encoder logic will inevitably move away from the computer mainframe, however. More digital transducers will be available than at present and analog units will more frequently be available with digital conversion built in. Local multiplexers will abound. Illustrating the technical feasibility of local multiplexing, conversion and encoding is the fact that a single MSI chip is available today that performs all the logic and control associated with an A/D converter for less than $100.00.[8]

Another example is a recently announced computer that operates from a *read only memory* (ROM), interfaces directly to a 120 V ac, signals and replaces interlocks, break detector systems and other relay control systems on an even cost basis.[9]

Packaging

ANOTHER major ramification of these developments will be their effect on the area of computer packaging and reliability. The hardware will be more compact allowing totally new packaging concepts. One result even now being offered in prototype is a hermetically sealed enclosure with a positive pressure inert gas atmosphere totally surrounding these new, small machines. The electronics are inherently immune to industrial temperatures and the packaging will render the machine proof against chemical and biological attack, since the process atmosphere is excluded. In addition, this packaging will obviate the need for special computer rooms allowing location close to the process under control. With ruggedisation and enclosures of this type high reliability is guaranteed, washdown hoses and forklift trucks not excepted. The first steps in this direction are indicated by several recent product announcements by major computer vendors.[10, 11] An artist's conception of one of our recent proposed configurations for applications in an area with a corrosive atmosphere is shown in Fig. 2.

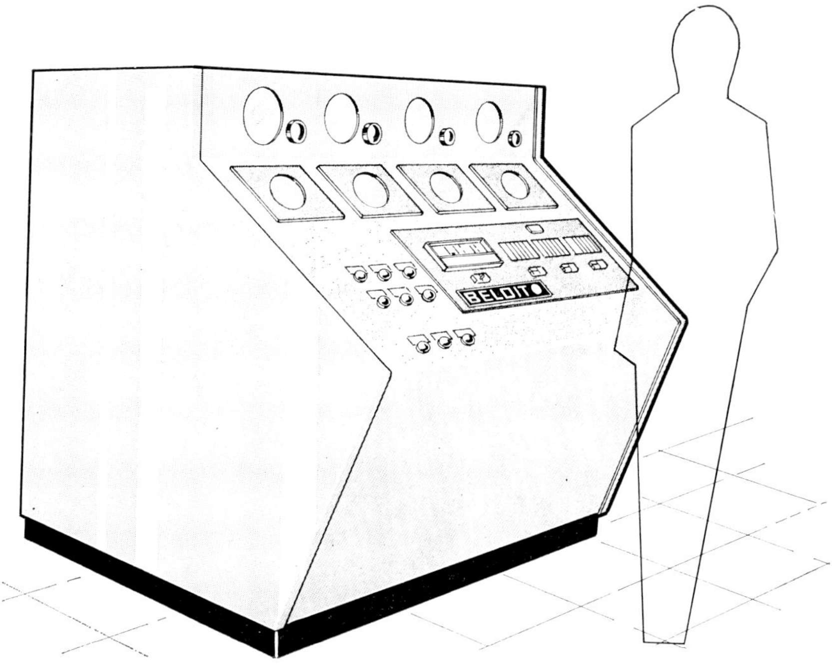

Fig 2—Environmentally protected small computer control system

System reliability

THE reliability of the computer will increase greatly, simply because of its MSI construction and concommitant lack of multi-level interconnections. Industrial type packaging as described will add to these inherent benefits. A side benefit of this reliability, in addition to less process down-time, reduced repair costs and simplified maintenance, will be gross reductions in manual or analog back-up necessary. With micro-computer replacement of present analog systems on new process installations 'back-up', as such will be a meaningless concept. This increase in reliability is not an unmixed blessing. Two problems will result, the first signs of which are evident in the highly computerised petrochemical industry, even in isolated pulp and paper installations. These include the gradual degradation of any analog back-up equipment installed through lack of use and maintenance plus the increasing unfamiliarity of the operators with the non-computer mode of operation, again through lack of use. The first problem can be prevented via an aggressive maintenance programme, the second with a periodic, scheduled half hour or so of process operation without the computer to enable the operating crews to maintain their old skills.

Hierarchical control

THE hierarchical or hierarchically distributed arrangement of computers is being made economically feasible by the advent of small low cost mini-computer and micro-computers and the improved packaging discussed above. Reasons for applying hierarchical control from an application standpoint are threefold. Firstly, wiring costs are reduced and very attractive savings may be realised by locating the computer close to the points being scanned and controlled. Secondly, the system design is simplified and reliability increased by the control and associated programming tasks into logical sub-units and sizing the computer appropriately. Thirdly, this hierarchical or distributed concept of control has the effect of greatly improving the system response time via parallelism, without having to support the cost of the equivalent reduction in computer cycle time. It is interesting at this point to note the recent announcements by major U.S. vendors of mini-computers to be marketed with their conventional large-scale process control computers and to act as remote peripheral controllers.

Digital telemetry

IN CONJUNCTION with and as an extension of this concept of distributed hierarchical control, increasing utilisation will be made of digital data transmission techniques and devices. Telemetry equipment has long been employed in the power and utility industries for remote control from a central location and will be employed in the paper industry in the near future. One U.S.A. user is currently planning to install telemetry equipment to control a bleaching plant several hundred miles distant, utilising excess capacity in the existing central control computer.

Alternatives to conventional hard-wired and microwave telemetry will become available through advances in other fields. One development that we are now examining with interest is the utilisation of lasers for digital data transmission. Laser links will have cost, reliability, noise and flexibility advantages over conventional methods on medium to intermediate ranges (that is, 0·25–10 miles).[12] A prototype installation is in operation at the Case Western Reserve University in Cleveland, Ohio, connecting a central scientific computer to a number of remote scientific terminals. The technology employed is directly transferable to the paper industry with suitable ruggedisation to meet the stringent requirements for industrial applications.

Graphic displays and the operator interface

ANOTHER area of great need where work is underway is that of operator communications and the graphic display of data. To enable an operator adequately to handle a centralised control system of vastly increased scope, two developments must occur. Extensive data reduction and decision making must be done within the framework of the computer hierarchy. Operators must be presented almost exclusively with information by exception. Unless specifically interrogated, the only data that will be displayed is that indicating a process upset, out of limit values or conditions that require his action. This data must also be presented in an easily interpreted graphical or pictorial format whenever possible. Several vendors have already taken the first step in this direction and with their nuclear gauging systems offer single colour

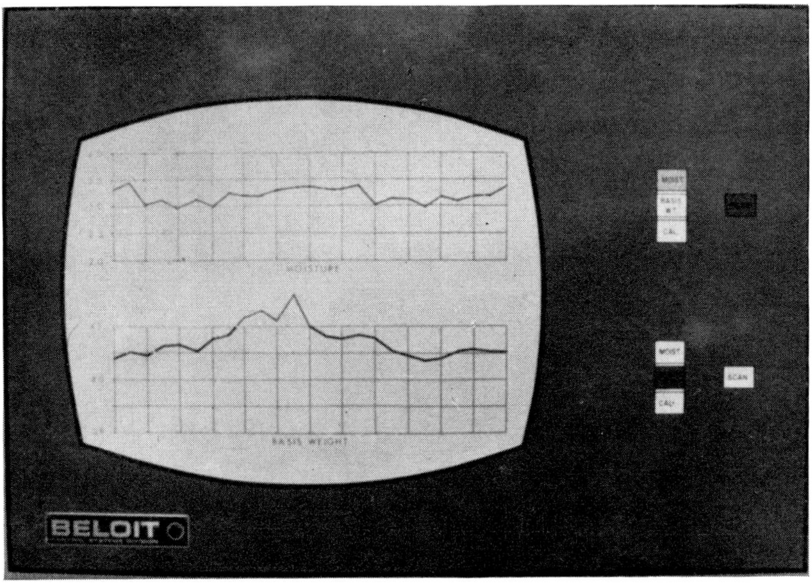

Fig. 3—Colour graphics console

graphic displays of cross-machine profiles (such as basis weight, moisture, caliper) as optional equipment. The next step is to extend the range and types of data presented graphically and to add to this colour-coded presentation to assist the operator in discriminating among the data presented (Fig. 3). Four and five colour graphic consoles will eventually be common in centralised control panels, presenting process conditions, production bar graphs, process schematics with variable data, etc. Much of this capability, with the prime exception of the colour graphics, exists today and is used in an embryonic form.[13] The power industry now makes use of computer-generated graphic displays of power grid schematics and operating conditions. Recent work in integrated circuitry and single gun multi-phosphor cathode ray tubes[14] will allow extension of single colour techniques at a price that will allow their use in industrial systems. A sketch of a console now under development incorporating a colour CRT system is shown in Fig. 4.

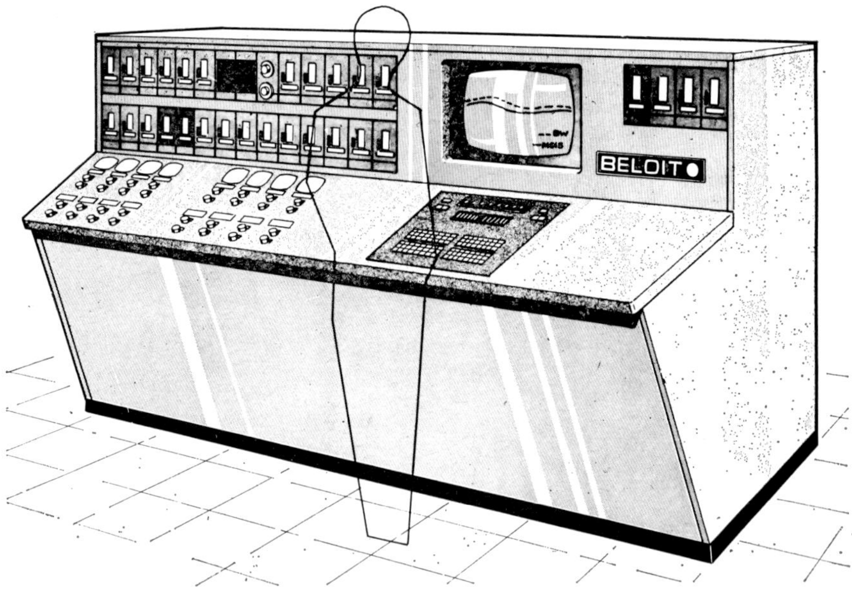

Fig. 4—Operator's control console utilising CRT-based multi-colour displays

Software and programming considerations

ONE OF the major drawbacks in the use of current mini-computers has been the limited number of programming languages available for a given machine (typically only USASI BASIC FORTRAN IV and ASSEMBLY language). Another problem of equivalent dimensions is the amount of time and effort involved in compiling or assembling programs in the languages that are available, since most mini-computer configurations are process-oriented and have few, if any, high-speed data processing peripherals (card or high-speed paper tape input, output, line printer, etc.). This means that, on a typical installation, a programmer will experience turn-around times of 1·5–2 h for programs of moderate length; thus, his productivity is greatly reduced.

Two solutions are beginning to appear for this problem. The first will be the availability of compilers and assemblers that generate object programs for mini-computer use, but that run on some other larger computer such as a scientific or EDP machine, which has the desired high-speed peripherals. We have taken this approach in our own installation, in which we have assemblers for a DEC PDP8, Varian 620/i, and Honeywell H316 and H516 computers running on several of our larger data processing machines. The other solution is indicated by the research presently being conducted[15, 16] in the area of higher level control programming languages and standardisation efforts on user and problem-oriented languages.[17]

An integrated system

As A summary overview of the preceding, Fig. 5 depicts a hypothetical integrated papermill with a hierarchically distributed total plant control system. The first level—or local control—incorporates eight physically small (20 in × 20 in × 60 in), freestanding (that is, not in a special computer room), dedicated computers. These first level 'area' computers perform either DDC or supervisory control (perhaps to local micro-computer controllers) as required, on the various process areas (such as bleaching plant, groundwood mill, stock preparation, chemical recovery, papermachine, power house). Graphic display capability is provided to meet each specific area's requirements. These computers are hierarchically linked to a much larger multi-programmed, time-sharing computer. This central machine co-ordinates the lesser computers and performs such functions as parameter optimisation, management and process information, reduction and presentation and on-line production scheduling.

The central computer may have additional small computers subservient to it that are used specifically as console and display drivers or for specialised input/output (I/O) tasks such as telemetry control. The optimisation of the total plant at this level may involve inventory or schedule control based on market considerations. Typical factors to be taken into account include order status, inventory position and current operating costs. The most profitable general conditions for each of the separate control areas may then be determined within observed operating constraints.

One final feature of our proposed system should be specifically noted. This is the presence of a separate monitoring and control computer allocated to the areas of water and waste treatment. This is in-line with our projection of the effects of public sensitivity, stricter legislation and responsible industry programs relative to environmental and pollution problems.

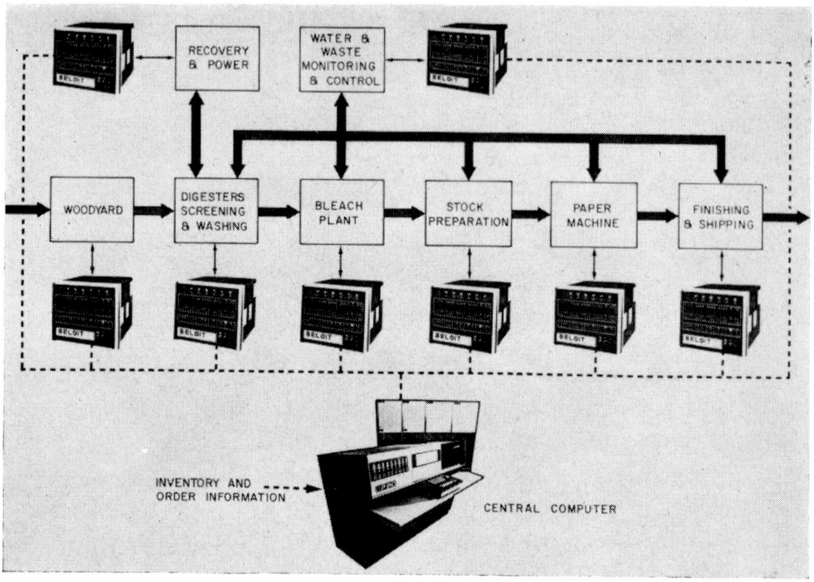

Fig. 5—Total plant hierarchically distributed control system

Application trends

THE expansion of hardware capabilities discussed in the first part of this paper will be utilised to apply techniques and strategies new to the paper industry that will improve current control system performance. Additional process parameters will also be controlled that are at present regulated manually.

We have attempted to cover the question of what will the functional appearance of control hardware be in the future, but this projection as yet has not met with the problems of application. Control and measurement problems are seldom more perplexing than those that we have in our industry. Almost every notable classified control difficulty is present to varying degrees in one pulp and paper area or another. Similarly, some of our important quality and process operating parameters still require inaccurate, slow, periodic, off-line laboratory test procedures.

With this ammunition, some critics of advanced control have yielded to the persuasive strains of 'our industry is not ready yet . . . no doubt the approach will be feasible someday'. Although this attitude undoubtedly serves to provide a reasonable amount of security in the face of new developments (whether controls or equipment), history has adequately demonstrated that it is not viable as a long-term approach. Fortunately, the pulp and paper industry as a whole is realistic enough to realise that, while perfect control does, in fact, require perfect measurements, our actual goal is *not* perfect control. Our goal is and must be an adequate return on invested capital. To achieve this, we must only provide *better* plant operation. Adequate gains can often be demonstrated with seemingly minor improvements in control easily within our present measurement capabilities.

Sensors

MEASUREMENT shortcomings nonetheless exist. Additional benefits would certainly be available if many of the lacking items were developed.

In general, two different approaches have been taken to the problem of inadequate or non-existent sensors. The first is the most straightforward—develop the necessary instrument. The saying is clearly easier than the doing. This approach has, through the combined efforts of industry, academic equipment suppliers and instrument manufacturers, provided us with an array of accurate on-line devices that we are only beginning to utilise fully. These include—

Basis weight	Gloss	Porosity
Moisture		Formation
Caliper		Pulp brightness
Colour		Pulp colour
Opacity		Spent liquor concentration

The future will bring more of these devices now undergoing development into the status of workable production tools. Research is now in progress on units for on-line measurement of modulus of elasticity (sonic) and sheet composition (X-ray analysis). Water and waste treatment control is also benefiting from new developments. Here, several continuous systems are now available for BOD via various inferential techniques, COD, turbidity, residual chlorine, dissolved oxygen, colour and zeta-potential. We are currently examining several of these units in the context of advanced waste treatment control systems. Analytical instruments of all types are being developed and will soon be standard items of instrumentation in every modern mill. In use now are systems for monitoring bleaching and pulping liquor strengths continuously. Automated wet-chemical equipment and methods abound.[18]

With these measurement systems, what is lacking? Probably the two items that come to mind easily are consistency and pjlp quality. We have been trying to obtain a precise measure of consistency for well over 100 years and it is certain that the papermaking wasp and the original Chinese practitioners of the papermaking art were also troubled by consistency control. We are only slightly closer to a final answer to this measurement problem today. Typical accuracy warrantees on consistency transmitters contain prose of the variety 'not to be calibrated in terms of absolute consistency measurements . . .'. Current work in this area includes examination of several potential new measurement principles, including nuclear magnetic resonance.

Pulp quality measurement is, if anything, harder yet in that the desired parameter has not yet been satisfactorily defined. The only universal consensus is that freeness in its various forms is not the right answer. There is even disagreement on what is to be expected of a pulp quality measurement. For example, should it predict the drainage characteristics on the machine, the runnability of the sheet or final paper strength factors? Without a statement of purpose even as broad as this, it is unlikely that any significant development in the framework of a useful operating tool will arise out of the present muddle in the near future.

The second or indirect approach to measurement problems utilises inductive measurement principles. Here a 'bad' measurement is accepted, provided the undesirable factors are largely deterministic. The measurement of consistency provides a classic example of the use of this approach. It is known that the output of all shear-based consistency transmitters is to an extent dependent upon flow, freeness, temperature, fibre blend and, hopefully, the true consistency. An illustration of this type of dependence is provided in Fig. 6.[15] The reduction of error in this type of measurement may be achieved by measuring (either continuously or in the laboratory) these additional variables and using this data to provide a corrected consistency via an empirically developed relationship of the form—

$$\hat{C} = a_0 + a_1 C + a_2 V_1 + a_3 V_2 + \ldots\ldots$$

where $\hat{C}$ = corrected consistency,
C = raw transmitter output,
V_1, V_2, etc. = additional measured factors,
a_0, a_1, etc. = empirical calibration constants.

This general technique has been further improved to limit residual measurement errors through stochastic feedback correction of the calibration equation.

It should be apparent that a digital computer is essential for the success of this approach, both in the initial equation development and in its implementation.

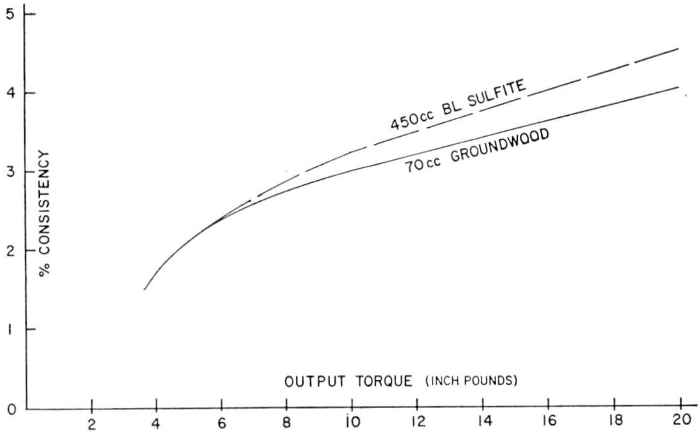

Fig. 6—Characteristic consistency transmitter response

While some specific examples of measurement problem areas admittedly tax the ingenuity of control systems engineers, it should be reasonably clear that the pulp and paper industry as a whole is no longer critically limited by lack of sensors. The majority of the burden now rests on all industry segments, both user and supplier, to utilise to the fullest extent the available measurement equipment.

Control techniques

THE mass of books, periodicals and miscellaneous publications relevant to modern control theory and control applications produced over the last few years staggers the imagination. The highly mathematical nature of the majority of this work renders it totally useless to the traditional instrument engineer as he exists today in the paper industry. The mathematics and technological level of this literature is certainly not going to get easier. TAPPI and other technical organisations have recognised the broad aspects of the re-education problem that exists because of technical advances and personnel obsolescence in all fields. Continuing education programs and various industry sponsored seminars are only part of the total solution to this problem. The ohter solution is indicated by the increasing reliance of industry on the special talents available through package or turnkey purchasing, whether the package is a power boiler or a process control system.

A review of the more noxious static and dynamic control aspects of pulp and paper processes will reveal how and where some of the current applications of advanced control principles arise. These characteristics are well-known and encompass—

1. The existence of large variable time delays.
2. Changing or unknown process dynamics.
3. Strong interaction between major control and quality parameters.
4. Non-linear static and dynamic plant characteristics.
5. A wide range of important process time constants (from approximately 0·02 s on some drive control systems to periods of days or weeks in some waste treatment systems).
6. The existence of many varied types of control problem, including continuous or conventional control, batch control (digesters and batch reactors), transitional control (grade change) and discrete control (sequencing, interlock systems).
7. The presence of significant process and signal noise.

As stated, control theory is an area of technology where theory has far outstripped applications. This is true to such an extent that a concentrated effort must be made to provide control at a level justified by the application expense and return on investment considerations. If this restraint is not exercised, 'over kill', excessive expense and late project schedules are guaranteed. A few examples of some typical approaches to the elicited problem areas will suffice.

Non-linear control

NON-LINEAR control may be applied when the process has some undesirable non-linear characteristic that non-linear control can alleviate, the actuator has some non-linear feature (such as backlash) or a non-linear response is desired. Applications of linearising cams or other elements to control systems is certainly not new. Valve manufacturers have for years supplied characterising mechanisms as an integral part of their valve positioners. The speeder-spring in a steam turbine mechanical governor is characterised in this fashion to provide more severe corrections for large deviations than for small. Modern analog and digital computer control systems simply allow easier implementation and some attractive extensions to these concepts.

Air-motors, hydraulic and electrical positioning systems on digester blow valves, refiner plate and plug position controls and horizontal or vertical slice adjustments have inherent on/off characteristics. Pulse, pulse duration or probability duty cycle control can be applied to these units, the optimum choice depending upon specific actuator characteristics in each case. The second technique has wide application in both digital computer and analog pulse duration control applications. A comprehensive computer aided design technique covering this type of control to a general dynamic process is the subject of a forthcoming technical paper by our staff.[20]

Dead time compensation

TIME lag or transport delay is present in significant amounts in water and waste treatment, digestion (continuous), consistency regulation and control of moisture and basis weight. Two approaches to the solution of this problem have been utilised. The first entails building a comprehensive feedforward model. This will allow a measure of control without the feedback stability problem. This typifies the approach initially taken with great labour and limited success to the control of bleaching plants. Unfortunately, no matter how comprehensive the model building techniques are, there will be model error.

The second method is to detune the feedback control system to the point where it is stable and operates acceptably under steady-state conditions, although generally not under set point changes. [21, 22] Prof. O. J. M. Smith of the University of California in 1958 proposed an analog feedback scheme for dead time compensation that allowed use of higher system gains, thus smaller errors. This scheme requires a simplified dynamic process model to provide synthetic feedback to the controller, but a simple model whose static and dynamic characteristics only crudely approximate to the actual plant is satisfactory for this purpose. Application of this technique to the control of basis weight pulp bleaching and a variety of other pulp and paper processes has been shown to be feasible by our Control Engineering Staff.[23]

Adaptive control

ADAPTIVE control is so vaguely defined that it led Eli Mishkin[24] to define an adaptive system as '. . . any physical system that has been designed with an adaptive viewpoint'. This definition is notably lacking in utility. For our purposes, adaptive control systems are those that embody the capability of making on-line modifications to their own control parameters to meet changing conditions. Fig. 7 and 8 indicate the improvements in control performance that may be obtained through the use of these techniques. Adaptive control complexity varies from an elementary application of the prior knowledge of how plant parameters vary with operating point to more complex schemes when spectral decomposition of the error signal or on-line plant identification is used.

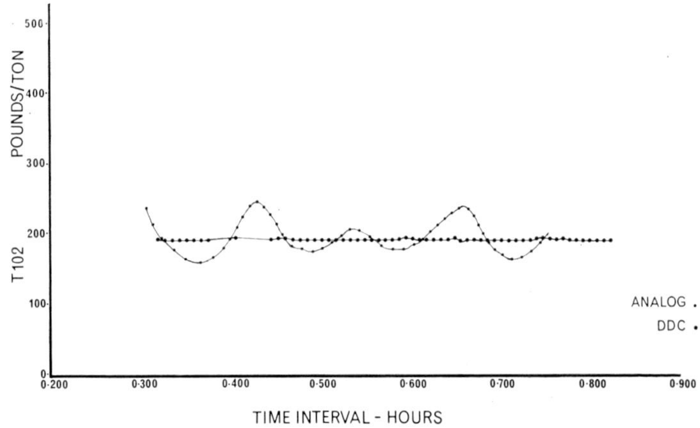

Fig. 7—Control comparison analog against adaptive DDC
Standard deviation analog 15.19, adaptive DDC 1.91

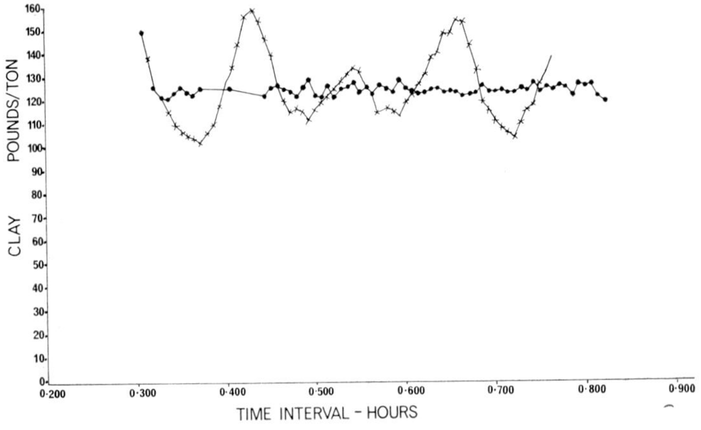

Fig. 8—Control comparison analog against adaptive DDC
Standard analog 23–69, adaptive DDC 1.13

Non-interacting control

FACTORS that interact both statically and dynamically include pulp viscosity and brightness in a bleaching plant; total head, level and head box consistency in the head box of a papermachine; and the classical papermachine variables of basis weight and moisture.

Decoupling techniques have been widely publicised by Gibson[25] for continuous systems. A technique utilised with success involves expressing the multi-variate plant and control characteristics as the product of two matrices of transfer functions. A decoupling controller is then derived via simple matrix algebra to provide independence of the various control parameters and the desired response characteristics. This method has been applied to basis weight and moisture control.[26] Its application is elementary with a digital process control computer system.

Range of dominant process dynamics

THE problems associated with the wide variety of dynamic characteristics encountered in pulp and paper practice are those of efficient utilisation of computational resources. There are at present a number of regulatory problems for which an analog system will still provide far superior dynamic response, although lower accuracy than is feasible with large-scale digital computers whose use is shared on a number of other tasks. These applications include drive system regulation where the bandwidth may typically be 40 radians/s or greater. To match analog performance on these systems, a digital unit would have to sample at a rate in excess of 30 times per second. A current answer to this problem is supervisory control of the analog subsystem.

Many drive suppliers today utilise special purpose digital techniques, for which both high accuracy and fast response are essential.[27] These applications include variable speed fan-pump drives where accuracies as high as $\pm 0{\cdot}00001$ per cent are claimed.[28] Within five years, however, the advent of general purpose LSI micro-computers will make the economics of such special purpose devices questionable This latter approach, as previously indicated, will simplify the integration of these systems into the complete hierarchical control structure.

Simulation

ALTHOUGH specific control techniques abound, very little in the way of simple pencil and paper methods for analysis or synthesis of complex control systems such as those that have been described is available. A technique widely utilised by control engineers for both complex process analysis and advanced control system design is that of simulation. The use of these methods originally was limited to those who both understood the electronics of the simulator (an analog computer) and the mathematical description of the process and control element static and dynamic properties. Development of digital and hybrid simulation languages that are user and use oriented has extended the use of simulation to a wider group of people and a wider class of problems.[29] A modern facility for the testing of advanced control algorithms, new process designs and complete real time software systems is illustrated in Fig. 9.

Fig. 9—A process control and simulation laboratory

Summary

An examination of available control technology indicates that no radical new developments are needed to solve our industry's control problems. The most complex of these systems is totally feasible with the process control systems of today. The majority of effort over the next five years will entail applications of existing technology and not, generally, from radical new development in the field.

Hierarchically distributed control concepts will achieve full acceptance during the next five years due to maturing application technology and massive reductions in hardware costs. With the realisation of these projects, the close integration of process control systems into plant and corporate management information and control systems will come to pass. The inescapable consequence of matching papermaking needs against modern digital technology will be the transition of the process control computer to its rightful status as an effective operating tool.

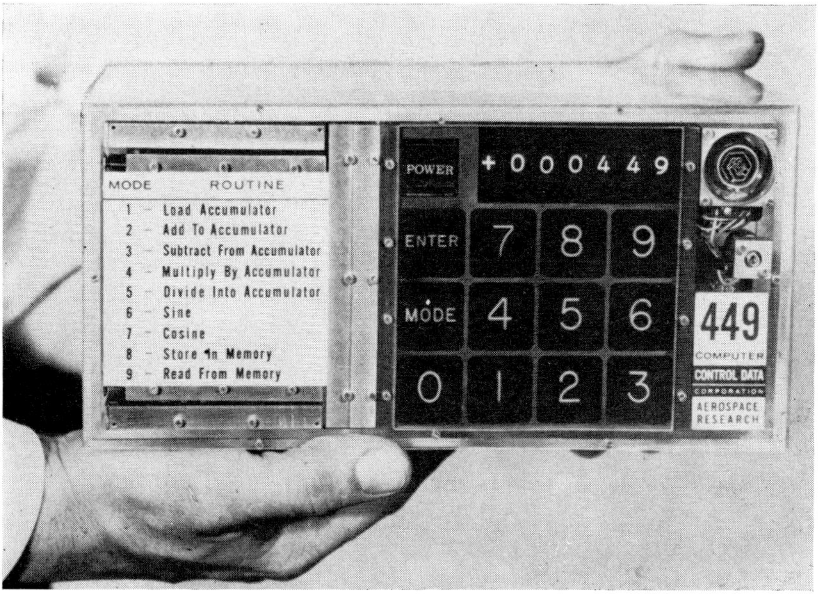

Fig. 10—Typical aerospace computer

References

1. Anon., 'Information processing': *Forbes Magazine*, January 1969, 181
2. *Control Data 449–1*, Micro-Electronic Computer System (Control Data Corporation product announcement)
3. Anon., 'New process makes smaller circuits': *Chem. Eng. News*, September 1969, 45
4. Anon., 'A trend report of electronic process controllers': *Control Eng.*, August 1969, 71–86
5. 'Mini-computers—What all the noise is about': *Control Eng.*, September 1968, 73–80
6. 'Digital proliferation—Mini-computers revisited': *Control Eng.*, November 1968, 72–74
7. 'Mini-computers—A look at the breadth of control applications': *Control Eng.*, January 1939, 128–132
8. *Data Sheet* 3751, 12 bit A/D converter (Fairchild Semi-conductor, Mountain View, California)
9. *Digital Equipment Corporation PDP–14 programmable controller* (product announcement)
10. *Ruggedised DDP–516 GP Computer* (Honeywell Computer Control Division product announcement)

11. *Dataspec—Ruggedised Varian 620/i* (Varian Data Machines product announcement)
12. Pratt, W., *Laser Communication Systems* (J. Wiley & Sons, New York, 1969)
13. Kazan, B. and Knoll, M., *Electronic Image Storage* (Academic Press, New York, 1968)
14. *Sylvania multi-colour display tube* (Sylvania SC–4876P226/P22R product announcement)
15. Temple, R. H. and Schoeffler, J. D., 'RTL—A real time language for computer control software': *Proc. IEEE*, January 1970
16. Temple, R. H., *RTL—A real time language for computer control software* (Ph.D. thesis, Case Western Reserve University)
17. Anon., Minutes of the *Workshop on standardisation of industrial computer languages* (Purdue University, 17th–21st February 1969)
18. Anon., *Automation in Analytical Chemistry* (Technicon Symposia 1965–68, Mediad Incorporated, New York)
19. DeZurik Corporation, Santell, Minnesota
20. Persik, S. R., *Pulse duration control of industrial processes* (Beloit Corporation, publication pending)
21. Persik, S. R., Process analysis report No. 19: *Design of a scan average feedback regulatory integral controller for direct digital control of machine-direction basis weight* (Beloit Corporation, internal report)
22. Persik, S. R., Process analysis report No. 20, *Feedforward basis weight control loop analysis* (Beloit Corporation, internal report)
23. Persik, S. R., Process analysis report No. 22, *Dead time compensation with the Smith linear predictor* (Beloit Corporation, internal report)
24. Mishkin, E. and Braun, L., *Adaptive Control Systems* (McGraw-Hill, New York, 1961), 4
25. Gibson, J. E., *Non-linear Automatic Control* (McGraw-Hill, London, 1963), 23–33
26. Dahlin, E. B., 'Integrated control of Fourdrinier machines': *Second International Symposium on Water Removal at the Press and Dryers* (Mont Gabriel, P.Q., Canada)
27. Keyes, M. A., Process analysis report No. 27, *Differential drive control analysis* (Beloit Corporation, internal report)
28. Strand, S. M., 'Frequency-based regulators lock dc machine speeds': *Control Eng.*, March 1967
29. Terhar, R. J., 'Simulation in the pulp and paper industry': *Pulp & Paper Mag. Can.*, 1967, **68** (6), T295–T300

Discussion

Mr S. Hem We had heard exclusively about the hardware side of your systems. As potential users of the computers, we are also interested in the control aspect of the problem. In what areas of the papermaking process can you see advanced control theory being used?

Mr M. A. Keyes All areas.

Dr H. Sandblom It has been very interesting to listen to this paper, for it is always of great interest to look into the future to see where computer technique is heading.

My company is of course very interested in the small, but powerful Data 449 computer control and has been studying it very carefully. In this connection, was the input/output equipment included in the basis weight figures (that is, its connections with the process)? If not, only very small computer systems were considered.

May I add a further comment. One of your diagrams shows the relationship between software and hardware costs. Even if systems are becoming increasingly involved and complex, with the use of better software packages like process input acquisition and better programming systems like AUTRAN, the software costs should not rise as steeply as you have shown.

Mr Keyes There are several vendors and systems houses such as ourselves who are introducing packaged software systems. The advantage of this approach is that these people are in a position to spread the development over a number of systems rather than on one specific application, thus reducing the user costs.

Dr S. W. Kingsnorth As you are head of the control division of a major machine manufacturer, could you comment on the likely future of papermachines in their ability to respond quickly and accurately to the information

Under the chairmanship of P. E. Wrist

that the control systems are feeding back to them on flow boxes, steam systems, calenders and everything else that is in the major hardware of papermaking?

Mr Keyes Well, of course there will be changes in the process and one of the things that is taken into account in a system design is the response capabilities, the controllability of process and machine equipment. There is a certain amount of literature on this co-ordinated design approach. I presented a paper at a TAPPI meeting in Boston some years ago, for example, on the integrated design of hardware systems for controllability.

The Chairman We will now proceed to the general discussion of the papers in this session and I would like first to say a few words in the context of control to review very briefly what these gentlemen seem to say and what people seem to hear them say.

Mr Jolliff raised questions about process interactions that are fundamental to solving the entire papermaking control problem. Mr Mardon and his colleagues raised questions about the control system learning curve and the return of investment curve. In particular, it was notable that the discussion centred around the psychological problems of training, man-machine communication and intergroup behaviour and communication. Messrs Cyprus & Attwood dealt with a computer control project, in particular with the economic aspects, organisational and interpersonal aspects, technical considerations, redundancy of control equipment and man-machine interface.

The two written contributions dealt again with man-machine problems. Mr Unthank talked about the conflict between production and development demands placed on a single computer and Mr Keyes brought out the question of good project management and continued with a review of some of the remarkable advances that continue to take place in digital computer equipment. He referred to software costs, hierarchical control, multi-connected computer control relative to single computer implementation of hierarchical control and again about man-machine communications.

In my opinion, a number of the pieces of the jigsaw of entire papermachine control have been discussed as has some of the glue (so to speak) that might stick these pieces together. I have a feeling that we really have not come to grips with the problem of entire papermachine control and that it is going to require some highly skilled artistry to complete this picture. I hope that we can deal with this in the discussion.

Dr H. K. Corte I am somewhat disturbed by the emancipation from papermaking of the development of electronic hardware in papermachine control, which seems to run far ahead of our papermaking knowledge. It is still the chief aim in the life of a papermaker to produce paper that performs to the satisfaction of the customer. The first word on Mr Keyes' first slide, giving reasons for computer control, was *quality*, but I have not heard one paper concerned with the control of quality. Basis weight and moisture control improve the quality of a reel or consignment, but, if the paper is found not to be dimensionally stable enough or if the *small-scale* basis weight distribution is unsatisfactory, what do you do then? Too much emphasis perhaps is given to those quantities that we happen to be able to control (such as flow rates of steam, water, etc.) and too little to those that we ought to control in order to make better paper, not only more money.

Mr Keyes It is readily admitted by most people who put in control systems that they do not pretend to have a final answer. Their justification of advanced control is measured rather in terms of improvement over standard operation than in achievement of ultimate performance.

Your second statement on pulp and paper quality interests me, for the industry as a whole has for a number of years been pursuing the elusive goal of a device that measures pulp quality. I have yet to see a satisfactory definition of just what pulp quality is, whether it is measured automatically or whether it is measured by manual means.

Mr J. Mardon I think we can draw an analogy from the rules of sport. If one referees a game, one is commonly expected to have a thorough knowledge of that game. You cannot referee a hockey match using football rules and such is the situation in which, in the application of advanced control systems, the companies are placing themselves. I absolutely agree with what you say, but much of this information is available. The point is that those implementing control systems are often not cognisant of it and do not attempt to use it. It is certainly true in any company thinking of applying an advanced control scheme that known papermaking technology should first be put to use; before applying an advanced control system, thoroughly investigate the process. If the available technology is inadequate—and, in certain areas of the manufacturing processes such as for pulp, the technology is clearly not adequate—one must find it out. This is one reason for our systems engineering group to be moving slowly. They are taking cognisance of the points you have raised and are making sure that they understand the rule book before they start to referee the game.

Mr G. F. Underhay Following on what Dr Corte said, I notice that, in stating the degree of accuracy in controlling basis weight, not one writer has said what size of paper he is concerned with when measuring basis weight. Square centimetre samples will exhibit far greater basis weight variations than square metre samples and, of course, the wider the paper, the worse will be the indicated weight variation, especially if small samples are used.

Mr M. I. MacLaurin Mr Jolliff remarked that the Georgetown project was directed from his New York office with no consultation at the mill. I expect this approach to raise problems.

Mr C. C. Jolliff The original computer installation was put in purely as a research project, without any planning or knowledge on the part of the individuals who were brought together to start the project. That was the last computer installed in that way, the others have been planned for and have been justified on the basis of expected improvements in product quality or productivity.

Mr R. G. Nagro Perhaps, to tie together the pieces of the jigsaw, we should look at the non-control aspects of the computers and see if we can get an analogy. Originally, when computers first came out, people installed them where they worked with well-defined, but limited objectives. We are still doing this with control computers today and I think we have reached the stage of some rather well-defined applications. We may not know exactly how well they pay off yet, but I assume in the next year or two that you will be able to assess them fairly well. Nevertheless, we are reaching the point of being on the verge of a major revolution in the management of papermaking companies, as we are in the management of *all* companies. What management is going to be faced with over the next two or three years is a major change in the techniques of doing business, not necessarily through the computer, but through a well-defined, very precise sort of system—all the elements of process from the receipt of the order to the shipping. That is the opportunity that the paper industry will have to grasp. Although some managements consider themselves being dragged into this, I am afraid that in one way or another we must all be involved. Maybe, we ought to look at the hardware and software we have been talking about and the definition of control in the context of how we really want to run our paper companies.

Mr G. D. Madeley My first point is that people have been very defeatist in their ideas—that, because we do not understand papermaking fully, we cannot attempt to control it. The way you learn to play football is just by playing it, then you start learning the rules. This is how we feel about it. Our computer installation highlighted our areas of ignorance: we rapidly learned to overcome them. We learned about papermaking—indeed, a great deal about it.

We are now getting a more consistent, uniform product. Perhaps we do not understand how to control the wet expansion of our paper, but at least it is uniform now. If we make a change, we obtain a uniform change.

My second point is on extending control throughout the whole of the management system. This is something we are very keenly interested in. At present, we have a computer on the papermachine by which we are hoping to control the actual papermaking. At the same time, we are introducing into the mill a planning system conceived in logic rather than just growing up like Topsy. This planning system, although essentially manual, is designed in such a way that automatic accounting machines—perhaps even computers—can be put into it with the minimum of problems. I am sure that this is the way papermill control will go.

Mr O. Alsholm Being in the initiation stage of the game, it is only natural that quite a bit of money has been spent on back-up instrumentation for computer control loops. Thus, it was interesting to hear from Mr Cyprus about the cheap 'degraded' controllers used at the Sittingbourne Mill.

Mr Keyes, you have been involved in two of the world's largest DDC installations, which have both almost 100 per cent back-up instrumentation. What of the future installations? For example, some people claim that one gains more than the cost of back-up installation by including several recorders during start-up. I cannot imagine that this will still be true in the future.

Mr Keyes Neither of these installations incorporates full back-up. Our experience has shown that the use of full back-up is detrimental both to reliability and to the human engineering aspects of the job. When a papermaker is confronted with a choice between a system that performs better, but is new and a system that performs not as well, but with which he is traditionally familiar, he will tend to use the familiar equipment with loss of performance. Secondly, in some of our installations, we found that one of the weakest links in the entire system has been the reliability of the back-up equipment.

Mr H. D. Cyprus The system at Sittingbourne has six bumpless controllers, which are quite expensive. To a large extent, this reflects what little lack of confidence we had when first putting this system together. Our plans for No. 17 machine include nothing like this extent of standby back-up; we would almost totally eliminate the need for the bumpless controller and standby system would therefore rely largely on the simple standby controllers (about £30 apiece). They share output lines with the normal computer outputs, so cabling, etc. is no additional cost. They have a very limited proportional band and fixed integral time, but both these are easily changeable from the nominal values that are set when the equipment is purchased. In any event, I believe a new improved range of equipment is at present available.

Mr B. W. Balls On this question of tying the machine together, I think a lot can be learned from a large number of installations that have been made in the chemical industry, where surprisingly enough the problems are similar. They talk about it in different terms, but the pattern that emerges is the importance of control enforcement, by which is meant making the system perform as it was designed to do with minimum operator interference. DDC provides a standard of control enforcement never before experienced in any control system. There is ample reference to this in the literature and, if you are able to talk to people who operate DDC systems, this will be said. Today, any machine of reasonable capacity can be tied together very easily and simply with a DDC system; I think it could be guaranteed that an upgrading of the whole performance of the machine would then be noted. One advantage of DDC that was not learned with some of the earlier supervisory computer systems is the need for total involvement of everybody in the project, from the management downwards. In this respect, I support what Mr Mardon says in his paper on the involvement of people. If the people are not involved, the computer project will fail.

AUTOMATION OF A PAPERMACHINE AND ITS ANCILLARIES*

A. L. RAMAZ, S. BAUDUIN, and D. MARCÉ, Centre Technique de l'Industrie des Papiers, Cartons et Cellulose, Grenoble, France

Synopsis A digital computer was installed on the Centre Technique du Papier's experimental machine in order to study the problems involved in the automation of a papermachine.

After a short description of the installation, this paper summarises the work done during the installation phase, in particular, control of the basis weight was designed to go into service as soon as the computer was installed. It uses a control algorithm compensating for the dead time from the point of view of loop stability. The initial results are encouraging.

The same principles will be extended to the control of the dryer section after development of an appropriate model.

Introduction

AN AGREEMENT on participation in a programme of research (No. 67.00.911) made it possible to install an Automateur MAT 01 computer built by the Automation Division of Télémécanique Electrique on the experimental papermachine at the Centre Technique du Papier.

This paper gives an account of the study up to the time when the computer was installed on site.

Aims of the study

THE general aim of the study was to investigate all the possibilities offered by the use of a digital computer in the control of a papermachine.

We feel it is necessary to emphasise the fact that the objective was not to automate the Centre's experimental machine, but to develop a general approach to the problem of papermachine automation.

We also intend to show that part of the programming can be done in

* This study was carried out by the Centre Technique du Papier under the terms of agreement for participation in a research programme (No. 67.00.911) with the Délégation Générale à la Recherche Scientifique et Technique with the participation of the firm Télémécanique Electrique, which supplied the computer and software

Under the chairmanship of P. E. Wrist

advance by means of approximate models so as to be able to put the computer to work on some jobs as soon as it is installed and thus reap some economic return almost immediately.

In this paper, we shall deal at any length only with the points already achieved or due to be achieved in the near future, the more long-term projects being merely roughly outlined.

Description of the pilot unit

THE investigation was carried out on the experimental machine at the Centre Technique du Papier at Gières.

Experimental machine

This machine was built by Ets. Neyrpic under licence to SICMA. Its main features (as far as this paper is concerned) are as follows—

Maximum speed	100 m/min
Trimmed width	50 cm

The stock prepared in the machine chest is sent to a constant level stuff box. The circuit includes a consistency regulation system.

Dilution takes place in a mixing chest combined with the centrifugal strainer. The stock and recirculated water are controlled by valves, the head on the valves being supplied by controlled level tanks.

The mixing chest is connected to a conventional head box, but is higher to provide a head. This box is placed under vacuum in order to keep the stock level higher than it would normally be for the pressure at the slice.

The Fourdrinier machine is equipped with wet boxes, table rolls, suction boxes and a suction couch roll. The water from the wet boxes and table rolls on the one hand and from the suction boxes and suction couch on the other hand is collected in two separate chests working as communicating vessels, which enables the wet boxes and table rolls water to be taken first for dilution purposes.

The press section has two plain presses—a couch press and a reversed press. The dryer part, which is equipped throughout with pocket ventilators, is split into two sections supplied by separate manifolds. All 12 sections of the machine are driven from a main shaft through tapered pulleys.

Instrumentation

It was planned from the start to provide quite comprehensive instrumentation so as to be able to meet all the requirements of the study as a whole and, if necessary, to investigate problems of paper technology not directly connected with the control machine.

Hence, it was planned to link up some 80 analog measurements to the computer. Among these, the items concerning us directly are the following—

Stock and dilution water flows and wire drainage flows (electro-magnetic flow meters).
Consistency at the above flows (consistency pick-ups and turbidity meters).
Head and level in the head box (differential pressure pick-ups).
Slice opening (potentiometer pick-ups).
Dry end and wet end basis weight and moisture content (beta-gauge and capacitance pick-up).
Steam pressure in the post-dryer section.

The dilute stock and whitewater consistencies and the value of the dry basis weight at the wet end will be taken into account for control purposes as soon as the pick-ups are correctly adjusted, which is proving difficult.

Seven separate controls are provided, with the possibility of future extension. They work through digital/analog converters, consisting of an inching control motor receiving the impulses from the computer and a force balance, which converts the position of the motor into a standardised 4·20 mA dc current. This signal is then transformed into a pressure of 0·2–1 bar actuating the control mechanisms.

The physical quantities controlled are—

Stock and whitewater flow rates.
Steam pressure at the post-dryer.
Slice opening.
Papermachine speed.
Degree of vacuum in head box.

In addition, the machine is equipped with a number of conventional pneumatically operated control systems—

The levels in the stuff boxes, which will probably remain as they are.
The consistency control, for which computer control is also being contemplated.

Computer and data input systems

Computer—The computer's main characteristics are—

8 K words of 21 bits (18 value bits).
Base cycle 5 μs.
Six interrupt levels.
Programming in machine or assembly languages.

The link with the operator is through a DDC console handling requests for display, print-out or modification of the parameters in the control loops. The messages are printed and punched by teleprinter. The teleprinter can be used also for program input from punched tape.

The link with the control process is either through relays (simple on/off relays) or a selector switch and analog/digital converter for the analog signals. All signals have been standardised at 50 mV and are amplified to 10 V before conversion.

The control operations are performed by relays in the case of on/off operations and by impulses in the case of the analog quantities, conversion being by means of the digital/analog converters described above.

Dry basis weight control

THE first regulation of the dry basis weight has been prepared before installation of the computer to enable it to be used as soon as the system is operational. This involved the development of a theoretical mathematical model of the papermachine.

Mathematical model

We need not go into a detailed description of the static model consisting of the steady state water balance and solids balance, which are expressed in the Sankey diagram well-known in the paper industry. The use of this model is nevertheless indispensable in order to establish the operating point on which the dynamic model will be based.

Development of the dynamic model—The development of models representing the dynamic behaviour of the papermachine is also well-known.[1-3] The method consists in breaking down the machine into separate parts, which can be represented by simple equations, then linking up the equations. In our case, we are basically dealing with volumes involving mixtures and piped transport.

There is no question of working on a model that describes the behaviour of the machine in response to all possible working conditions. We have confined ourselves to building up a model that can express mathematically the relationships between the control quantities and the output quantities under certain operating assumptions.

In the present case, the control quantities are the stock and recirculation water flow rates. The output is the dry basis weight calculated from the data provided by the dry end combined metering system (basis weight and moisture content measurements).

The operating assumptions are essentially as follows—

1. Proportionality between the dry basis weight and the quantity of solids leaving the wet end. This hypothesis enables the presses and dryers to be replaced by a dead time.

2. Constant total flow in the head box—the first model developed in fact showed (and practical tests confirmed) that possible variations in stock flow rate with respect to total flow rate cannot be ignored in an open head box from either hydraulic operating characteristics or response in dry basis weight points of view and it was therefore decided to control both flow rates simultaneously so as to work at a constant total rate of flow.
3. Constant volumes—this hypothesis is justified in view of the constant rate of flow in the head box and mixing box and the constant levels in the other tanks.
4. Retention varying linearly with the quantity of solids on the wire. (This latter assumption, although at first sight questionable, checks fairly satisfactorily in the case of small variations; an increase in the solids content on the wire is accompanied by a decrease in the solids content in the table roll water and that of the suction box water rises as a result of the transfer of part of the drainage from the table rolls to the suction boxes.)

On these assumptions, the wet end diagram shown in Fig. 1 may be expressed in the form of the block diagram in Fig. 2, which describes the dynamic behaviour of the quantiites of matter θ on the machine in the case of small variations.

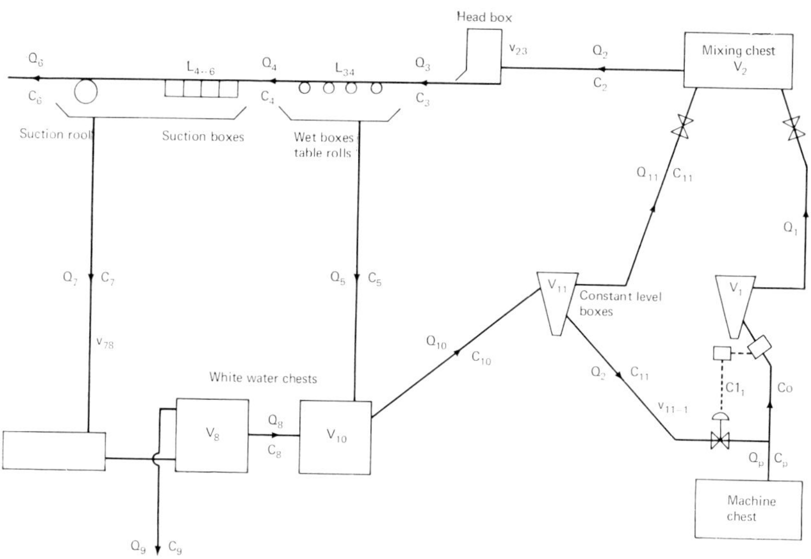

Fig. 1—Wet end diagram of the papermachine

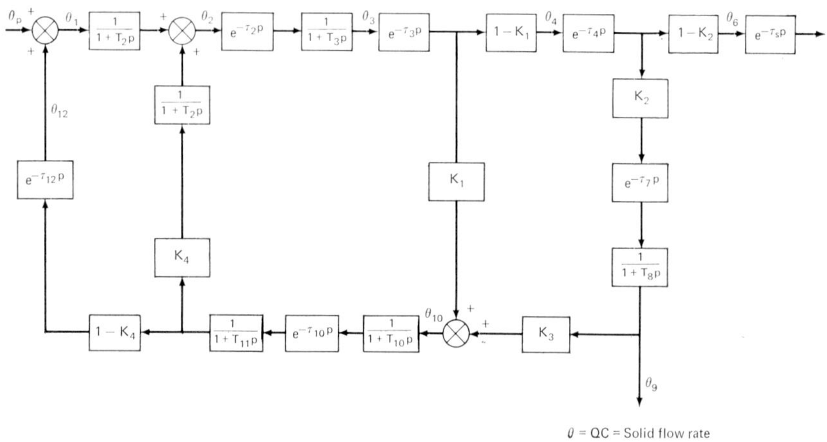

Fig. 2—Block diagram of the wet end

Using the following notation, we have in the case of the mixing chest, for example—

Q_1, C_1, θ_1	Stock flow rate, consistency and solids throughout, respectively.
Q_{11}, C_{11}, θ_{11}	Ditto for the whitewater flow.
Q_2, C_2, θ_2	Ditto for the diluted stock flow.
V_2	Volume of the mixing chest.

We then have the following in the mixing chest—

$$Q_1C_1+Q_{11}C_{11} = Q_2C_2+V_2\frac{dC_2}{dt}$$

If we introduce a variation (small letter)—

$$q_1 = -q_{11} = q$$

it becomes—

$$q(C_1-C_{11}) = Q_2C_2+V_2\frac{dC_2}{dt}$$

or in Laplacian notation—

$$q(C_1-C_{11}) = (Q_2+V_2p)C_2 = \left(1+\frac{V_2}{Q_2}p\right)Q_2C_2$$

and by putting $\theta_2 = Q_2C_2$

$$\frac{\theta_2}{q} = (C_1-C_{11})\Big/\left(1+\frac{V_2}{Q_2}p\right)$$

and, if in addition, we make—

$$\theta_1 = C_1q \text{ and put } V_2/Q_2 = T_2$$

we get—

$$\frac{\theta_2}{\theta_1} = \frac{C_1 - C_{11}}{C_1} \cdot \frac{1}{1+T_2p}$$

We thus replace the mixing chest by an in-line mixing followed by a time constant.

The model assembled from the various elements was studied using two different approaches.

Frequency aspects—We calculated the transmittance of this model in gain and in phase and compared it to the transmittance of the direct flow circuit (without recirculation).

These two transmittances are shown in graphical form in Fig. 3 in the Bode diagram (ignoring the dead time in the dryers). The curves are slightly different at very low frequencies and practically identical above 2×10^{-3} cycles.

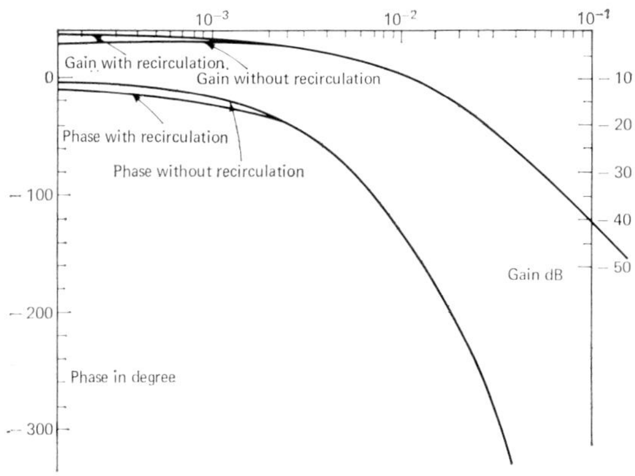

Fig. 3—Bode diagram of the wet end

As an example, Fig. 4 gives the comparison between the theoretical response (three time constants) and the response actually recorded on the unit. These tests were done before simultaneous valve control was installed and provide a good illustration of the effects of variations in total flow rate (first part of the curve). From this, it was concluded that the process may be fairly approximated by a system involving three time constants and a dead time.

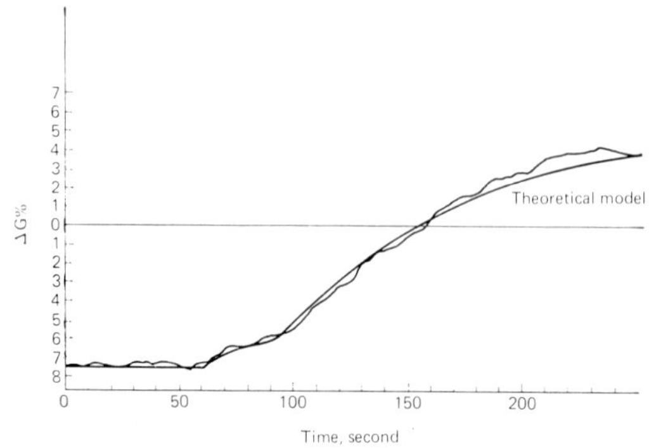

Fig. 4—Basic weight response to a step of stock flow rate

The orders of magnitude of these parameters are—

Mixing chest $V_2 = 250$ litre, $T_2 = 40-130$ s
Head box $V_3 = 40$ litre, $T_3 \approx 6-20$ s
Filter on measuring instrument $T_s = 3$ or 10 s
Dead time in pipe 6–20 s; on machine 40–120 s.

Allowing for the fact that T_2 is quite a good deal larger than T_3 and T_s, it is even possible to envisage a model with one time constant and one dead time; this case has also been examined, as will be seen later.

Time aspect—Chronologically, this study was in fact done before the frequency study.

The first model developed was intended to be of a more general character than the one described above. In particular, it was desired to find out the stabilisation time for the various chests at the start of a manufacturing run, but before the machine was actually started.

In view of the complexity of the phenomena involved, also for reasons of convenience, it was decided right from the start to use a numerical simulation approach.

To do this, the equations for the mixtures in the stock chests were put in finite difference form and the dead times taken as simple time decrements.

In addition, numerical simulation makes it possible to account for non-linearities, which are difficult to deal with using a frequency approach. Thus, the first program, written in ALGOL for processing on an IBM 7044, allowed for flow and level variations and the separations on the wire were represented by various sets of fairly complex equations taken from the literature on the subject.

The stabilisation times derived from simulation were found to give good agreement with practical results. The response of this model to small variations showed on the one hand the all-important role of the direct flow circuit (without recirculation) and, on the other, the need for simultaneous control of both head flow rates to maintain a constant total discharge. These two observations led us to simplify the original model. This program was transposed into FORTRAN for processing on an IBM 360 computer and it is being used for an initial verification of the control algorithms (with adaptation of the coefficients, if necessary) before they are applied on the machine.

Feedback control

Originally, we asked the supplier to provide the software for logging and for classical three-mode control (proportional, integral and derivative), as well as the computer hardware. In particular, we intended to obtain exact performance data on the three-mode control in the difficult case of basis weight in order to compare it with the performance of any other control of better design.

Since the frequency study and simulation test showed that this type of control system could not be expected to work well, it was abandoned in favour of a study of a regulation system enabling stability to be maintained despite dead time.

Dead time compensation principle—There are now many publications available on control systems using dead time compensation, which should not be confused with predictive control. In the first case, the dead time is compensated for as far as the stability of the loop is concerned, but the dead time remains in the actual process. The principle is shown schematically in Fig. 5. A control algorithm $C(p)$ is designed such that, the transmittance of the process being $G(p)e^{-\tau p}$, scheme No. 1 is equivalent to scheme No. 2, in which $H(p)$ is chosen such that the response has a predetermined form.

Identification of the two transfer functions—

$$\frac{C(p)G(p)e^{-\tau p}}{1+C(p)G(p)\text{-}e^{-\tau p}} \text{ and } \frac{H(p)e^{-\tau p}}{1+H(p)}$$

enables $C(p)$ to be computed by identification—

$$C(p) = \frac{H(p)}{G(p)}[1+H(1-e)^{-\tau p})]^{-1}$$

A control algorithm of this type, involving the dead time in a process, can easily be applied on a digital computer. On the other hand, the use of this computer involves sampling and it is better to study the system using the z-transform.

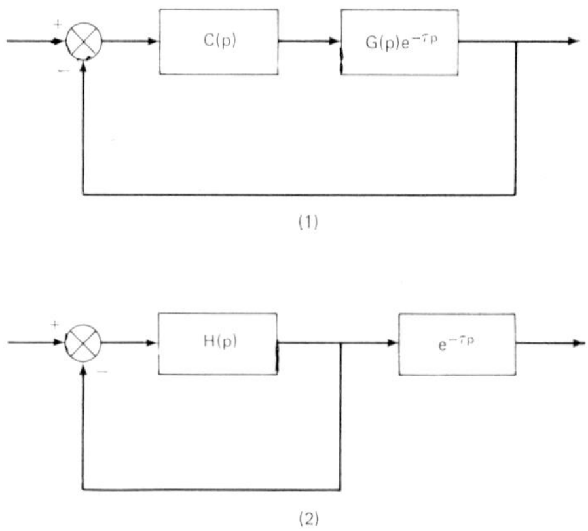

Fig. 5—Principle of dead time compensation

Numerical correction of dry basis weight—This is a cascade control system: the stock and whitewater flow rates are held to given settings that can themselves be modified by the main loop from the output of the end of machine combined metering unit (Fig. 6).

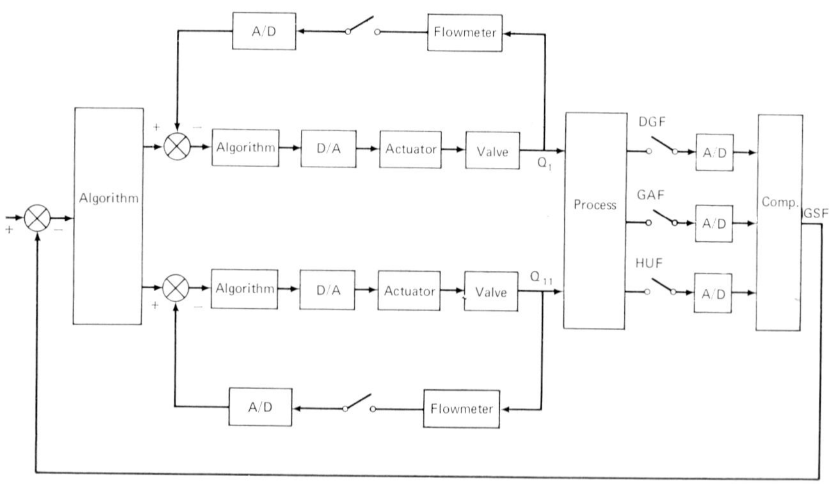

Fig. 6—Numerical correction for dry basis weight

A 5 s sampling period was adopted for the measurements and control operations.

For the theoretical aspect, the reader is referred to Volgnine.[5]

Automatic flow control—The computer controls the valves from the indications given by the flow meters. Since the set points of these loops are to be driven by the computer, it is advisable to take a minimum prototype whose response is complete by the time the next sample is taken.

Furthermore, since the digital analog converter/current pressure transducer/control valve/flow meter system, considered as a whole, has a response much shorter than the sampling period (less than 1 s) we have a controller of the single mode type.

The problem is thus reduced to determining the gain and the whole control system may be considered as a clamp.

Main control loop—In view of the results obtained when designing the model, we are trying to develop a control strategy for a process involving three time constants and dead time. Considering the values of these time constants, the 5 s sampling period satisfies, at least approximately, Shannon's theorem.

For the first application, we have chosen a control algorithm that is a good compromise between a very good response to a one-step change in set point and a wave-free response.

The expression for the control algorithm defined by the above method, but using the z-transform gives—

$$S_i = v_o S_{i-(n+1)} + v_1 S_{i-(n+2)} + v_2 S_{i-(n+3)} + \frac{a_o}{K}[\varepsilon_i - \lambda_1 \varepsilon_{i-1} + \lambda_2 \varepsilon_{i-2} - \lambda_3 \varepsilon_{i-3}]$$

in which S_i is the output at time i,
ε_i is the deviation at time i,
n is process dead time expressed in number of sampling periods,
a_o, v, λ are coefficients determined from time constants for the process,
K is process gain.

For the regulation study, we decided to start by investigating the response to a single thick stock consistency control step.

In theory, the output z-transform can be obtained in terms of the input z-transform, the actual output values being obtained by means of a polynomial division program.

Furthermore, we have applied this type of control algorithm on a simulation on the full model in order to ascertain the effect of the whitewater return line.

Fig. 7 shows the two responses obtained. An explanation has not yet been found for the fact that the response of the full model appears at first sight to be better than the theoretical response. The practical tests are in progress and will be presented at the conference.

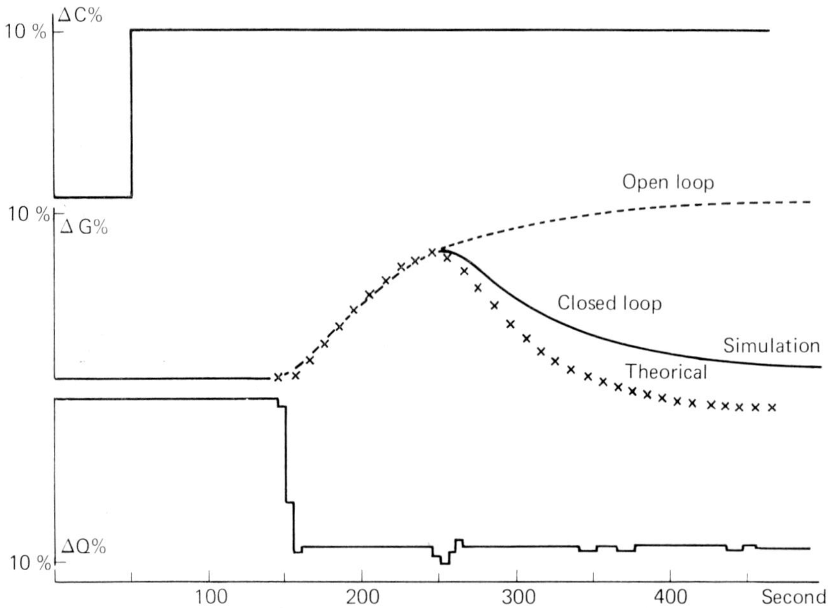

Fig. 7—Basis weight response to a step consistency (three time constants model)

If the process is approximated to one time constant plus a dead time, the control algorithm is simpler—

$$S_i = S_{i-(n+1)} + a_1 \varepsilon_i - a_2 \varepsilon_{i-1}$$

The responses are given in Fig. 8.

It is this control algorithm that was programmed at first, but, since adjustment of the coefficients was found to be a tricky business, it has been discarded in practice—at least for the time being—in favour of the earlier one. Anyhow, the use of a more complex algorithm is no problem when a computer is available.

A few practical considerations

Automatic flow control—Owing to the non-linearity of the valves, the optimum gain of the control system is variable, depending on gate opening. This problem is fairly easily solved. The computer selects the gain itself from the actual flow rate and the valve calibration curve stored in the memory.

The slack in the valves corresponds to about 10 control impulses (for a total travel of 1 000 impulses). This problem has not yet been fully solved, since the compensation tests carried out up to now have yielded poorer results than with no compensation at all. As a result, the response in certain cases is obtained only in the second sampling period, instead of in the first.

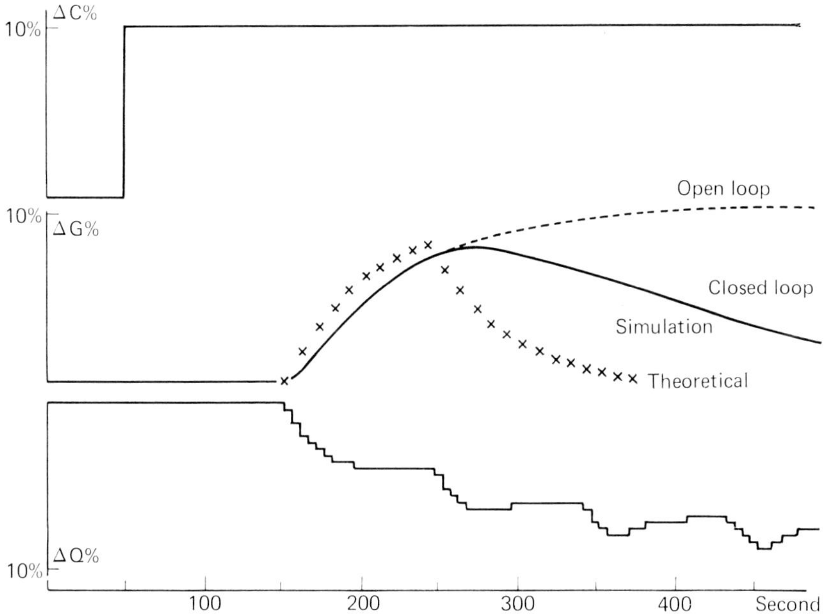

Fig. 8—Basis weight response to a step consistency (one time constant model)

Dry basis weight—An industrial pick-up is used, which sends out three signals—

GAF Total basis weight setting.
DGF Percentage deviations on the above.
HUF Moisture content related to the indicated basis weight (the maker assumes that the capacitance pick-up gives a picture of the basis weight in water and divides this signal by the indicated basis weight).

A 20 points linearisation is applied to GAF, which has a logarithmic variation law and the dry basis weight is then computed—

$$GSF = GAF\left(1+\frac{DGF}{100}-\frac{HUF}{100}\right)$$

The accuracy of this computed value governs the whole regulation system, which is thus highly dependent on the correct operation and calibration of the combined metering unit, in particular the moisture content meter.

Case of an industrial machine—In the case of a laboratory papermachine with a width of 50 cm, the combined metering unit works in a set position and measurements are continuously available.

On an industrial papermachine, it is debatable whether the pick-up should be kept in one position or whether it should scan the sheet in order to give an average reading. In the latter case, the signal is available only at much longer intervals (of the order of a minute). In this instance, a correction of the type used in the flow control system must be used, which is much less efficient with relatively rapid perturbations. The final choice between the two solutions—a single scanner type pick-up or one fixed pick-up for control purposes and a scanner for monitoring across the sheet—depends on the operating characteristics of the machine.

Short-term projects

THESE include simultaneously the progress of a simulation study and of experiences on the machine.

The work is now oriented towards a study with a control algorithm of the type defined above for a certain number of products (grades) so as to be able to check performance in the face of various perturbations (retention, adjustment of suction pressure in the wet end, for example), also towards a study of the dryers, in which it is intended to use basically the identification method.

In the corresponding experiments, we expect to be able to achieve—

1. Finer calibration of the wet end and dry end moisture and basis weight combined metering units, as well as turbidity meters.
2. A basis weight control system that, in a first stage during the first half of 1969, will take into account the indications from the wet end metering unit, with further refinements later in the form of flow and consistency measurements around the wire.

We have also started investigating changes in machine operating conditions, the practical application of the theoretical results being due during the second half of the year.

Conclusions

THE first part of this study, lasting over some 18 months, has been mainly preparation—definition of the problems, installation of the equipment and familiarisation with it.

The theoretical study carried on concurrently with the practical installation work has enabled a sound foundation to be laid for a satisfactory start when the computer was brought in.

In particular, the model study of the wet end showed that the effect of whitewater recirculation may be ignored, around an operating point. This simplification enabled a type of dry basis weight control to be devised, which was applied right from the start.

We are now entering a new phase of the study, the addition of a computer enabling our working methods to be modified.

We are in fact going to put into practice the first control systems considered and make any necessary additions. The powerful instrumentation facilities now at our command will enable us to deepen our knowledge of the unit with a view ultimately of developing more sophisticated control systems and procedures for production changes.

References

1. Sullivan, P. R. and Schoeffler, J. D., 'Dynamic simulation and control of the Fourdrinier papermaking process': International Federation of Automatic Control Congress, London, June 1966
2. Smethurst, J. A., 'Direct computer production changes on a simulated quality papermachine': *ibid.*
3. Terhar, R. J., 'Simulation in the pulp and paper industry': Seventh Pulp & Paper Instrument Society of America Symposium, Augusta, U.S.A., April 1966
4. Nougaret, M., *Commande par calculateur en ligne de la composition de tête d'une colonne de distillation pilote* (Thesis for Doctor in Engineering, Science Faculty, University of Grenoble, August 1966)
5. Volgnine, L. N., *Programmation des calculatrices numériques de commande* (Bibliothèque de l'automaticien, Dunod, Paris, 1966, translation into French by J. Coeuillet)

HEAD BOX CONTROL

H. H. CHAO, Development Engineer and I. B. SANBORN, Process Control Manager, Consolidated Papers Incorporated, Wisconsin Rapids, U.S.A.

Synopsis Head box control requirements have been reviewed and the equipment required to achieve control has been discussed briefly. In addition, methods that have been used to model the head box successfully are described and discussed from both an analytic and practical point of view. Finally, six head box control schemes varying in complexity from a simple analog level control to a complex decoupled 3×3 controller (controlling head box total head, liquid level and flow from the box) are described and fully discussed. The paper is closed by a brief discussion of work that ought to be done in the relatively near future.

Introduction

THE head box is of central importance to the papermaker, for with it he generates the thin jet of stock subsequently deposited on the Fourdrinier wire and formed into paper. In the past, this operation has required considerable papermaking art, but engineers have recently made significant progress in systemising the means by which head boxes are operated and controlled. The result has been not only to make head boxes easier to operate, but to increase the efficiency and accuracy with which they can be run.

Head box requirements

THOUGHTFUL consideration indicates that the optimum control system for an air-loaded head box should be able to meet six criteria—

1. A constant liquid level must be maintained in the box for practical reasons.
2. The total head within the box must be controlled to close tolerances to ensure that a constant difference (drag) between wire speed and jet velocity is maintained, despite changes in wire speed.
3. The flow through the box must be controlled to allow adjustments for changes in drainage, retention, formation and the like.
4. A constant jet escape angle (as defined in Fig. 9) should be maintained to achieve delivery of stock to the Fourdrinier machine in a controlled manner.

Under the chairmanship of P. E. Wrist

5. The mass flow rate of fibre across the head box slice must be held constant to ensure a flat basis weight profile in the finished paper.
6. Turbulence within the head box must be controlled closely to ensure that stock of a reproducible flock size is delivered to the slice and formed into a sheet showing good formation as well as stable strength and optical properties.

At present, it is practicable to meet only the first three of the above criteria on a closed loop basis. The three most convenient and economic manipulated variables to accomplish such control are air valves stem position, water valve stem position and slice opening. Unfortunately, the adjustment of any one of these variables causes a change in all three controlled variables (liquid level, total head and liquid flow); hence, the trick is to know how and when to adjust the valves and slice. This problem will be discussed in quite some detail later.

Control of escape angle, basis weight profile and jet flock size, though theoretically possible, is difficult on a practical basis for several reasons. First, in the case of escape angle and jet flock size, there is no good means of measuring the variable to be controlled. Second, there is not enough known about the effect of convenient manipulated variables on these controlled variables. Third, very little control can be exercised over the degree of activity on a Fourdrinier wire; hence, it is very possible that the beneficial effects of proper head box operation can be completely masked by unfortunate conditions on the wire. Lastly, there is only very limited evidence about how profound an effect the last three variables have on final sheet properties (perhaps because of variations in wire activity); and it is difficult to justify work in this area on a strictly economic basis.

Such difficulties should not be unduly discouraging, however, for until recently the benefits of total head and flow control were not generally recognised. Now, anyone having such controls will attest to their practical and economic worth. As a result, some hypothesising and discussion on how closed loop control of escape angle, basis weight profile and slice flock size can be achieved will be included at the end of the paper.

The head box as a system

THE physical limitation of a system are invariably the factors that determine its controllability. Hence, system controllability is affected not only by shortcoming in the mechanical design of the head box, but also by deficiencies in the equipment used to sense and adjust controlled and manipulated variables. Such drawbacks are best appreciated by studying Fig. 1, a schematic sketch of a typical, air-loaded head box system designed to operate in the 1 000–1 500 ft/min speed range. No controllers are included in the sketch to simplify matters and concentrate attention on process flows, manipulated and controlled variables.

Two flows are of interest in the system—

1. The flow of stock through the system and out the slice.
2. The flow of air to and from the head box air pad.

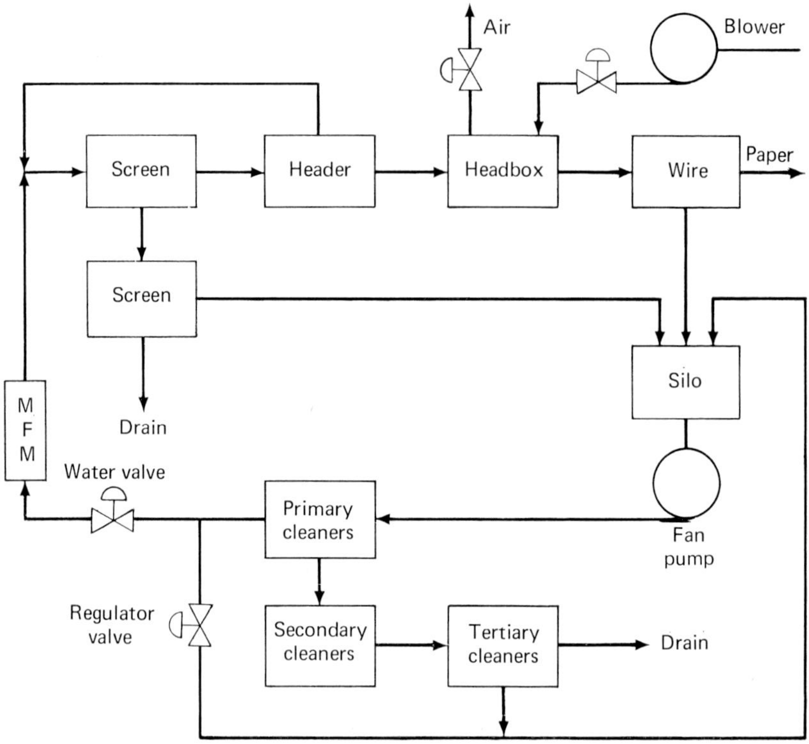

Fig. 1—Wet end system

Stock flow originates at the fan pump (a constant speed centrifugal pump), the prime mover for the entire whitewater system. From there, is passes to a bank of primary centrifugal cleaners, where approximately 80 per cent of the flow (3 500–4 500 gal/min at 0·5–0·8 per cent consistencies) is accepted and sent on through the system. Rejects are recirculated to the silo (by way of the secondary and tertiary cleaners depicted in Fig. 1) and from there to the fan pump again. On the exit side of the cleaners, the accepts stream encounters a pressure regulating valve, which serves to maintain constant pressure on the upstream side of the systems water valve or stream flow valve by recirculating part of the accepts back to the silo. The function of the water valve is to control the entire flow rate of stock in the whitewater system. In addition, it is one of the manipulated variables used by the head box control system.

Surprising as it seems, in at least one instance, this manipulated variable is a conventional 90° butterfly valve (without the benefit of a 'fishtail') equipped with a pneumatic stem positioner controlled by an electrically driven precision pressure regulator. Initially, it was felt that a small by-pass valve with good control characteristics would be required to achieve the accuracy of adjustment demanded by the head box control system, but subsequent practical experience proved this not to be the case and it has been found that a simple digital controller capable of gain adjustment when load changes require it is sufficient for the purpose. Feedback of valve stem position or pressure is possible, but not used, because of a lack of accuracy. Instead, a difference type algorithm is used in the digital controller, doing away with the need for a feedback signal from the manipulated variable.

Immediately after the water valve is located a 14 in magnetic flow meter used to measure the total flow to the head box. As previously mentioned, liquid flow to the head box is a controlled variable and, in this instance, it can be sensed with an accuracy and reproducibility of 1 per cent and 0·5 per cent of full scale, respectively. Since the principle of operation of the magnetic flow meter is well understood, it will not be discussed here. Suffice it to say that this meter generates an electric signal that makes it quite compatible with modern control systems hardware.

Just before it enters the centrifugal screen, the main stock stream is joined by a recirculated stream coming from the exit side of the head box distribution header. This recirculated stream amounts to about 10 per cent of the whole flow and enters the centrifugal screen along with the main stream, whereupon fibre bundles and the like are removed and sent to a vibratory screen for further purification before being either accepted and returned to the silo or rejected to the sewer. Accepted stock at the screen flows on directly to the head box distribution header. At the distribution header, the stock stream is converted into a slowly moving mass of stock of roughly rectangular shape before its narrowing into the narrow rectangular jet subsequently deposited on the wire. Excess stock at the end of the header is recirculated to the centrifugal screen as previously mentioned.

The two other controlled variables in the head box system are sensed at the head box proper. These variables are head box liquid level and head box total head. Both are essentially pressure measurements made with the aid of differential pressure (DP) cells equipped with a diaphragm, the position of which is sensed by a lever system that in turn converts mechanical movement into a pneumatic signal by a flapper-nozzle arrangement. In the case of liquid level, the variable is sensed by attaching one side of the cell to a tap in the bottom of the head box and the other side to a tap connected to the air pad of the box. Total head, on the other hand, is sensed by connecting one side of the cell to the atmosphere and the other to another tap located at the bottom of the head box. The total accuracy of DP cells is usually about 0·5 per cent of range, but usually the pneumatic signal put out must be converted to its electric analog via a PI converter, which has an accuracy and repeatability in the order of 0·2 per cent and 0·10 per cent of range, respectively. Time lags generated in sensing these signals are not substantial, though it must be admitted that nothing is stated in the instrument specification about this matter.

Slice opening is a head box manipulated variable and is adjusted by means of a conventional electric motor (with all winding sealed from the atmosphere to prevent problems caused by the hostile environment) controlled by an ordinary open/close contact, which may or may not be remotely driven. Slice position is not sensed, since experience has shown that it is difficult to obtain an accurate signal. Hence, slice adjustments are calculated with the aid of a difference type digital controller, which eliminates the need for manipulated variable feedback.

Modern head box control systems also require a knowledge of wire speed to allow drag calculations to be made and total head set points to be adjusted accordingly. Normally, the papermaker wishes to control drag at a given wire speed. Hence, he is not interested in total head as a control variable directly, but rather in maintaining a drag set point. This means that, although total head is a true control variable, it very frequently is the slave of a cascaded loop in which drag is specified wire speed measured and total head set point calculated. Wire speeds may be sensed by a variety of methods. The method used in our particular instance was to attach a magnetic pulse generator to the couch roll of the Fourdrinier machine and use an electronic digital computer to count the pulses generated and to calculate wire speed accordingly. The accuracy and reproducibility seems very good, amounting to perhaps 0·1 per cent of the range.

The second flow of interest in the study of the head box as a system is the flow of air to and from the air pad. Compressed air for this purpose is provided by a small water-sealed, centrifugal blower. Air passes from the blower through a control valve and into the head box, from which it subsequently escapes via an outlet equipped with a second control valve. These valves taken as a pair are the final manipulated variables used by the head box control system. They are usually mounted in a push/pull arrangement, which allows the very rapid adjustment of air pad pressure with an air supply and air lines of minimum size. The valves themselves are usually pneumatically operated by stem positioners that have remotely adjustable electric pneumatic converters controlling their stem positions. In this case, a serious time lag (in the order of

5 s or more) results whenever these valves must be moved, but such lags are unavoidable at the present time and may be compensated for quite adequately if controllers are properly tuned.

In summary, it should be noted that there are two flows, three manipulated variables, four controlled variables and wire speed that are of interest in a head box system. The flows of course are liquid flow through the box and air flow through the air pad. The manipulated variables are water valve stem position, slice opening and air valves stem position. Controlled variables are liquid flow through the box, total head, liquid level and drag. Wire speed information is required to allow total head set points to be calculated from a knowledge of the drag desired.

Model development

As previously mentioned, three manipulated variables (air valves stem position, water valve stem position and slice opening) are used to maintain control of the head box systems. Unfortunately, each of the system's controlled variables is affected by all three of its manipulated variables. For example, if the air valves are adjusted, they not only affect total head by changing air pad pressure, but also upset liquid level and stock flow rate as well. Interactions can make the design and tuning of a control system quite tedious. Hence, it is essential to establish the dynamic response of all the controlled variables to adjustments in each of the manipulated variables as a first step in developing a suitable control scheme for a head box system.

TABLE 1

No.	*Basic equation*	*Laplace transform upon linearisation*	*Description*
1	$\frac{dM}{dt} = M_a$	$M = \frac{M_a}{s}$	The rate of air mass change in air pad is equal to the net air flow
2	$P_a = \frac{M}{V} RT$ $V = V_a - Ah$	$P_a = C_1 M + C_2 h$	The pressure in air pad is related to air mass and volume by a gas equation, where V_o is total head box volume and Ah is the volume occupied by water
3	$H = h \times \frac{P_a - 14{\cdot}7}{e}$	$H = h + C_3 P_a$	Total head is equal to the water head plus pressure head
4	$Q_o = CdWY\sqrt{2gcH}$	$Q_o = C_4 H + C_5 Y$	The jet velocity is $\sqrt{2gcH}$, the flow rate through slice is the product of jet velocity, cross-section of slice and flow coefficient
5	$\frac{dh}{dt} = \frac{1}{A}(Q_i - Q_o)$	$h = \frac{1}{A_s}(Q_i - Q_o)$	The rate of liquid level change is equal to the difference between incoming and outgoing flows divided by cross-section of head box

A list of nomenclature is given in Appendix 2

In 1960, Mardon *et al.*[1] developed the basic equations describing the head box system (see the first column in Table 1) and linearised them so that they could be Laplace transformed and used to characterise the system in block diagram form (Fig. 2). A typical example of how this was accomplished is described below for equation (*2*) of Table 1.

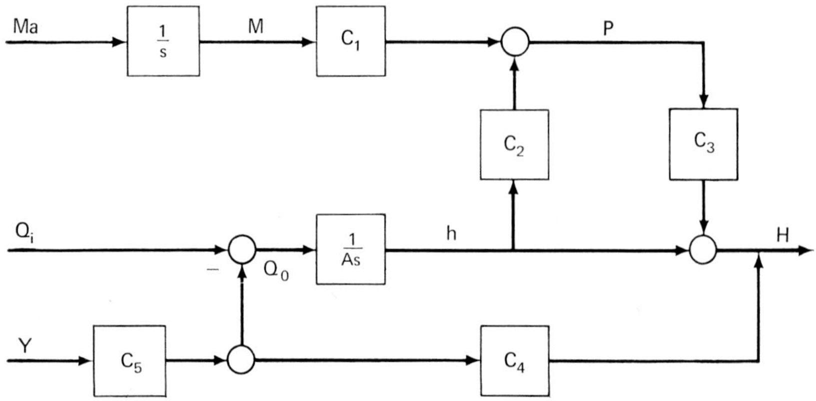

Fig. 2—Block diagram of head box

Consider the pressure P in the air pad of a typical head box. If it is assumed that variations in this pressure are characterisable by the perfect gas law, then—

$$P = M.RT'/V \qquad (1)$$

Now if—

$$V = V_o - Ah$$

then equation (*1*) becomes—

$$P = \frac{M}{V_o - Ah} RT' \qquad (2)$$

If M and h are independent variables, then—

$$dP = \frac{\partial P}{\partial M} dM + \frac{\partial h}{\partial P} dh \qquad (3)$$

where $\dfrac{\partial P}{\partial M} = \dfrac{RT'}{V_o - Ah}$; $\dfrac{\partial P}{\partial h} = \dfrac{-AMRT'}{(V_o - Ah)^2}$

If the perturbations of M and h are small, then both partial derivatives will be a constant. Hence—

$$\Delta P = C_1 \Delta M + C_2 \Delta h \qquad . \quad . \quad . \quad . \quad (4)$$

where $\dfrac{\partial P}{\partial M} = \dfrac{RT'}{V_o - Ah} = C_1 \qquad \dfrac{\partial P}{\partial h} = \dfrac{-AMRT'}{(V_o - Ah)^2} = C_2$

and the Δ operator signifies the magnitude of the deviations from an original or reference value. Taking the Laplace transform of equation (*4*), we obtain—

$$P(s) = C_1 M(s) + C_2 h(s) \qquad . \quad . \quad . \quad . \quad (5)$$

which is equivalent to the expression listed in column 2 of Table 1.

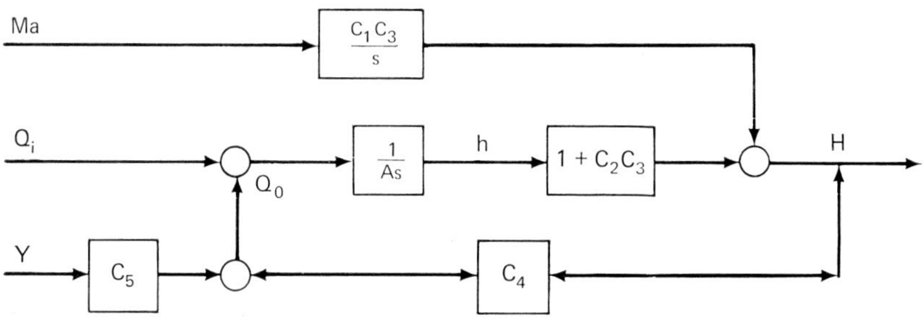

Fig. 3—Condensed block diagram of head box

The block diagram of the head box system shown in Fig. 2 may be reduced to a simpler form as shown in Fig. 3. Before the development of a system model is discussed further, however, it should be pointed out that the manipulated variables assumed in the equations of Table 1 are not the same as discussed previously. Slice position is assumed to be a manipulated variable in each case, but stock flow to the box and the mass of gas in the air pad in the theoretical model are taken as manipulated variables instead of air and water valves stem position, which are the true variables. Assuming these intermediate variables to be manipulated variables leads to a considerable simplification of the modelling problem, however, without any serious consequences to model validity (as will be shown in a subsequent section) and is justified on practical grounds.

From Fig. 3, it is possible to obtain the general transfer functions of system-controlled variables (H, h and Q_0) with respect to system-manipulated variables (Q_i, M_a and Y). For instance, the transfer function between total head H and flow to the box Q_i is of the form—

$$\frac{H}{Q_i}=\frac{\frac{1}{A_s}(1+C_2C_3)}{1+\frac{1}{A_s}(1+C_2C_3)C_4}=\frac{1+C_2C_3}{A_s+(1+C_2C_3)C_4} \quad . \quad . \quad (6)$$

Nine transfer functions are required completely to characterise system dynamics. The other eight functions along with the one developed here are presented in Table 2, from which it may be seen that seven of the nine transfer functions are simple first order lags. This is of considerable import, as will be evident later when practical identification methods are discussed.

The two transfer functions that differed from the rest were G_{22} and G_{33}. G_{22} relates h to M_a and is of the form $K/S(Ts+1)$. Hence, it contains a first order lag, but includes an integrator as well. The function G_{33}, which relates Q_o and Y, is unique, because, although it also contains a first order lag, derivative action is present as well. These last two types of transfer function present somewhat different identification requirements.

Before proceeding further with a discussion of practical identification procedures, one additional complication should be discussed. As has already been mentioned, the transfer functions contained in Table 2 represent the dynamics between intermediate variables and system-controlled variables. Hence, the dynamics relating the intent (on the part of a controller) to move the water valve, the air valves or the slice and the intermediate variables of Table 2 must be discussed.

TABLE 2

Value	*Total head* (*H*)	*Liquid level* (*h*)	*Outlet liquid flow* (Q_o)
Inlet liquid flow (Q_i)	$G_{11}=\dfrac{1+C_2C_3}{A_s+C(1+C_2C_3)}$	$G_{12}=\dfrac{1}{A_s+C_4(1C_2C_3)}$	$G_{13}=\dfrac{(1+C_2C_3)C_4}{A_s+(1+C_2C_3)C_4}$
Net air flow (M_a)	$G_{21}=\dfrac{AC_1C_3}{A_s+(1+C_2C_3)C_4}$	$G_{22}=\dfrac{1}{s}\dfrac{C_1C_3C_4}{A_s+(1+C_2C_3)C_4}$	$G_{23}=\dfrac{AC_1C_3C_4}{A_s+(1+C_2C_3)C_4}$
Slice position (Y)	$G_{31}=\dfrac{-(1+C_2C_3)C_5}{A_s+(1+C_2C_3)C_4}$	$G_{32}=\dfrac{-C_5}{A_s+(1+C_2C_3)C_4}$	$G_{33}=\dfrac{AC_5s}{A_s+(1+C_2C_3)C_4}$

Consider the case of a controller being used to cause a change in flow into the head box via an adjustment in water valve stem position. At least three distinct transfer functions are involved in achieving this objective (Fig. 4).

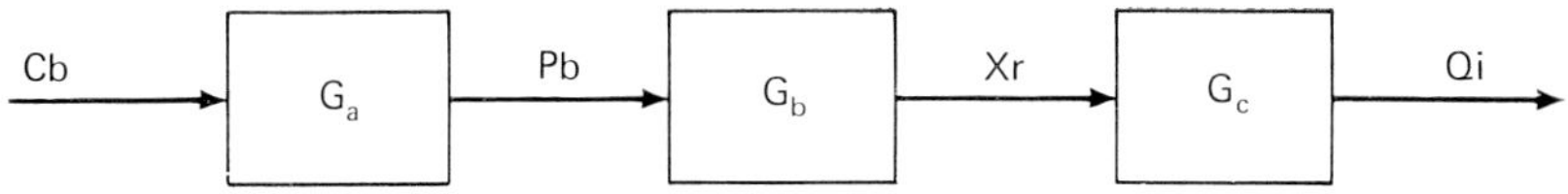

Fig. 4—Water valve transfer function

First, the controller signal $C(s)$ must be converted to bonnet pressure P_b, which in turn is converted to a new valve stem position X_r by means of the stem positioner. The change in stem position causes a change in the flow into the head box Q_i. Each of these transfer functions G_a, G_b and G_c are at best a first order lag, with a decided possibility that G_b could be a second order system. In addition, dead time may exist in all three transfer functions as a result of slippage and the like. The net effect of this is to give a general transfer function of at least third and probably fourth order with dead time. Nevertheless, the situation is not as formidable as might be expected, for such actuating systems are invariably set up to be decidedly overdamped. The net result is that the complete dynamics of the actuation system can be characterised quite adequately via a first order lag with dead time. The applicability of this simplifying assumption has profound beneficial effects for a practical identification procedures.

In summary, it may be said that adequate analytical models are currently available to relate the various manipulated variables to their controlled variable counterparts. In addition, the dynamics of the various actuation systems adjusting manipulated variables have been studied and found to be characterisable by a first order lag plus dead time. With the form of the general transfer functions thus established, the task remaining is to describe the methods used to estimate the specific numerical values of the parameters appearing in said models.

System identification

A DETAILED description of the identification procedure used to characterise model parameters is uncalled for in this text, since it has already been described in detail elsewhere.[(2)] Suffice it to say here that the technique used consists of—

1. Choosing a model of a form capable of characterising system dynamics adequately.
2. Inducing a rather prolonged, closed pulse in each manipulated variable of interest (water valve position, air valves position and slice opening in this case) while taking particular pains to note the response of each controlled variable to said upset.
3. Employing a rather specific mathematical search technique to estimate the model parameters giving the best fit to the data at hand.

Fig. 5–7 contain plots of the response of total head and liquid level to pulse upsets in air valves stem position, water valve position and slice opening, respectively. The magnitude and duration of these pulses are such that undue upsets in the process are avoided. Hence, it is possible to identify transfer functions for the head box system at will with considerable convenience. The first requirement is that the control loops on each controlled variable be

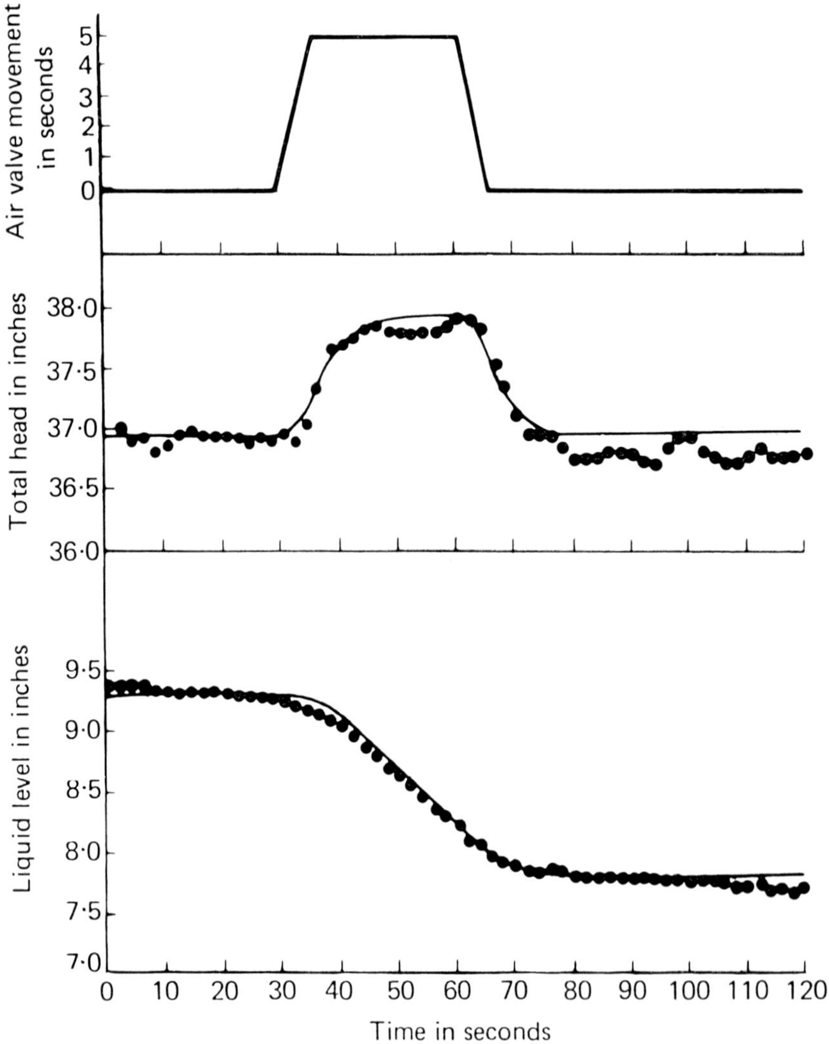

Fig. 5—Air valve identification data

opened shortly before the identification is to begin. Subsequently, the manipulated variable of interest is pulsed as required, with pains being taken that all appropriate data is obtained (usually automatically). The net effect is to put quite a convenient method of determining complete head box dynamics at the disposal of the engineer.

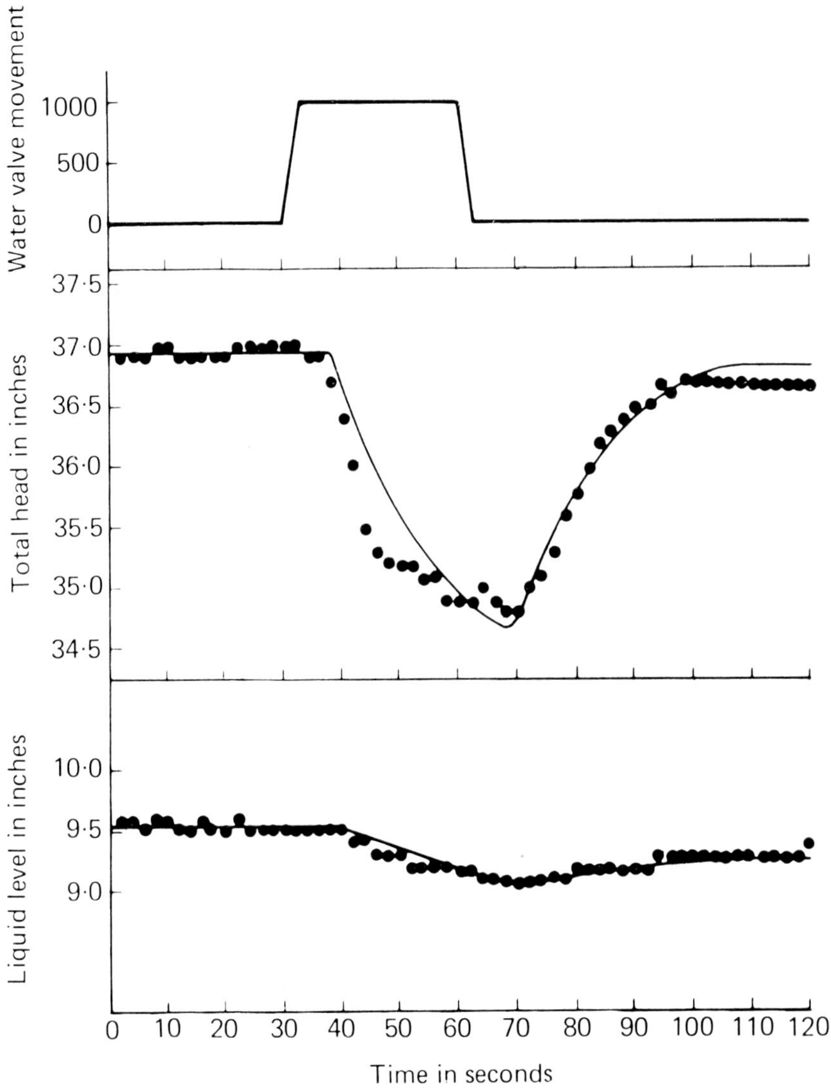

Fig. 6—Water valve identification data

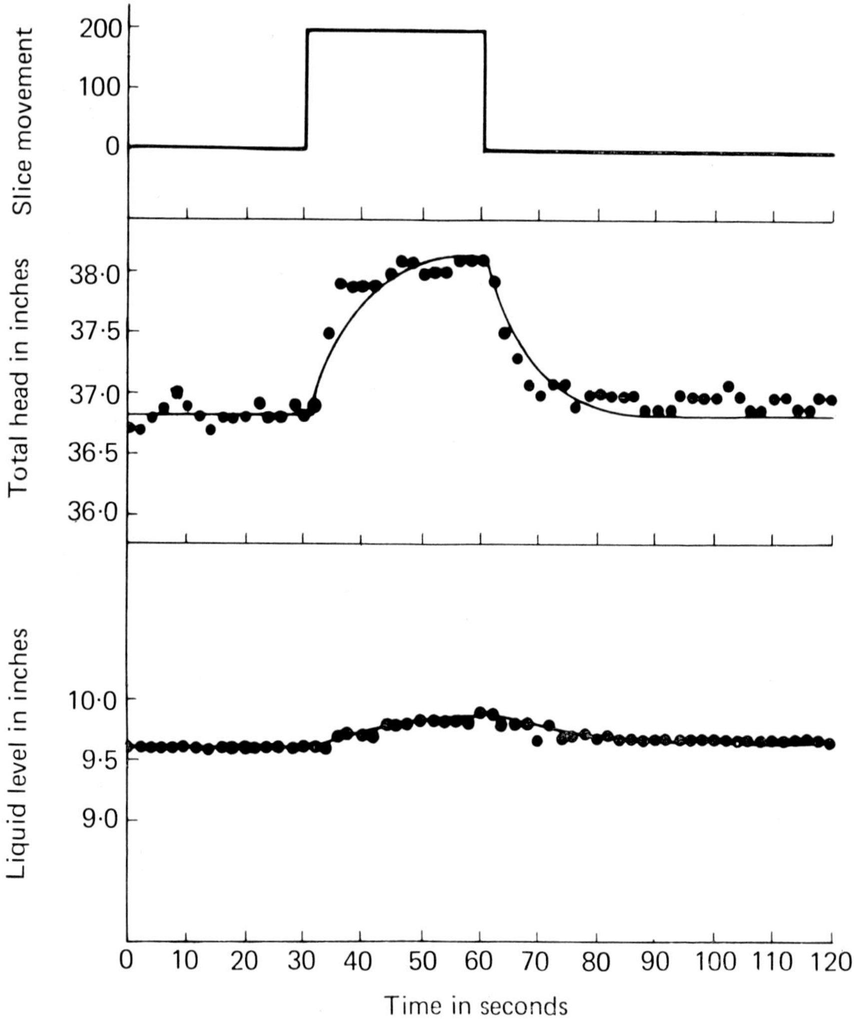

Fig. 7—Slice opening identification data

The closed circles in Fig. 5–7 represent actual data points with the continuous lines (in the case of liquid level and total head) being the response expected of the system if it were in fact completely characterised by the process model subsequently identified by our search procedure as being best. Needless to say, the fit in all six cases is quite good. The transfer functions identified are presented in Table 3. In each case, a dead time as well as a time constant and gain was identified. These dead times can probably best be ascribed to time delays in the actuation systems for the various manipulated variables; yet time constants and gains are undoubtedly the result of a combination of actuation and process dynamics. The integrator appearing in P_{22}, on the other hand, is completely the result of process dynamics as already discussed.

TABLE 3

Value	*Total head* (H)	*Liquid level* (h)
Water valve position (W)	$P_{11}(s) = \dfrac{0{\cdot}262e^{-8s}}{14{\cdot}62s+1}$	$P_{12}(s) = \dfrac{0{\cdot}142e^{-8s}}{75{\cdot}3s+1}$
Air valve position (X)	$P_{21}(s) = \dfrac{0{\cdot}02e^{-1{\cdot}5s}}{4{\cdot}46s+1}$	$P_{22}(s) = \dfrac{0{\cdot}001}{s}\,\dfrac{e^{-1{\cdot}5s}}{4{\cdot}46s+1}$
Slice position (Y)	$P_{31}(s) = \dfrac{0{\cdot}667e^{-s}}{6{\cdot}95s+1}$	$P_{32}(s) = \dfrac{-0{\cdot}21e^{-s}}{27{\cdot}5s+1}$

System non-linearities

As PREVIOUSLY mentioned, the basic equations defining the head box system are non-linear and need to be linearised via a perturbation technique. Experience has shown that this approach has worked well for head boxes operated close to one standard set of operating conditions, but that difficulties are encountered when appreciable changes are made in system operating points. Studies of the situation have shown that the difficulties experienced are a result of changes generally in process dynamics reflected by differences in identified process gains and time constants. Dead times for the most part have remained relatively constant, indicating that they are strongly related to actuator rather than to process dynamics.

TABLE 4

Type of transfer function	*At 50 lb/ream grade*	*At 80 lb/ream grade*	*At 100 lb/ream grade*
$\dfrac{H}{W}(s)$	$\dfrac{0{\cdot}354e^{-5{\cdot}5s}}{20{\cdot}6s+1}$	$\dfrac{0{\cdot}262e^{-8s}}{14{\cdot}6s+1}$	$\dfrac{0{\cdot}132e^{-8{\cdot}5s}}{16{\cdot}2s+1}$
$\dfrac{H}{Y}(s)$	$\dfrac{0{\cdot}749e^{-s}}{6{\cdot}9s+1}$	$\dfrac{0{\cdot}669e^{-s}}{6{\cdot}95s+1}$	$\dfrac{0{\cdot}628e^{-s}}{4{\cdot}5s+1}$

The data presented in Table 3 was obtained while producing an 80 lb/ream (3 300 ft^2) coated sheet at a speed of approximately 800 ft/min. To appreciate the effect that changes in operating conditions can have on system dynamics, refer to Table 4, which contains the transfer functions relating total head to water valve position and slice opening for three sets of operating conditions—the manufacture of a 50 lb/ream sheet at 1 100 ft/min, an 80 lb/ream sheet at 800 ft/min and a 100 lb/ream sheet at 600 ft/min. From Table 4, it is apparent that there are significant differences in process dynamics, indicating that system dynamics form a continuum varying continuously across the entire spectrum of operating conditions. During normal operations (at a fixed set of points), this fact causes no great difficulties, provided the appropriate system dynamics are used in calculating controller parameters. Occasionally, most papermachines must be moved from one set of operating conditions to another. This grade change presents particular difficulties from a control standpoint as a result of the head box non-linearities now being discussed. Steps taken to alleviate the problems encountered will be discussed at some length later under the heading *Controller design—part 5.*

Controller design—part 1

THE transfer functions given in Table 2 can be used to determine the effect of changes in any manipulated variable (Q_i, M_a or Y) on any one or all three of the controlled variables (H, h or Q_o). To be specific, consider the magnitude of the steady state error in total head resulting from a step change (of unit size) in liquid flow to the box. If the final value theory is applied, then we have—

$$\lim_{t \to \infty} H_{(t)} = \lim_{s \to o} (s)\left(\frac{1}{s}\right) G_{11}(s) = \lim_{s \to o} \frac{1 + C_2 C_3}{As + C_4(1 + C_2 C_3)}$$

$$\text{Lim} H_{(t)} = 1/C_4 \quad . \quad . \quad . \quad . \quad . \quad (7)$$

This means that, if a unit step in flow to the box takes place, a constant offset equal to $1/C_4$ will be experienced in total head. Following the same procedure, it can be shown that all the controlled variables save one will suffer a constant offset as a result of a step change in any one of the system's manipulated variables. The one exception to this rule is head box liquid level during a step change in air flow. Applying the final value theory in this instance shows that—

$$\lim_{t \to \infty} h_{(t)} = \lim_{s \to o} (s)\left(\frac{1}{s}\right) G_{22}(s) = \lim_{s \to o} \frac{1}{s} \frac{C_1 C_3 C_4}{As + (1 + C_2 C_3) C_4}$$

$$\lim_{t \to o} h_{(t)} = \infty \quad . \quad . \quad . \quad . \quad . \quad . \quad (8)$$

Hence, if there is an upset in air flow, the head box will eventually either completely fill or empty itself of stock, because of the integrator that appears in G_{22}.

As a result of the characteristic just discussed, it is an absolute necessity to provide each and every air-loaded head box with some type of liquid level controller. Originally, this controller was the Hornbostel hole; but now analog pneumatic controllers with liquid level as a controlled variable and air valves stem position as the manipulated variable are usually provided for the purpose. An air-loaded box with any less of a control system would be inherently unstable. Hence, at the very least, liquid level control must be provided as back-up for every head box system, regardless of its sophistication and complexity.

Further use of the final value theory will show that it is possible to stabilise the head box system with a simple proportional liquid level controller. If offsets are to be avoided, however, both proportional and integral action must be provided. As a result, modern head boxes are invariably supplied with at least a PI liquid level controller as standard equipment.

Generally speaking, tuning of simple liquid level controllers is no problem, for only one loop is involved and the interaction between manipulated and controlled variables is intuitively easy to grasp. As a result, any competent instrument man can tune the controller in the field using trial and error methods, thereby eliminating the need for formal identification and tuning procedures. It should be pointed out, however, that in all cases the controller used must be tuned with its parameters set so that the closed loop system will be stable under all possible operating conditions. The system is therefore usually somewhat sluggish over most of its operating range in order to avoid instability when process gains are at a maximum (usually when lightweight sheets are being made at high speeds).

Controller design—part 2

THOUGH only liquid level control is an absolute necessity for an air-loaded head box, systems are now supplied with both liquid level and total head control provided. The reasons for supplying the more sophisticated system are numerous, but boil down to the fact that it makes a head box (and hence its associated papermachine) easier to operate. For example, a total head/liquid level controller simplifies grade change procedures considerably. In addition, if a total head/liquid level controller is combined with means for automatically sensing wire speed, all the ingredients are available for providing automatic drag control on a Fourdrinier machine.

Fig. 8 contains a block diagram of a typical total head/drag control loop, from which it may be seen that—

1. The operator is allowed to specify either total head or drag for the sake of convenience.

2. Total head set point can be calculated directly from a knowledge of wire speed and drag set point.

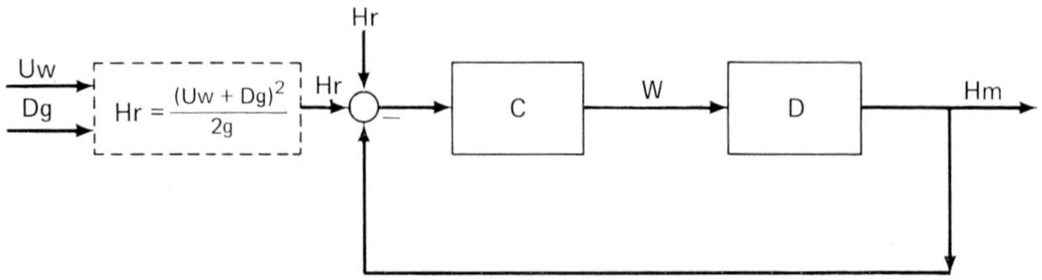

Fig. 8—Block diagram of drag controller

The main advantage of such a drag control feature is that it allows an important sheet forming condition to be held constant with time. This leads to more stable operations on the wet end and allows the manufacture of paper of better uniformity and quality.

Mardon *et al.*[1] have studied two specific controller configurations with which total head and liquid level can be controlled. Both proportional and integral action were required in all loops, which made one configuration very difficult to tune and the other moderately so. Nevertheless, the systems were studied extensively via analog simulation and it was concluded that the configuration shown in Fig. 9 was the best for stability and tuning generally. Unfortunately, no performance data for an operating head box was reported by Mardon *et al.*; but experience on the part of the authors indicates that such a system is very workable in a practical sense. Not only is it stable and dependable, but has the added benefit that, should the total head controller fail, liquid level will still be maintained via the conventional approach outlined in part 1. This latter feature has been of considerable practical benefit when it has been necessary to go on and off total head control.

Control according to Fig. 9 is usually implemented via analog hardware and, though tuning is possible by trial-and-error methods, it is difficult. Recently, Lee *et al.*[3] described a tuning method using a root contour approach that can successfully choose advantageous starting points for controller parameters. This greatly reduces the amount of 'witchcraft' involved in tuning such a multi-loop, interacting system as the head box. In addition, the analysis involved leads to a better appreciation for factors important to system stability.

In summary, it is possible to control both total head and liquid level with conventional analog hardware attached to a head box in a straightforward manner. In addition, experience shows that the control system diagrammed in

Fig. 9 works best on a practical basis. Tuning methods are also available to give good starting points for controller parameters. This simplifies system tuning considerably.

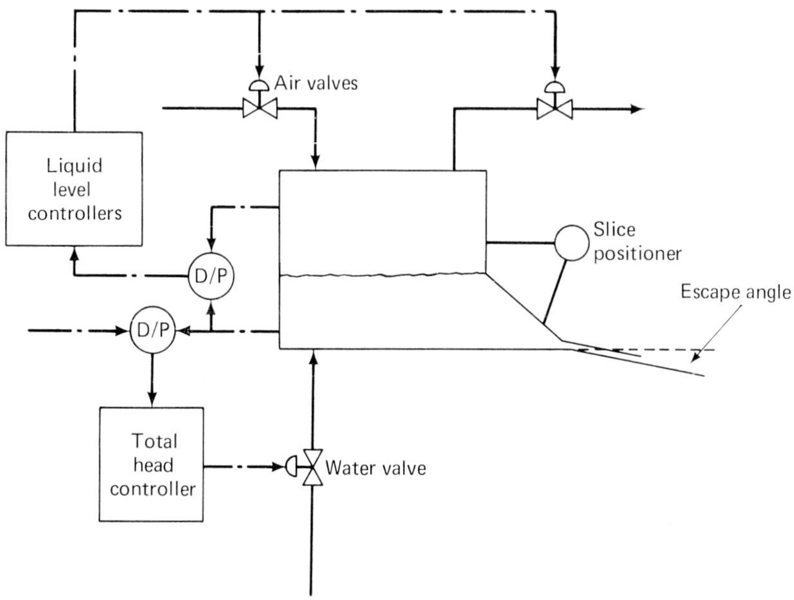

Fig. 9—Schematic diagram for best analog controller configuration

Controller design—part 3

HEAD box dynamic parameters are affected by changes in operating point (Table 4). This means that a control system tuned is an optimum under one set of circumstances may be useless for a second set. To illustrate the point, consider the response curves shown in Fig. 10. These data were obtained via simulation (on a digital computer), but should be quite representative of the response shown by a full-scale operating head box, because the model and identified parameters previously reported in Tables 3 and 4 were used. (A block diagram of the simulation program and an explanation of it are included in Appendix A.) In addition, the use of analog controller hardware with invariant parameters was assumed. The upset studied was a 1 in change in total head set point and the controllers were set to give relatively good response for a 100 lb/ream sheet.

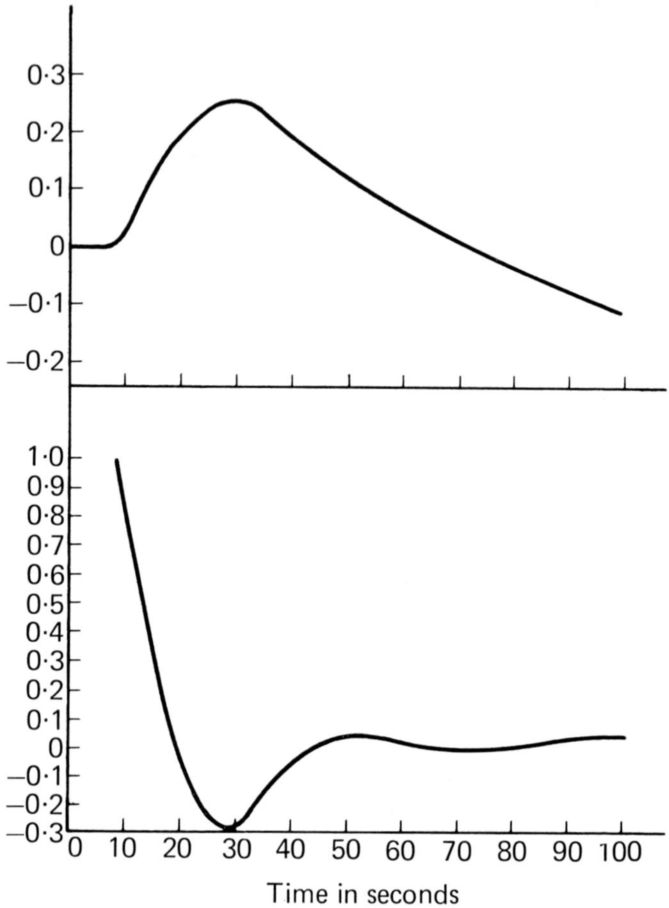

Fig. 10a—Simulated response of an analog control system of 100 lb/ream grade

If settling time to a 10 per cent error and amount of overshoot are taken as performance criterion, it may be seen that—

1. Total head control though somewhat oscillatory (overshoot 27 per cent, settling time 37 s) is adequate in the case of 100 lb/ream paper manufacture (Fig. 10*a*).
2. The total head controller is definitely too active (overshoot 61 per cent, settling time 47 s) in the case of a 50 lb/ream sheet (Fig. 10*b*).

3. Though upsets in liquid level were minor in each case, liquid level showed a greater deviation and slower period of oscillation in the case of 100 lb/ream paper manufacture than for a 50 lb/ream sheet.

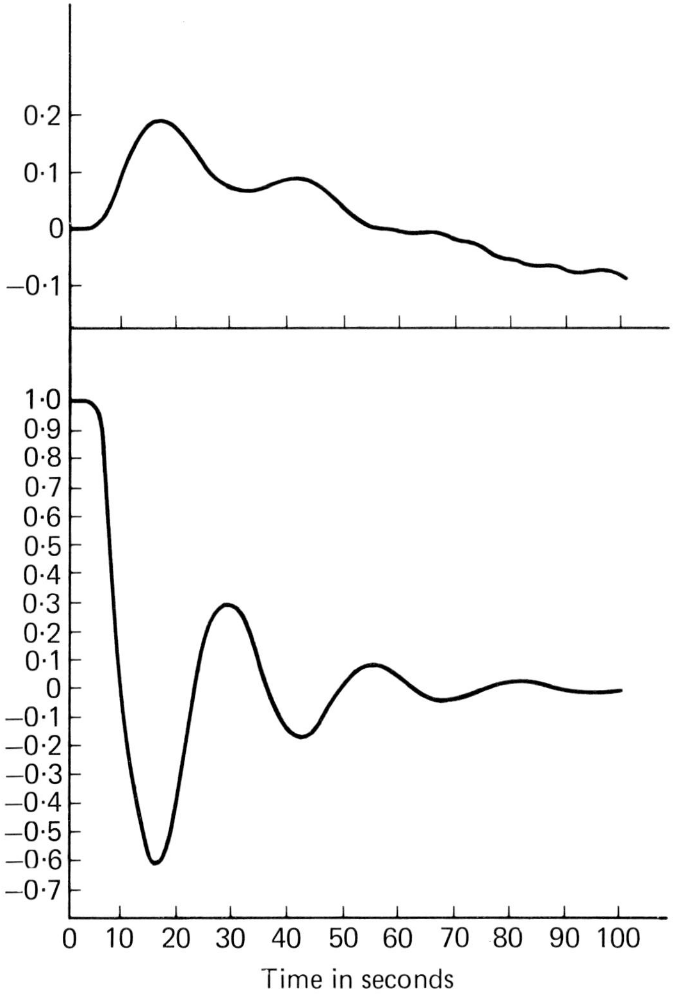

Fig. 10b—Simulated response of an analog control system at 50 lb/ream grade

This sacrifice of liquid level control undoubtedly helped the system in the case of the 100 lb/ream sheet, but it still must be recognised that the primary cause of the trouble in the case of the 50 lb/ream sheet was the change in head box dynamics.

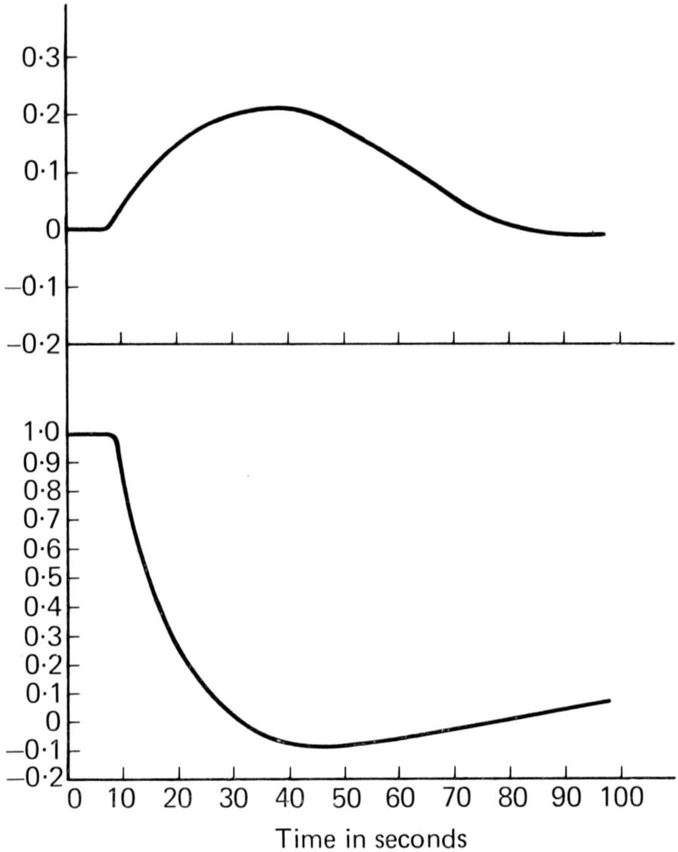

Fig. 11a—Simulated response of a digital control system at 100 lb/ream grade

Fig. 11 contains the same type of simulation results for a head box equipped with a digital, adjustable parameter, total head controller and analog liquid level control. Two major advantages of the digital controller are illustrated by these data. First, comparison of the response curves in Fig. 10(*a*) and 11(*a*) (manufacture of a 100 lb/ream sheet) shows that the digital controller (9 per cent overshoot and a 25 s settling time) out performed the analog controller, despite the fact that they were both tuned to give the same initial response to the total head upset. This was the result of the digital system's better ability to deal with the 8 s dead time in the head box.

The second advantage of the digital controller is best illustrated by comparing Fig. 10(*b*) and 11(*b*) (manufacture of a 50 lb/ream sheet). Here the digital system's advantage lies in its adjustable parameters and the result is far better performance (no overshoot and a 30 s settling time) than the analog system. In addition, it should be noted that less of a load was placed on the analog liquid level controller when a digital total head control was used and this resulted in less liquid level disturbance and a slower period of oscillation.

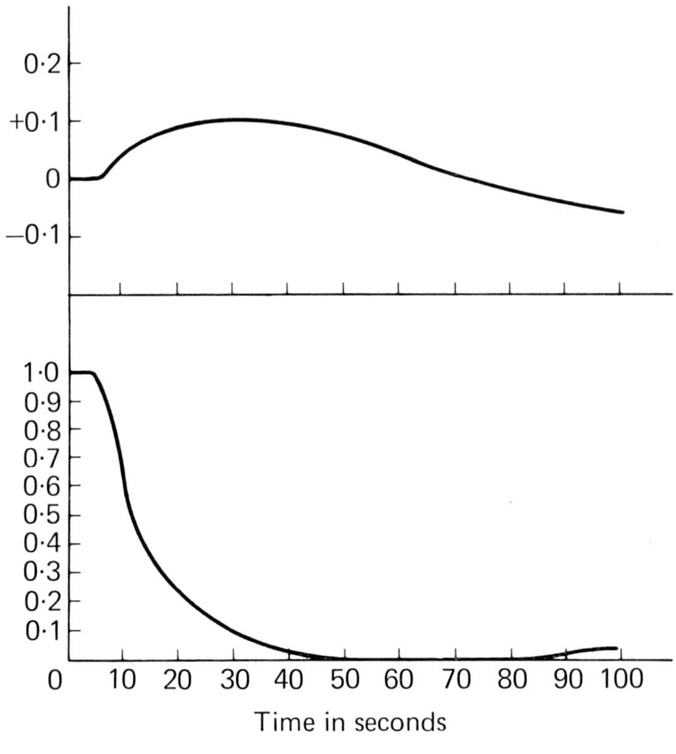

Fig. 11b—Simulated response of a digital control system at 50 lb/ream grade

The digital controller used in the above simulation was designed by a cancellation compensation procedure reported by Dahlin *et al.*[4] The approach used was to first specify the closed loop response desired, then derive the form of the controller required algebraically from a previous knowledge of the process' open loop response. For example, consider the system block diagram in Fig. 12. The general transfer function is of the form—

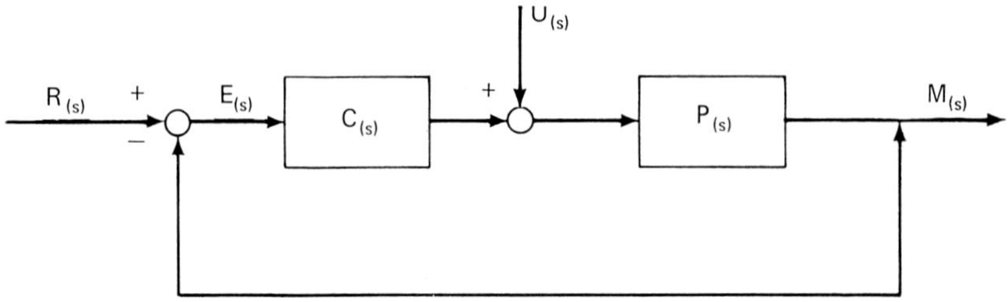

Fig. 12—Block diagram for generalised control loop

$$\frac{M(s)}{R(s)} = \frac{P(s)C(s)}{1+P(s)C(s)} \qquad (9)$$

where $P(s)$ and $C(s)$ are the open loop transfer function of the process and the controller, respectively. If we specify that—

$$\frac{M(s)}{R(s)} = F(s) \qquad (10)$$

then $C(s)$ is defined and may be solved for algebraically.

$$C(s) = \frac{F(s)}{1-F(s)} P(s)^{-1}. \qquad (11)$$

Now if, as in the case of the head box, $P(s)$ is a first order system with time constant T, gain K and dead time D, then it is most practical to specify that—

$$F(s) = \frac{\lambda}{s+\lambda} e^{-sD} \qquad (12)$$

where λ is the reciprocal of the system's closed loop time constant and is referred to as its cut-off frequency. Making the substitution indicated in equation (*11*), we obtain—

$$C(s) = \frac{1}{K} \cdot \frac{\Upsilon s+1}{\lambda s+1-\lambda e^{-sD}} \qquad (13)$$

Such a transfer function is difficult if not impossible to implement via conventional analog hardware. Nevertheless, it can be easily implemented digitally as evidenced by the fact that, if the z-transform of equation (*13*) is taken, of we obtain—

$$C(z) = \frac{1-\theta}{K(1-\eta)} \frac{1-\eta z^{-1}}{1-\theta z^{-1}-(1-\theta)z^{-d-1}} \qquad (14)$$

where T is the sampling interval and—

$$\theta = e^{-\lambda T},\ \eta = e^{-T/T},\ d = D/T,$$

but, since $C(z)$ denotes the transfer function between the manipulated and controlled variables of the system, we may say that—

$$C(z) = \frac{W(z)}{\Delta H(z)} \qquad (15)$$

which, if substituted in equation (*14*), gives—

$$W(z) = \frac{(1-\theta}{K(1-\eta)}(\Delta H - \eta\,\Delta Hz^{-1}) + \theta W(z)z^{-1} + (1-\theta)W(z)z^{-d-1} \qquad (16)$$

which when reduced to its recursive form becomes—

$$W_i = \frac{(1-\theta)}{K(1-\eta)}(\Delta H_i - \eta\Delta H_{i-1}) + \theta W_{i-1} + (1-\theta)W_{i-d-1} \qquad (17)$$

the digital algorithm required to achieve the closed loop response $\lambda e^{-sD}/s+\lambda$.

Such an algorithm has been used to control the total head in the head box of an operating machine for some two years now. Typical responses for this box to upsets in total head and slice opening are illustrated in Fig. 13, from which it may be seen that response is somewhat sluggish, but dependable. This response is by design, since an entire closed loop time constant of 25 s for the digitally controlled total head loop has been chosen. This allows a sampling interval (T) of 8 s to be used, which greatly simplifies programming problems and allows one computer to be spread further while still providing adequate control.

In summary of this section, a digital total head controller holds certain advantages over an analog system—namely, better dead time compensation and easy parameter adjustment. In addition, cancellation compensation design techniques taking advantage of the versatility of digital algorithms are available and one design procedure was reviewed. Finally, the actual performance of an operating head box was illustrated and shown to be satisfactory from a practical standpoint.

Controller design—part 4

THUS far, the controls of only total head and liquid level have been discussed, but occasionally the papermaker wishes to adjust the flow of stock from his box as well. Under normal steady-state conditions, the effect of this is to change the consistency of the stock in the head box. Such a change is one of the most direct methods by which a machine tender can adjust the paper's formation. Hence, though there is a definite need for flow control, flow set

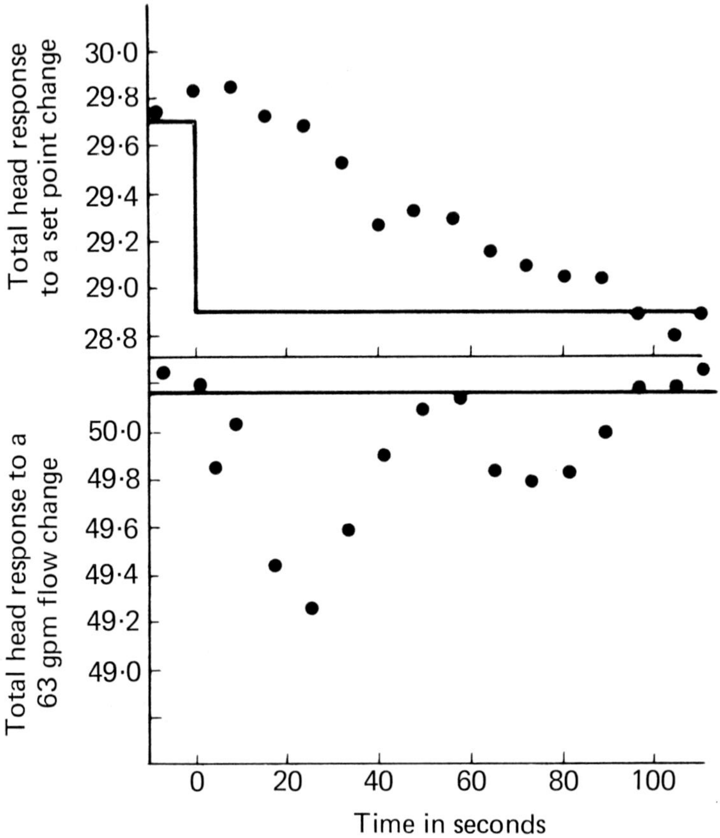

Fig. 13—Actual digital controller

point is adjusted infrequently, because the papermaker does not usually wish to toy with a property as elusive as formation.

As already mentioned, the manipulated variables used in gaining control of total head and liquid level were water valve and air valves stem position. The only other variable that can be conveniently manipulated in the head box system is slice position. Consequently, it is used somewhat indirectly to achieve control of this third controlled variable—flow from the box. In other words, in the control system found to be most convenient by these authors, water valve stem position, air valves stem position and slice opening were the variables used to control total head, liquid level and flow from the box, respectively. Other combinations of manipulated and controlled variables can be devised, of course, but this arrangement has seemed to give the best general head box performance.

The ideal feedback signal for a head box flow controller is flow from the box, which should be returned to the controller via a set point comparator. Unfortunately, it is not possible to measure flow from the box directly, but, under steady state conditions, it is a fact that flow to the box must equal flow from the box. Hence, it is possible to use a measure of flow to the box (which can be measured) as an estimate of flow from the box, but this complicates the control system considerably (as will be illustrated in the next section) by virtue of the fact that the total head controller's dynamics are inserted into the feedback path of the flow control loop. Under normal (non-grade change) circumstances, however, this complication is not particularly restrictive, because an active flow controller is not particularly desirable (the papermaker does not appreciate having adjustments made to 'his' slice however small and for whatever reason). This means that the controller is intentionally designed to be sluggish (low gain) and a dead band of appreciable size (2 per cent of the mean flow) is incorporated to avoid unnecessary slice movements.

Experience has shown proportional action to be quite adequate for this application and a digital algorithm was used to close the loop because of the ease with which a dead band could be incorporated and the fact that the necessary hardware was at hand. Under normal circumstances, however, it is doubtful if the extra cost of digital adjustment could be justified. Finally, it should be stated that such a controller has been in operation for some two years now with no mishaps of wire flooding and the like. Set points are entered in terms of gal/min of stock desired and so far the machine tenders involved seem well pleased by the results.

Controller design—part 5

ALTHOUGH an intelligent combination of the control approaches described in parts 3 and 4 of this paper seem best at present for normal operating conditions, the requirements during a grade change are markedly different. For example, it might take as long as 20 min for the head box controller to settle down after a grade change of substantial magnitude (300 ft/min speed change or the like) due to interactions between the total head and flow from the box control loops. Such performance would be intolerable in that it would delay grade changes inordinately, as well as increasing the danger of a wet end break.

The best way to achieve fast, accurate response in both the total head and flow loops during grade change is to use a decoupled controller that takes the above-mentioned interactions into account. Decoupling in this context requires that—

1. When only a total head set point change is imposed, both the water valve and slice are used to achieve said change while holding the flow constant.
2. When only a flow change is required, again use both the water valve and slice to achieve it while holding total head constant.
3. If set point changes in both total head and flow are called for, use both manipulated variables to achieve the change while simultaneously taking into account interactions and correcting for them.

Such a procedure will insure not only fast, but accurate changes to reach the new operating point.

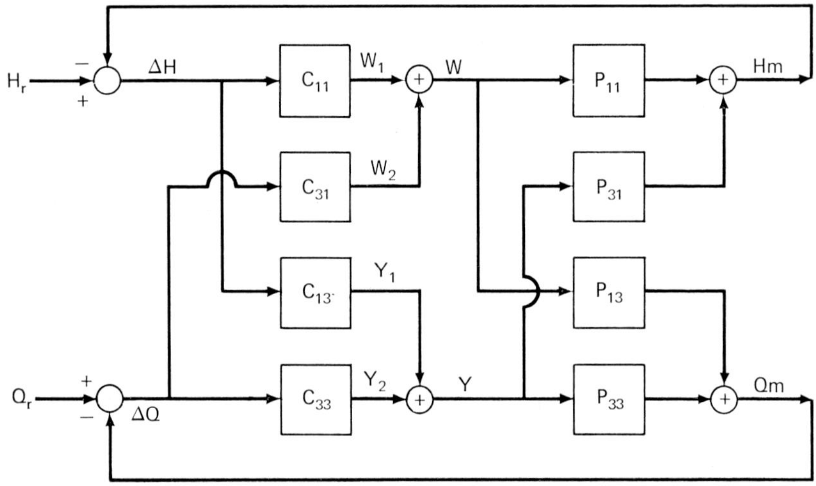

Fig. 14—Block diagram for the 2×2 decoupled controller

A block diagram of the approach proposed is contained in Fig. 14 with the C values and P values denoting the appropriate transfer functions of the controllers and the process, respectively. From Fig. 14, it can be seen that—

$$W = C_{11}\Delta H + C_{13}\Delta Q \qquad (18a)$$

$$Y = C_{31}\Delta H + C_{33}\Delta Q \qquad (18b)$$

$$H_m = P_{11}W + P_{31}Y \qquad (19a)$$

$$Q_m = P_{13}W + P_{33}Y \qquad (19b)$$

Now, if equations (*18a*) and (*18b*) are substituted into equations (*19a*) and (*19b*)—

$$\begin{aligned} H_m &= P_{11}(C_{11}\Delta H + C_{13}\Delta Q) + P_{13}(C_{31}\Delta H + C_{33}\Delta Q) \\ &= (P_{11}C_{11} + P_{13}C_{31})\Delta H + (P_{11}C_{13} + P_{13}C_{33})\Delta Q \qquad (20a) \end{aligned}$$

$$\begin{aligned} Q_m &= P_{31}(C_{11}\Delta H + C_{13}\Delta Q) + P_{33}(C_{31}\Delta H + C_{33}\Delta Q) \\ &= (P_{31}C_{11} + P_{33}C_{31})\Delta H + (P_{31}C_{13} + P_{33}C_{33})\Delta Q \qquad (20b) \end{aligned}$$

where $\Delta H = H_r - H_m$ and $\Delta Q = Q_r - Q_m$.

From the first definition of decoupling, it can be stated that—

$$Q_m = 0 \text{ when } \Delta H = 0$$

which is the necessary and sufficient condition for—

$$P_{31}C_{11}+P_{33}C_{31} = 0 \quad . \quad . \quad . \quad . \quad . \quad (21a)$$

Likewise, from the second definition, it may be shown that—

$$P_{11}C_{13}+P_{13}C_{33} = 0 \quad . \quad . \quad . \quad . \quad . \quad (21b)$$

Now, if equations (*21*) are substituted into equations (*20*), we have—

$$H_m = (P_{11}C_{11}+P_{13}C_{31})(H_r-H_m) \quad . \quad . \quad . \quad (22a)$$
$$Q_m = (P_{31}C_{13}+P_{33}C_{33})(Q_r-Q_m) \quad . \quad . \quad . \quad (22b)$$

If the closed loop transfer functions are now specified to be—

$$\frac{H_m}{H_r} = \frac{\lambda_1}{s+\lambda_1}e^{-sD1} \quad . \quad . \quad . \quad . \quad . \quad . \quad . \quad . \quad . \quad (23a)$$

$$Q_m = \frac{\lambda_3 e^{-sD3}}{s+\lambda_3-\lambda_3 e^{-sD3}}(Q_r-Q_m) \quad . \quad . \quad . \quad . \quad (24b)$$

Hence, by comparing equations (*22*) and (*24*), it is obvious that—

$$P_{11}C_{11}+P_{33}C_{31} = \frac{\lambda_1 e^{-sD1}}{s+\lambda_1-\lambda_1 e^{-sD_3}} \quad . \quad . \quad . \quad (25a)$$

$$P_{31}C_{13}+P_{33}C_{33} = \frac{\lambda_3 e^{-sD3}}{s+\lambda_3-\lambda_3 e^{-sD3}} \quad . \quad . \quad . \quad (25b)$$

Combining equations (*21*) and (*25*) and expressing the result via matrix notation, one obtains—

$$\begin{bmatrix} C_{11} & C_{13} \\ C_{31} & C_{33} \end{bmatrix}\begin{bmatrix} P_{11} & P_{13} \\ P_{31} & P_{33} \end{bmatrix} = \begin{bmatrix} \dfrac{\lambda_1 e^{-sD1}}{s+\lambda_1-\lambda_1 e^{-sD1}} & 0 \\ 0 & \dfrac{\lambda_3 e^{-sD3}}{s+\lambda_3-\lambda_3 e^{-sD3}} \end{bmatrix} \quad . \quad (26)$$

If the four controller functions are then solved for, one obtains—

$$\begin{bmatrix} C_{11} & C_{13} \\ C_{31} & C_{33} \end{bmatrix} = \begin{bmatrix} \dfrac{\lambda_1 e^{-sD1}}{s+\lambda_1-\lambda_1 e^{-sD1}} & 0 \\ 0 & \dfrac{\lambda_3 e^{-sD3}}{s+\lambda_3-\lambda_3 e^{-sD3}} \end{bmatrix}\begin{bmatrix} P_{11} & P_{13} \\ P_{31} & P_{33} \end{bmatrix}^{-1} \quad . \quad (27)$$

The process transfer functions required in equation (*27*) are contained in Table 5 (for both the *s* and *z* domains), from which it should be noted that P_{33} (the transfer function between slice changes and flow to the box) is equal to zero, showing that flow to the box is not substantially affected by changes in slice opening. Hence, the P matrix in equation (*27*) may be inverted in the following manner—

$$\begin{bmatrix} P_{11} & P_{13} \\ P_{31} & P_{33} \end{bmatrix}^{-1} = \begin{bmatrix} P_{11} & P_{13} \\ P_{31} & 0 \end{bmatrix}^{-1} = \begin{bmatrix} 0 & 1/P_{31} \\ 1/P_{13} & -P_{11}/P_{13}P_{31} \end{bmatrix}$$

By definition—

$$P_{11} = H/W \text{ and } P_{31} = Q/W$$

therefore—

$$\frac{P_{11}}{P_{31}} = \frac{H}{Q}$$

Since H/Q can be identified directly, this is the method used to estimate it rather than by taking the ratio of P_{11} to P_{31}, which are also identified.

TABLE 5

Transfer function	*s domain*	*z domain*
$\frac{H}{W} = P_{11}$	$\frac{a_1 g_1}{s+a_1} e^{-sD_1}$	$g_1(1-\eta_1)\frac{z-d_1-1}{1-\eta_1 z-1}$
$\frac{Q}{W} = P_{13}$	$\frac{a_2 g_2}{s+a_2} e^{-sD_1}$	$g_2(1-\eta_2)\frac{z-d_1-1}{1-\eta_2 z-1}$
$\frac{H}{Y} = P_{31}$	$-\frac{a_3 g_3}{s+a_3}$	$-g_3(1-\eta_3)\frac{z-1}{1-\eta_3 z-1}$
$\frac{Q}{Y} = P_{33}$	0	0
$\frac{H}{Q} = \frac{P_{11}}{P_{13}}$	$\frac{a_4 g_4}{s+a_4}$	$g_4(1-\eta_4)\frac{z-1}{1-\eta_4 z-1}$

Remark: $\eta_i = \exp(-a_i T)$, $d_i = D_i/T$

Using the notation of Table 5, the P matrix becomes—

$$\begin{bmatrix} 0 & \frac{s+a_2}{a_2 g_2} e^{sD_1} \\ \frac{-s+a_3}{a_3 g_3} & \frac{a_4 g_4}{s+a_4} \cdot \frac{s+a_3}{a_3 g_3} \end{bmatrix}$$

Actually, a_3 and a_4 are essentially equal, because they are the reciprocal time constants of total head to flow from and to the box, respectively. Hence, the P matrix may be reduced further to the form—

$$\begin{bmatrix} 0 & \dfrac{s+a_3}{a_2 g_2} e^{sD1} \\ \dfrac{-s+a_3}{a_3 g_3} & \dfrac{g_4}{g_3} \end{bmatrix}$$

which, when substituted in equation (*27*), gives—

$$\begin{bmatrix} C_{11} & C_{13} \\ C_{31} & C_{33} \end{bmatrix}$$

$$= \begin{bmatrix} \dfrac{\lambda_1 e^{-sD1}}{s+\lambda_1-\lambda_1 e^{-sD1}} & 0 \\ 0 & \dfrac{\lambda_3 e^{-sD3}}{s+\lambda_3-\lambda_3 e^{-sD3}} \end{bmatrix} \begin{bmatrix} 0 & \dfrac{s+a_3 e^{-sD1}}{a_2 g_2} \\ \dfrac{-s+a_3}{a_3 g_3} & \dfrac{g_4}{g_3} \end{bmatrix} \quad (28)$$

Expressions equivalent in the z domain as well as the appropriate recursive formulas are shown in Table 6.

TABLE 6

Value	*z domain*	*Recursion formula*
$\frac{W}{H} C_{11}$	0	$W_1 = 0$
$\frac{W}{Q} C_{31}$	$\dfrac{(1-\theta)(1-\eta_2 Z^{-1})}{g_2(1-\eta_2)[1-\theta Z^{-1}-(1-\theta)Z^{-\Gamma-1}]}$	$W_2^i = \dfrac{(1-\theta)}{g(1-\eta_2)}(Q^i-\eta_2 Q^{i-1})+$ $W_2{}^{i-1}+(1-\theta)W_2{}^{i-\Gamma-1}$
$\frac{Y}{H} C_{13}$	$\dfrac{-(1-\theta)(1-\eta_3 Z^{-1})Z^{-\Gamma}}{g_3(1-\eta_3)[1-\theta Z^{-1}-(1-\theta)Z^{-\Gamma-1}]}$	$Y_2^i = \dfrac{-(1-\theta)}{g_3(1-\eta_3)}(H^{i-\Gamma}-\eta_3 H^{i-\Gamma-1})+$ $\theta Y^{i-1}+(1-\theta)Y^{i-\Gamma-1}$
$\frac{Y}{Q} C_{33}$	$\dfrac{g_4(1-\theta)Z^{-\Gamma-1}}{g_3[1-\theta Z^{-1}-(1-\theta)Z^{-\Gamma-1}]}$	$Y_2^i = \dfrac{g_4(1-\theta)}{g_3} Q^{i-\Gamma-1}+$ $\theta Y_2{}^{i-1}(1-\theta)Y^{i-\Gamma-1}$

Needless to say, the algorithms required by this approach are considerably more complex than any of those discussed previously. In addition, considerably more identification work is required because interactions between loops must be quantified. Nevertheless, the added effort is well repaid as evidenced by the excellent response shown by total head and flow to the box during a typical grade change (Fig. 15).

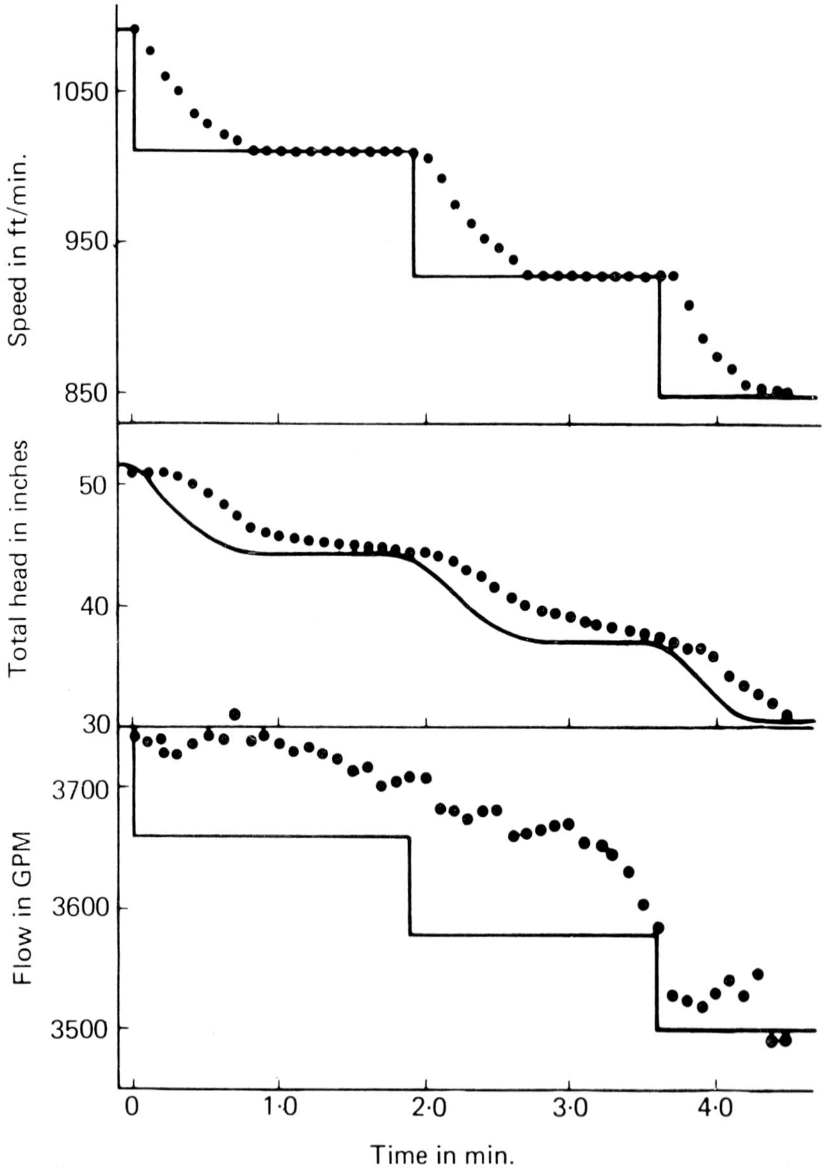

Fig. 15—Head box response during a grade change

The grade change program serviced by the head box control system in this instance is built around a series of large steps (80 ft/min) in machine speed, which are used to drive the papermachine from one operating point to another. The net effect of this is to impose a series of step-like changes in total head set point (as calculated by the head box drag controller) and step adjustments in flow rate set point. The job of the head box controller is to respond to these set point changes in minimum time and with minimum overshoot. Fig. 15 shows that the 2×2 decoupled controller described there seems well able to meet these requirements.

The changes in total head and flow operating points were 19·9 in and 250 gal/min, respectively. The maximum error encountered in each of the two loops were 3·7 in and 127 gal/min, which is very good considering that the 80 ft/min speed changes engendered 6·5 in and 84 gal/min total head and flow set point changes each time they occurred. Experience so far has shown that the head box system is the one limiting the speed of grade change. Hence, the next speed step is not allowed to occur until the total head error is reduced to 0·5 in. Nevertheless, the 240 ft/min change was accomplished in a mere 4·5 min, which explains the importance of having an active, accurate head box control system.

In summary, it is possible to design a 2×2 decoupled controller for total head and flow from the box via Dahlin's method of cancellation compensation. Liquid level control was ignored in the derivation, since its set point is seldom, if ever, changed and it is felt that a conventional analog controller arranged as described in part 1 would provide adequate performance.

Additional identification work was required to quantify all the interactions in the 2×2 controller, but this effort was well repaid via a decided improvement in head box control without the necessity of resorting to exotic manipulated variables or controlled variable sensing devises. Needless to say, the 2×2 algorithms were implemented digitally because of their complexity. Level control, on the other hand, continued to be via a conventional analog PI controller. Finally, it should be noted that adjustments to the 2×2 controller's parameters have been found to be necessary to retain good operating performance over the entire head box's operating range. When such changes are necessary during a grade change they are made automatically via the digital computer used to implement the algorithms.

Controller design—part 6

PART 5 having just outlined the advantages of a 2×2 total head/flow from the box, decoupled control, one might ask, 'If a 2×2 decoupled controller is superior to the simpler systems described in parts 3 and 4, would not a 3×3

decoupled controller be still better?' The answer is, 'Yes, it is in certain circumstances, but a 3×3 controller also places a substantial load on the digital data acquisition and control systems.' Nevertheless, let us proceed with a description of how such a system can be designed and worry about its limitations later.

The first thing to realise when designing a 3×3 controller for the head box is that complete decoupling is not required or even desirable. To be specific, complete decoupling is desired between total head and liquid level, whereas only partial decoupling is needed between total head and flow and no decoupling is required between liquid level and flow. The practical reasons for this are best understood by referring to Fig. 16, from which it may be seen that—

1. Total head and liquid level are completely decoupled as evidenced by the fact that controllers C_{11}, C_{12}, C_{21} and C_{22} exist. This design is justified by the fact

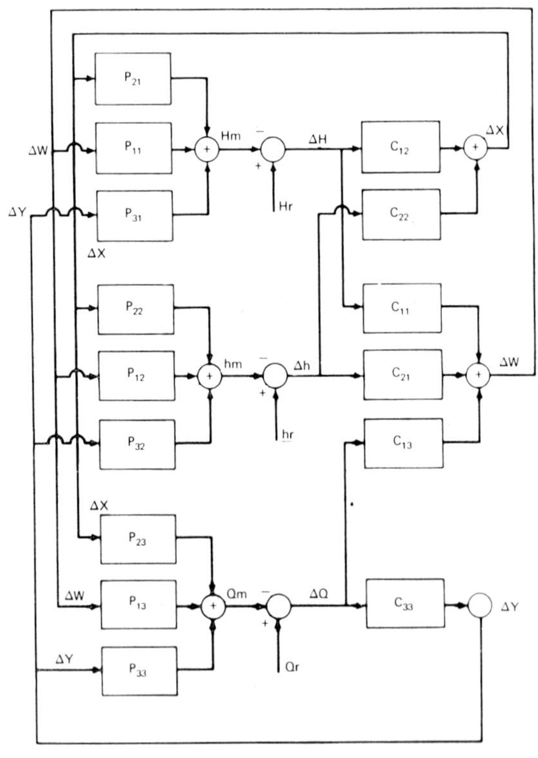

Fig. 16—Block diagram for the 3×3 decoupled controller

that it does give optimum total head and liquid level control and the papermaker does not object to rapid and/or extensive manipulation of the head box air and water valves.

2. Total head and flow from the box are only partially decoupled, because the papermaker objects to excessive adjustment of his slice. Hence, controller C_{13} is included to minimise the effect of slice adjustments on total head and liquid level, but no controller C_{31} is provided to allow slice adjustments when a total head error exists.
3. Liquid level and flow from the box are not decoupled at all, because the papermaker again objects to excessive slice manipulation, thereby eliminating controller C_{32}; controller C_{23} was left out, because air valve manipulation was considered to be inconsequential when dealing with flow errors.

Mathematically, the relationships depicted in Fig. 16 can be expressed as follows—

$$\begin{aligned} H_m &= P_{11}\Delta W + P_{12}\Delta X + P_{13}\Delta Y \\ h_m &= P_{21}\Delta W + P_{22}\Delta X + P_{23}\Delta Y \\ Q_m &= P_{31}\Delta W + P_{32}\Delta X + P_{33}\Delta Y \end{aligned} \qquad (29)$$

Actually, however, P_{32} and P_{33} turn out to be very near to zero, because changes in air valve position and slice opening have very little effect on the flow of stock to the box the variable being measured. Hence, equation (*29*) reduces to—

$$Q_m = P_{31}\Delta W$$

Also from Fig. 16, it may be seen that—

$$\begin{aligned} \Delta W &= C_{11}\Delta H + C_{12}\Delta h + C_{13}\Delta Q \\ \Delta X &= C_{21}\Delta H + C_{22}\Delta h \\ \Delta Y &= C_{33}\Delta Q \end{aligned} \qquad (30)$$

Reducing equations (*29*) and (*30*) to matrix form, we have—

$$\begin{bmatrix} H_m \\ h_m \\ Q_m \end{bmatrix} = \begin{bmatrix} P_{11} & P_{12} & P_{13} \\ P_{21} & P_{22} & P_{23} \\ P_{31} & 0 & 0 \end{bmatrix} \begin{bmatrix} C_{11} & C_{12} & C_{13} \\ C_{21} & C_{22} & 0 \\ 0 & 0 & C_{33} \end{bmatrix} \begin{bmatrix} \Delta H \\ \Delta h \\ \Delta Q \end{bmatrix}$$

which may be reduced to—

$$\begin{bmatrix} H_m \\ h_m \\ Q_m \end{bmatrix} = \begin{bmatrix} P_{11}C_{11}+P_{12}C_{21} & P_{11}C_{12}+P_{12}C_{22} & P_{11}C_{13}+P_{13}C_{33} \\ P_{21}C_{11}+P_{22}C_{21} & P_{21}C_{12}+P_{22}C_{22} & P_{21}C_{13}+P_{23}C_{33} \\ P_{31}C_{21} & P_{31}C_{22} & P_{31}C_{13} \end{bmatrix} \begin{bmatrix} \Delta H \\ \Delta h \\ \Delta Q \end{bmatrix} \qquad (31)$$

From the decoupling restrictions discussed above, it can be shown that—

$$P_{11}C_{12}+P_{12}C_{22}=0 \quad (32a)$$
$$P_{21}C_{11}+P_{22}C_{21}=0 \quad (32b)$$
$$P_{11}C_{13}+P_{13}C_{33}=0 \quad (32c)$$
$$P_{21}C_{23}+P_{23}C_{33}=0 \quad (32d)$$

where equations (*32a*) and (*32b*) imply complete decoupling between total head and liquid level. Equation (*32c*), on the other hand, expresses the partial decoupling between flow and total head. This also forces a partial decoupling between flow and liquid level—see equation (*32d*)—because, in order to achieve the decoupling specified in equation (*32c*), the water valve is adjusted before the slice so that flow in and out of the box are held in equilibrium. This not only eliminates bumps in total head, but in liquid level as well. As a result of the above, equation (*31*) reduces to—

$$\begin{bmatrix} H_m \\ h_m \\ Q_m \end{bmatrix} = \begin{bmatrix} P_{11}C_{11}+P_{12}C_{21} & 0 & 0 \\ 0 & P_{21}C_{12}+P_{22}C_{22} & 0 \\ P_{31}C_{21} & P_{31}C_{22} & P_{31}C_{13} \end{bmatrix} \begin{bmatrix} \Delta H \\ \Delta h \\ \Delta Q \end{bmatrix} \quad (33)$$

which may be stated as a series of algebraic equations in the following manner—

$$H_m = (P_{11}C_{11}+P_{12}C_{21})\Delta H \quad (34a)$$
$$h_m = (P_{21}C_{12}+P_{22}C_{22})\Delta h \quad (34b)$$
$$Q_m = P_{31}(C_{21}\Delta H+C_{22}\Delta h+C_{13}\Delta Q) \quad (34c)$$

Equations (*34a*) and (*34b*) show that total head and liquid level will only be affected by changes in their set points, whereas flow to the box is affected by changes in any of the three variables—see equation (*34c*). Therefore, it is necessary to have a dead band around flow set point or the decoupling so laboriously avoided in the case of slice adjustments will be negated.

To complete the design of the six controllers required (Fig. 16), it is necessary to specify the response desired for each of the loops. Mathematically, this is accomplished via equations (*35*)—

$$\frac{H_m}{H_r} = \frac{\lambda_1}{s+\lambda_1}e^{-sD1} \quad (35a)$$

$$\frac{h_m}{h_r} = \frac{\lambda_2}{s+\lambda_2} e^{-sD2} \qquad (35b)$$

$$\frac{Q_m}{Q_r} = \frac{\lambda_3}{s+\lambda_3} e^{-sD3} \qquad (35c)$$

which state that each loop is expected to show a closed loop response equivalent to a first-order system with dead time. If equations (*35*) are now combined with equations (*32*) and (*34*), the result is a matrix analogous to equation (*26*) in part 5. This expression can in turn be manipulated as shown previously to obtain an equation analogous to equation (*28*) (in part 5) that specifies the controllers required to achieve the performance specified by equation (*35*) and Fig. 16.

Needless to say, the control system required to achieve 3×3 decoupling of the type specified is complex, but it has been implemented and shown to work on a practical basis. Such a system does not show substantially superior performance to the system described in parts 3 and 4 however. Hence, its utility is rather severely limited, because it requires a considerably more complex algorithm and identification structure than previous approaches. In addition, this approach requires a considerably faster sampling rate (once per second or less) than other systems, because the transfer functions between air valves position and both total head and liquid level show a 3·5 s time constant (Table 3) plus an integrator in the case of liquid level. This integrator is the worst offender, since it tends to make liquid level oscillate unless given continual attention. Hence, it has been the authors' experience that the use of an analog controller on the liquid level system is best. Decoupling between the air valves and total head is sacrificed, but this has been found to be a small price to pay for the greater stability and lower sampling rates achieved.

To summarise, it is possible to design and implement a 3×3 digital control system for the head box. Such a system is not substantially superior in performance to the systems described in parts 3, 4 and 5, however and the increase in sampling and algorithm complexity required to implement it are not justifiable.

Conclusions

FROM the above, it should be obvious that it is possible to characterise the head box as a control system, to identify its dynamic parameters as a function of operating point and to design a variety of control systems with which to control it with varying degrees of facility. In addition, it should be obvious that though many Fourdrinier machines are still equipped only with head box liquid level control systems, there is no reason that the papermaker must tolerate such inconvenience. A 2×2 digital head box controller with analog control of liquid level is recommended as the best system currently available. Finally, it should be noted that no modern head box should be installed without means for sensing not only liquid level and total head, but flow to the box as well. Such sensors are required to implement modern control systems, which it should also be remembered require an ability to adjust head box water valve, air valves and slice opening automatically. In concluding, the reader is again reminded that all of the work reported has been carried out at speeds of 1 100 ft/min or less. It is expected, however, that the insights gained should be transferable to Fourdrinier machines operating at considerably greater speeds.

Future work

So FAR, attention has been directed almost entirely toward gaining good total head, liquid level and flow control for the head box and relatively efficient means for accomplishing this have been achieved. Other factors associated with the head box do have a significant effect on Fourdrinier machine operation, however and should be brought under control. For example, industrial experience shows that the basis weight profile of a sheet of paper can be obtained by merely averaging several successive beta-gauge traces. In addition, the authors have determined that this profile can be manipulated via slice screw adjustments without encountering the anomalous side effects reported in papermaking lore and fable. As a result, it would seem quite possible to achieve true basis weight profile control in the relatively near future by merely designing a matrix controller, using the basis weight profile as a controlled variable and slice screws position as the manipulated variables. The algorithm structure required will not be simple, particularly if decoupling between adjacent screws is required, but it is feasible with existing equipment and should be implemented as soon as adequate means for controlling moisture profile are also available. We at Consolidated Papers have already shown that control to within a range of 0·3 lb/ream is attainable with a simple supervisory type controller.

Another factor having an important bearing on sheetforming is the jet escape angle from the head box. No good method is currently available for sensing this angle directly, but it can be estimated using the data developed by Nelson[5] if slice opening and lower lip extension are known. Using these data as feedback, it is then possible to adjust lower lip position when making slice adjustments on the head box to ensure delivery of the jet to the wire at a fixed point. This in turn should aid in stabilising drainage on the machine substantially.

Finally, a third factor having a decided effect upon sheet formation and drainage is the degree of flocculation existing in the jet as it is delivered to the wire. Many factors influence this property and it is not at present possible to measure it on-line in a practical manner. Nevertheless, it is known that holey roll position and rotational speed as well as stock velocity in the box affect flocculation. Hence, there should be variables that can be manipulated to control the parameter, provided it is subsequently found to be of sufficient importance to merit closed loop control.

Appendix 1—Head box simulation program

SIMULATION methods have been used extensively throughout the work reported here. Fortunately, the authors had a modest sized IBM 360 computer available to them and were able to modify IBM's continuous systems modelling program[6] to operate on this computer. The material presented in the following consists of a block diagram for the head box equipped with analog controllers (Fig. A–1) and the CSMP diagram (Fig. A–2) corresponding to it. Tables A–1 to A–3 contain the configuration and specifications for the analog controlled head box depicted in Fig. A–2, whereas Tables A–4 to A–6 do the same for a digitally controlled box.

TABLE A–1 CONFIGURATION SPECIFICATION OF HEAD BOX MODEL WITH ANALOG CONTROLLERS

Output name	*Block number*	*Type*	*Input 1*	*Input 2*	*Input 3*	*Description*
Air valve movement	1	1	21			Dead time block
Water valve movement	2	3	31			Dead time block
Total head error	3	t	11	−12		Summer
Liquid level error	4	W	6	8		Weighted summer
	5	I		1	5	First order lag
	6	I	5			Integrator
	7	I		9	7	First order lag
	8	I		9	8	First order lag
	9	t	2	10		Summer
Disturbance	10	K				Constant block
Total head set point	11	K				Constant block
Total head response	12	W	13	5	7	Weighted summer
Initial total head	13	K				Constant block
Analog liquid level controller	21	W	4	22		Weighted summer
	22	I	4			Integrator
Analog total head controller	31	W	3	32		Weighted summer
	32	I	3			Integrator

TABLE A–2 NUMERICAL VALUES OF HEAD BOX MODEL AT 50 lb/ream GRADE

Parameter name	*Block number*	*Parameter of input 1*	*Parameter of input 2*	*Parameter of input 3*
Dead time, D_2	1		1·0	
Dead time, D_1	2		5·5	
Process gain, $-g_2$, g_4	4	−0·001	0·142	
a_1, $-a_1$	5		0·224	−0·224
a_3, $-a_3$	7		0·0484	−0·0484
a_4, $-a_4$	8		0·0133	−0·0133
Total head set point	11	31·0		
Process gains, g_1 and g_3	12	1·0	0·02	0·354
Gain and reset, p_1 and i_1	21	20·0	1·0	
Gain and reset, p_2 and i_2	31	11·1	0·555	
Initial total head	13	30·0		

TABLE A–3 NUMERICAL VALUES OF HEAD BOX MODEL AT 100 lb/ream GRADE

Parameter name	*Block number*	*Parameter of input 1*	*Parameter of input 2*	*Parameter of input 3*
Dead time, D_2	1		1·0	
Deaɑ time, D_1	2		8·5	
Process gain, $-g_1$ and g_4	4	−0·001	0·142	
a_1 and $-a_1$	5		0·224	−0·224
a_3 and $-a_3$	7		0·0619	−0·0619
a_4 and $-a_4$	8		0·0091	−0·0091
Total head set point	11	18·0		
Total gains	12	1·0	0·02	0·132
Initial total head	13	17·0		
Gain and reset, p_1 and i_1	21	20·0	1·0	
Gain and reset, p_2 and i_2	31	11·1	0·555	

TABLE A–4 CONFIGURATION SPECIFICATION OF HEAD BOX MODEL WITH DIGITAL TOTAL HEAD CONTROLLER AND ANALOG LIQUID LEVEL CONTROLLER

Output name	*Block number*	*Type*	*Input 1*	*Input 2*	*Input 3*	*Description*
Air valve movement	1	1	21			Dead time block
Water valve movement	2	3	31			Dead time block
Total head error	3	t	11	−12		Summer
Liquid level error	4	W			5	Weighted summer
	5	I		1	5	First order lag
	6	I	5		7	Integrator
	7	I		9	7	First order lag
	8	I		9	8	First order lag
	9	t	2	10		Summer
Disturbance	10	K				Constant
Total head set point	11	K				Constant
Total head response	12	W	13	5	7	Weighted summer
Initial total head	13	K				Constant
Analog liquid level controller	21	W	4	22		Weighted summer
	22	I	4			Integrator
Digital total head controller, W_i	31	W	32	37	39	Weighted summer
	32	W	35	33		Weighted summer
H_{i-1}	33	Z	34	40		Zero order holder
	34	U	35			Unit delay
H_i	35	Z	3	40		Zero holder order
	36	U	31			Unit delay
W_{i-1}	37	Z	36			Zero order holder
	38	U	37			Unit delay
W_{i-2}	39	Z	38	40		Zero order holder
	40	t	76			Time trigger
	76					Time

TABLE A–5 NUMERICAL VALUES OF HEAD BOX MODEL WITH DIGITAL CONTROLLER AT 50 lb/ream GRADE

Parameter name	*Block number*	*Parameter of input 1*	*Parameter of input 2*	*Parameter of input 3*
Dead time, D_2	1		1·0	
Dead time, D_1	2		5·5	
Process gain, $-g_2$, g_4	4	−0·001	0·142	
a_1, $-a_1$	5		0·224	−0·224
a_3, $-a_3$	7		0·0484	−0·0484
a_4, $-a_4$	8		0·0133	−0·0133
Total head set point	11	1·0		
Process gains, g_1 and g_3	12	1·0	0·02	0·354
Gain and reset, p_1 and i_1	21	20·0	1·0	
	31	5·223	0·428	0·572
	32	1·0	−0·691	
Sampling interval	40	8·0		

TABLE A–6 NUMERICAL VALUES OF HEAD BOX MODEL WITH DIGITAL CONTROLLER AT 100 lb/ream GRADE

Parameter name	*Block number*	*Parameter of input 1*	*Parameter of input 2*	*Parameter of input 3*
Dead time, D_2	1		1·0	
Dead time, D_1	2		8·5	
Process gain, $-g_2$, g_4	4	−0·001	0·142	
a_1, $-a_1$	5		0·224	−0·224
a_3, $-a_3$	7		0·0619	−0·0619
a_4, $-a_4$	8		0·0091	−0·0091
Total head set point	11	18·0		
Process gains, g_1 and g_3	12	1·0		
Initial total head	13	17·0		
Gain and reset, p_1 and i_1	21	20·0	1·0	
	31	11·1	0·428	
	32	1·0	−0·61	
Sampling interval	40	8·0		

Needless to say, such a simulation tool has been invaluable to the authors and has been found to be far more convenient to use than a conventional analog computer, because scaling problems are virtually eliminated. The program used does require more or less easy access to a modest sized computer, but, if this condition can be met, it is the authors' opinion that one would be foolish to use an analog machine in preference to it.

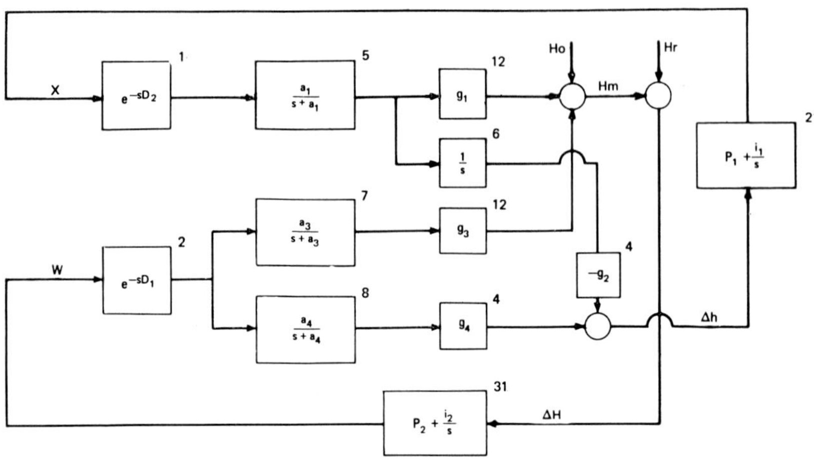

Fig. A1—Block diagram of head box model with analog controllers

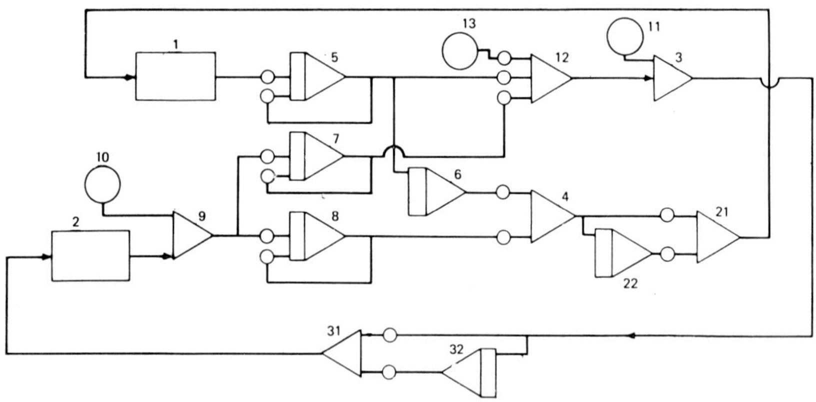

Fig. A2—CSMP block diagram

Appendix 2—Table of nomenclature

A	Cross-section of head box
a	Reciprocal of time constant in process transfer function
$C(s)$	Generalised transfer function for a controller
C_1–C_4	Physical constants of head box
C_{11}	Controller transfer function between water valve and total head

C_{12}	Controller transfer function between air valve stem positions and total head
C_{13}	Controller transfer function between slice position and total head
C_{21}	Controller transfer function between water valve and liquid level
C_{22}	Controller transfer function between air valve stem positions and liquid level
C_{23}	Controller transfer function between slice position and liquid level
C_{31}	Controller transfer function between water valve and flow
C_{32}	Controller transfer function between air valve stem positions and flow
C_{33}	Controller transfer function between slice and flow
D	Dead time
d	Integral of dead time to sampling interval ratio
d_g	Drag (in feet)
e	Natural lag base
$E(s)$	Generalised error
$F(s)$	Generalised close loop transfer function
G	Theoretical transfer function
g	Gravitational constant
g_1–g_4	Gain in process transfer function
H	Total head (in inches)
h	Liquid level (in inches)
K	Gain
M	The mass of air pad
M_a	The net air flow into head box
$M(s)$	Generalised output
P	Air pad pressure
$P(s)$	Generalised transfer function for a process
P_{11}	Process transfer function between total head and water valve
P_{12}	Process transfer function between liquid level and water valve
P_{13}	Process transfer function between flow and water valve
P_{21}	Process transfer function between total head and air valve stem positions
P_{22}	Process transfer function between liquid level and air valve stem position
P_{23}	Process transfer function between flow and air valve stem position
P_{31}	Process transfer function between total head and slice
P_{32}	Process transfer function between liquid level and slice
P_{33}	Process transfer function between flow and slice
Q_i	Inlet liquid flow
Q_o or Q	Outlet liquid flow
R	Gas constant
$R(s)$	Generalised set point
s	Laplace operator
T'	Absolute temperature
T	Sampling interval
t	Time

$U(s)$	Generalised disturbance
U_w	Wire speed
V	Volume of air pad
V_o	Total volume of head box
W	Water valve movement (in seconds)
X	Air valve stem movement (in seconds)
Y	Slice movement (in seconds)
z	z-transform operator
Υ	Time constant
λ_1	System cut off frequency for total head box
λ_2	System cut off frequency for liquid level loop
λ_3	System cut off frequency for flow loop
Γ	Dead time, same as d
η	Dummy variable
θ	Dummy variable

Subscripts—

i	Flow to head box
m	Measured response
o	Flow from head box
r	Set point

References

1. Mardon, J. *et al.*, 'A theoretical and experimental investigation into the stability and control of papermachine head boxes'—part 1: *Paper & Timber (Finland)*, 1966, **48** (1), 3–14; part 2: (5), 301–310; part 3: 1967, (4a), 189–197
2. Sanborn, I. B. and Chao, H., 'The identification of process dynamics': *unpublished* (presented at International Symposium on Pulp and Paper Process Control, Vancouver, B.C., April 1969)
3. Lee, J. K. *et al.*, 'The design of a head box control system'—part 1, Analytical consideration: *unpublished* (presented at Fourth International Federation of Automatic Control Congress, June 1969)
4. Dahlin, E.B. *et al.*, 'Designing and tuning digital controllers': *Instrument & Control Systems*, 1968, (June), 77–83
5. Nelson, H. C., 'Papermachine inlet performance in relation to the Fourdrinier wire': *Tappi*, 1960, **43** (4), 330–342
6. *IBM 1130 Continuous system modelling program* (system manual, 1969)

Discussion

The Chairman It seems that the day of the overflowing head box on grade change or start-up has gone for ever.

Mr J. Mardon First of all, I am very grateful for your courtesy. I suppose Dr Sanborn that you took the total head by a dp cell, converted it to digital form and matched in the digital form.

Dr I. B. Sanborn Yes.

Mr H. B. Carter Please clarify if the fan pump runs at the same speed with all these various grades?

Dr Sanborn Yes, it does. We have not experimented in terms of variable speed fan pumps to achieve total head control. Admittedly, this is a definite possibility, but we have no experience.

Mr P. A. A. Talvio Dr Sanborn's model includes the constants C_2 and C_3, but they are not in fact constants, as they depend on time and frequency. The simplification you made causes quite a large error at low total head values.

All parts of the head box model can be readily calculated on paper by hand. The equations are given, for instance, in Dr B. W. Smith's paper at this symposium and in the paper I presented at the IFAC conference, London 1966.

Did you Dr Sanborn ever try to calculate the complete transfer functions instead of making identification tests?

Dr Sanborn Yes, we did try to calculate some equations analytically. Initially, reasonable agreement was found between measured and actual values, but the trouble is that it is almost impossible to predict what the dynamics of valves, etc. will be. Since we have implemented this control, we have eliminated identification in terms of actual operation. Now, we usually identify when we first start a system up at some specific set of operating conditions. We are then able to calculate how the systems parameter will change with operating conditions. As a result, we have what might be called an adaptive controller.

On C_2 and C_3, all I can say is that we have had no difficulty with controllability. This may be because we have never run our machines below about 600 ft/min.

Dr D. B. Brewster In the grade you showed, I was not clear whether the total head set point was being determined by the speed set point or by the actual speed.

Dr Sanborn The total head set point is determined by the actual speed. The way that we have head box control set up is that the operator enters the drag at which he wants to operate; we then measure wire speed, calculate the total head required and set the total head set point accordingly.

Mr E. Justus In controlling total head box head on these particular installations, do you have to show the difference between head and set head—setting the valve in a fixed position mechanically, then trying to identify what the variables are?

Dr Sanborn The only information we have is the performance in terms of drag variations for constant speed with the total head controller loop closed and open. With it closed, the standard deviation of the drag was $\pm 2{\cdot}5$ ft/min. About half of that variation is due to wire speed fluctuations. Hence, only half of it can be ascribed to the total head controller. Without total head control, a standard deviation of ± 5 ft/min was noted. This shows that we had a well-behaved system before we implemented the controller.

Mr B. W. Wells (written comment) This comment refers to the problems of the need for different flow box control constants during different paper grade conditions. Surely, when using a computer there would be no more difficulties involved in using a supervisory program during grade changes to change the control constants of a DDC three-term control algorithm than changing the control constants in a digital compensator (as mentioned under *Control design—part 3*)?

Regarding the problem of 8 s dead times associated with the flow box, in our experience, such dead times would inevitably impair the response characteristics of any control system and would justify effort to remove them before serious thought could be given to designing a sophisticated control system.

Prepared discussion contributions

An Example of Head Box Control

W. T. WHIGHT, Empire Paper Mills and R. E. JONES, AEI-Elliott Automation Ltd.

THE head box on No. 1 papermaking machine at Empire Paper Mills is an Escher Wyss pressurised type, which is run by direct digital control from the digital control computer. Control outputs are to air valves, thin stock by-pass valves and slice.

The head box stock level is maintained at a mean value by the air pressure within. The main stream flow into the head box by a coarse/fine by-pass system under DDC from the main stream magnetic flow meter. The total head now takes up some value as a dependent variable.

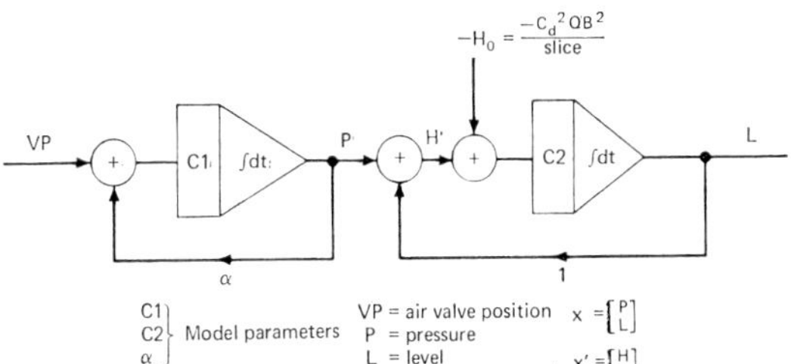

Fig. 1—Simplified small-signal head box model

Fig. 1 shows a simplified small-signal model of the head box. The structure of this model and modern control theory each suggest that the optimum controller will take the form of proportional feedback from head and level transmitters, both channels adding into the set point for the air valve position DDC. The relative importance attached to head or level variations can be traded off by the choice of proportional gains in each channel. Fig. 2 shows the scheme implemented with CONRAD.

Under the chairmanship of P. E. Wrist

On grade, a linear combination of head and level deviations is used to determine the air valve position. The control parameters have been adjusted to give preferential attenuation of head fluctuations at the expense of level. Resulting small level transients have no effect on papermaking performance. Large head transients that occur during deliberate transient conditions (such as grade or production changes) are not recognised by the system, as we have introduced a *velocity saturation* in the head box channel (not illustrated).

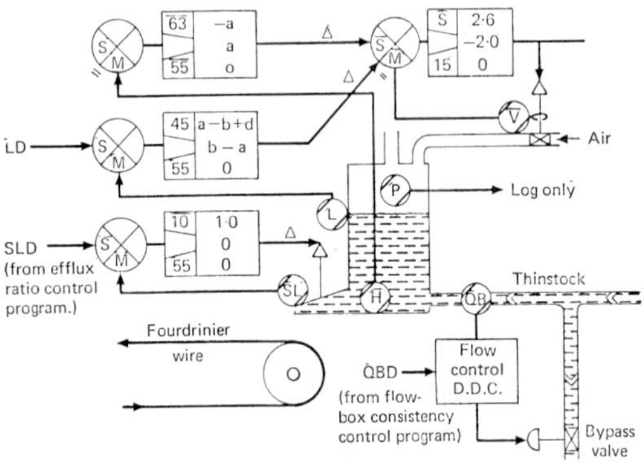

Fig. 2—Basic head box control using CONRAD

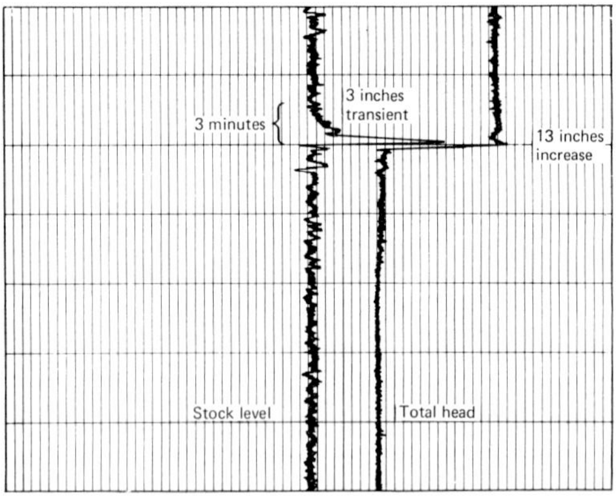

Fig. 3—Recorder chart trace

Since CONRAD works in the velocity mode, purely proportional controllers contain no set point information. In order to retain the level set point, this channel includes a small integral term δ. No such term must appear in the head channel, since we do not want a specific head set point, head being dependent on slice and flow. We want to transmit only head changes to the air valve position DDC loop.

Before this scheme was introduced, the flow box was controlled by two analog controllers—level to air valves and head to by-pass valves. Performance was then similar to Fig. 3.

With the DDC scheme described, however, the head overshoot has not been observed at grade change and on-grade head variability is now better than $\pm$ 0·1 in.

The result of this control system is an improvement in performance over the level to air valve schemes by a factor of 5. Thus, we have removed any on-grade basis weight variations owing to transients in total head.

The Design and Operation of a Computer-Controlled Flow Box

B. W. WELLS, Wiggins Teape Ltd., Dartford Paper Mill, Kent

Introduction

One of the problems encountered during the computerisation of No. 6 machine at Dartford was the need for a flow box control system to replace the existing Hornbostel hole arrangement. The circumstances were such that we required a control system implemented by the computer program (DDC) and the facilities for bumpless transfer to an analog controller arrangement in the event of computer failure.

To attain reliability and simplicity, we limited our investigations to the manipulation of the stock and air pressure supply valves to effect automatic control. Our control objectives were—

1. The minimisation of level and total head variability.
2. The maintenance of stable control at all times.
3. The inclusion of efflux ratio control, in the computer controller, to reduce general variability and aid weight changes.

Simulation

We proceeded with an analog computer simulation based upon the flow box equations previously deduced by J. Mardon *et al.*[1] and P. A. A. Talvio.[2] Owing to the lack of instrumentation on our papermachine at the time, we were compelled to deduce the constants for the analog computer equations by measuring the flow box and calculating the stock flow from a knowledge of the slice opening and total head. It was our intention to use the simulation to investigate alternative multi-loop three-term control systems to those developed by Mardon and Talvio.

We used a step input of stock flow as a disturbance in the simulation and overshoot plus settling time as a criterion for comparisons. After investigating 10 multi-loop control systems, the simulation results showed that a control system very similar to the one proposed by Talvio gave the best results,

Under the chairmanship of P. E. Wrist

whereas the system originally proposed by Mardon was the best and simplest alternative for use as an analog standby control system. We therefore decided to go ahead with—

1. Total head control by stock flow and level control by air pressure—for our standby control system.
2. A level controller cascading on to an air pressure controller and a total head controller cascading on to a stock flow control loop—for our computer control system.

The simulation predicted a 3:1 reduction in variability from the computer system compared with the standby system and about a 30:1 reduction in variability from either of the instrumented systems compared with the Hornbostel hole.

During the computerisation of No. 6 machine, the validity of our choice of flow box control systems was checked by R. Fletcher of Ferranti Automation Systems Division, using a digital simulation.

Results

Both of the control systems mentioned have now been in use for a year. Only a small reduction in total head variability could be attributed to the computer system compared with the standby system, although both of these systems were substantially better than the Hornbostel hole.

Tuning the cascaded loop computer control system was not easy and the controller parameters currently being used, pending further investigation, are only really suitable for the top three quarters of our normal total head operating range. Much difficulty was experienced while attempting to use the computer control system at machine startups. We have therefore tuned the standby control system to give a satisfactory response for machine start-ups, which also gives a reasonable response for steady running conditions. The procedure we currently use is to start up on standby control and switch to computer control as soon as the large start-up transients have disappeared.

We have experienced no problems while using our computer-controlled flow box during large basis weight changes. The level merely remains fairly steady, whereas the efflux ratio control ramps the total head.

Fig. 1 shows an example of total head/level responses during a combined machine speed and weight change.

Conclusions

Our flow box simulation work and subsequent operating experience has highlighted two main problem areas—

1. The necessity for very reliable control system hardware.
2. The necessity for wider investigations at the simulation stage.

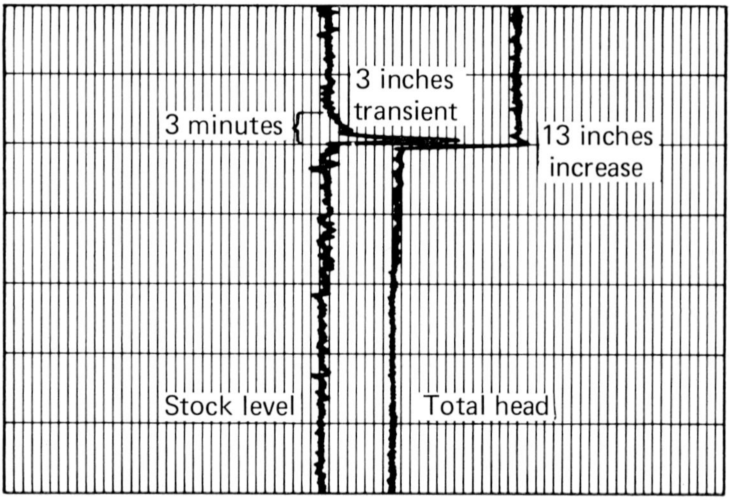

Fig. 1

Desirable improvements are—

1. More accurate values for flow box parameters.
2. Better representation of the dynamic characteristics of measurement and control hardware.
3. More realistic criterion for comparison of simulated control systems.

For example, a digital computer could be used to data log the measured flow into the flow box and statistical techniques employed to produce a 'noise model' of the flow. Then, using a digital simulation of the flow box, representative flow input disturbances could be applied to the various flow box control systems. A suitable criterion for controllability would then be the reduction of the slice flow variability.

References

1. Mardon, J., Monahan, R. E., Mehaffey, W. H. and Dahlin, E. B., 'A theoretical and experimental investigation into the stability and control of papermachine head boxes': *Paper & Timber (Finland)*, 1966, **48** (1), 3–14; (5), 301–303, 305–310
2. Talvio, P. A. A., 'A study of papermachine head box control systems with linear transfer functions' (IFAC Conference, London, June 1966)

Non-Interacting Control of Pressurised Flow Boxes

R. E. JOHNSTON, University of Manchester Institute of Science & Technology

THE dynamics of a pressurised flow box can be reasonably well described by simple transfer functions, as Dr Smith has shown. Fig. 1 represents the system as multi-variable (two input/two output).

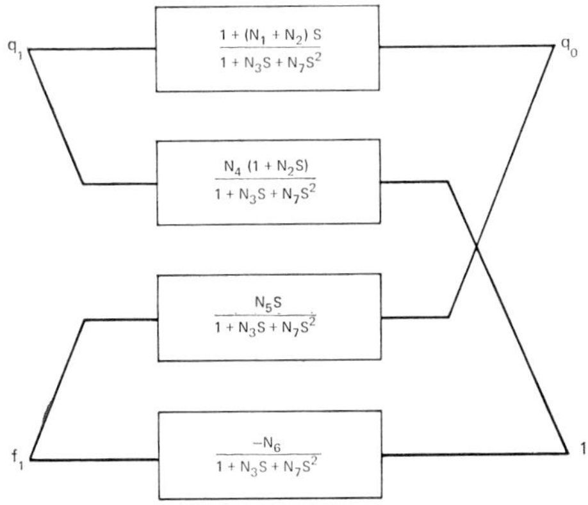

Using Smith's terminology, the values of the parameters N are given by—

$$N_1 = PAM_w/kRT$$
$$N_2 = VM_w/kRT$$
$$N_3 = 2HA/Q_o + N_1 + N_2$$
$$N_4 = 2H/L$$
$$N_5 = F_1A/kQ_o$$
$$N_6 = F_1/kL$$
$$N_7 = 2VAM_wH/kRTQ_o$$

Under the chairmanship of P. E. Wrist

(These apply to a flow box without a hornbostel hole or an internal overflow.)

It is clear that simple feedback controls (for example, from l to f_1 and from q_0 to q_1) will be interacting. Prof. Rosenbrock* at Manchester has developed a technique for tackling the problem of interaction and an application of the technique to this particular system throws some light on to the problem of designing a non-interacting controller. Using this method, it is not necessary to diagonalise the system completely and simpler compensating networks consequently arise.

The transfer matrix in this case is given by—

$$G(s) = \frac{1}{1+N_3s+N_7{}^2}\begin{pmatrix} 1+(N_1+N_2)s & N_5s \\ N_4(1+N_2s) & -N_6 \end{pmatrix}$$

and the inverse of this (using Rosenbrock's symbols) is—

$$\hat{G}(s) = \begin{pmatrix} 1 & (N_5/N_6)s \\ (N_4/N_6)(1+N_2s) & -(1/N_6)(1+(N_1+N_2)s) \end{pmatrix}$$

Working on the inverse, Rosenbrock shows that, if the moduli of the off-diagonal elements are made less than the moduli of the diagonal elements, then the stability of the system can be assured, provided the diagonal elements indicate stability. With luck, the transformations can be carried out by premultiplication by constant matrices.

Premultiplication by G(O) will eliminate zero-frequency interaction, thus—

$$\hat{Q}(s) = \begin{pmatrix} 1 & 0 \\ N_4 & -N_6 \end{pmatrix}\begin{pmatrix} 1 & (N_5/N_6)s \\ (N_4/N_6)(1+N_2s) & -(1/N_6)(1+(N+N_2)s) \end{pmatrix}$$

$$\hat{Q}(s) = \begin{pmatrix} 1 & (N_5/N_6)s \\ -N_2N_4s & 1+N_3s \end{pmatrix}$$

(N_4 and N_6 are easily found from the steady state gains in simple experiments.)

Generally, $N_2N_4 < N_3$, hence loop (*ii*) satisfies the non-interaction criteria. To eliminate high frequency interaction in loop (*i*), subtract (N_5/N_3N_6) times the bottom from row the top—that is, premultiply by—

$$\begin{pmatrix} 1 & -(N_5/N_3N_6) \\ 0 & 1 \end{pmatrix}$$

* U.M.I.S.T. Control Systems Centre Report No. 48 (February 1969)

Thus $\hat{Q}(s)$ becomes—

$$\begin{pmatrix} 1+(N_7/N_3)s & -(N_5/N_3N_6) \\ -N_2N_4s) & 1+N_3s \end{pmatrix}$$

There is no interaction provided that $(N_5/N_3N_6)<1$ and the two 'external' controllers can be tuned independently.

(For complete elimination of high frequency interaction, (N_5/N_3N_6) should be known to within $\pm 100\ N_2N_4/N_3$ per cent of the true value. This should be possible through on-machine studies, together with theoretical estimates, although one value may not be adequate over the whole working range.)

Thus, the complete 'correcting matrix' is given by—

$$K = \left[\begin{pmatrix} 1 & -(N_5/N_3N_6) \\ 0 & 1 \end{pmatrix} \begin{pmatrix} 1 & 0 \\ N_4 & -N_6 \end{pmatrix}\right]$$

$$K = \frac{1}{N_6}\begin{pmatrix} N_6 & (N_5/N_3) \\ N_4 & (N_5N_4/N_3N_6)-1 \end{pmatrix}$$

The main benefits can be seen from an example using proportional control only—

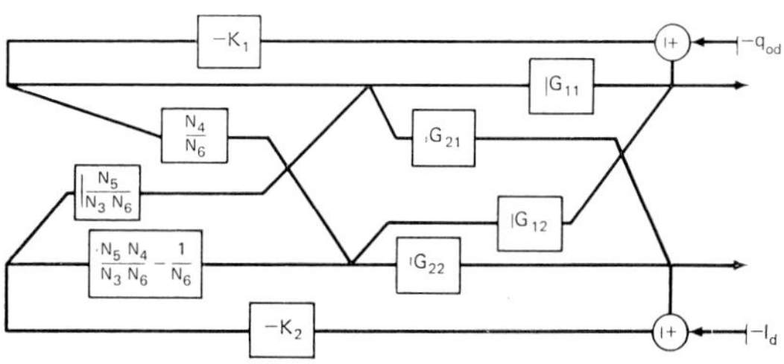

For a flow box similar to the one discussed by Smith with parameters $N_1 = 1$, $N_2 = 0{\cdot}04$, $N_3 = 1{\cdot}7$, $N_4 = 4$, $N_5 = 0{\cdot}1$, $N_6 = 0{\cdot}6$, $N_7 = 0{\cdot}04$, the closed loop response matrix (for $k_1 = 10$, $k_2 = 10$) is approximately—

$$R = \begin{pmatrix} 0{\cdot}91/(1+0{\cdot}002s) & 0{\cdot}008/(1+0{\cdot}156s) \\ 0{\cdot}013/(1+0{\cdot}156s) & 0{\cdot}91/(1+0{\cdot}156s) \end{pmatrix}$$

This is far better than the closed loop response matrix that would arise from an 'uncorrected' system. For example, direct proportional control between

$q_0 - q_1$ and $1 - f_1$ (with $k_1 = 10$, $k_2 = -10$) leads to a response matrix approximating—

$$R = \begin{pmatrix} \dfrac{0{\cdot}91(1+0{\cdot}15s)}{(1+0{\cdot}05s)(1+0{\cdot}11s)} & \dfrac{0{\cdot}013s}{(1+0{\cdot}05s)(1+0{\cdot}11s)} \\ \dfrac{0{\cdot}52}{(1+0{\cdot}11s)} & \dfrac{0{\cdot}86}{(1+0{\cdot}05s)(1+0{\cdot}11s)} \end{pmatrix}$$

Sensitivity of the control performance to errors in the system parameters can be determined analytically as can the frequency responses to disturbances.

Using this approach, it is also easy to determine the likelihood of instability in the two alternative 'uncorrected' control systems—that is, either $q_0 - q_1$ and $1 - f_1$, or $q_0 - f_1$ and $1 - q_1$.

Comments on Head Box Control

K. G. A. GUNNARSON, ASEA, Dept. FAP, Vasteras, Sweden

THE PAPERS on this subject have described total head controlled by pneumatic valves as actuators. In the Chao & Sanborn paper, the authors mention that a time lag of 5 s or more must be tolerated when using these actuators. I suggest that somewhat more rapid control can be arranged by using a modern electronic controller.

Because pneumatic valves are slow acting, I wish to report on the ASEA head box control system for papermachines in which such valves are not used for control purposes. The system that has been marketed for the last two years (to date 25 September 1969) employs the fan pump as a rapid and accurate actuator for level control.

The level controller was previously combined with a conventional control system for maintaining constant air pressure in the head box. If the short settling time of the fan pump control is to be fully utilised, corresponding demands must be imposed on the control equipment for the air pressure. Furthermore, the characteristics of the level and air pressure controllers must be matched with each other, since the process quantities involved greatly influence one another.

The system has been further developed to reduce the settling time of the air control and eliminate the disadvantages of having two different types of controller. With this new system for integrated total head and level control (patents pending)—

1. The air pressure (total head) is controlled with the aid of a speed-controlled Roots blower.
2. The controller for the air pressure (total head) is of the same type as the level controller and built together with this.

Fig. 1 shows the basic principle of this system.

In this new system, the level is controlled as previously with the aid of a speed-controlled fan pump and it is measured in the overflow channel. If the head box is not provided with an overflow channel, the level is measured

Under the chairmanship of P. E. Wrist

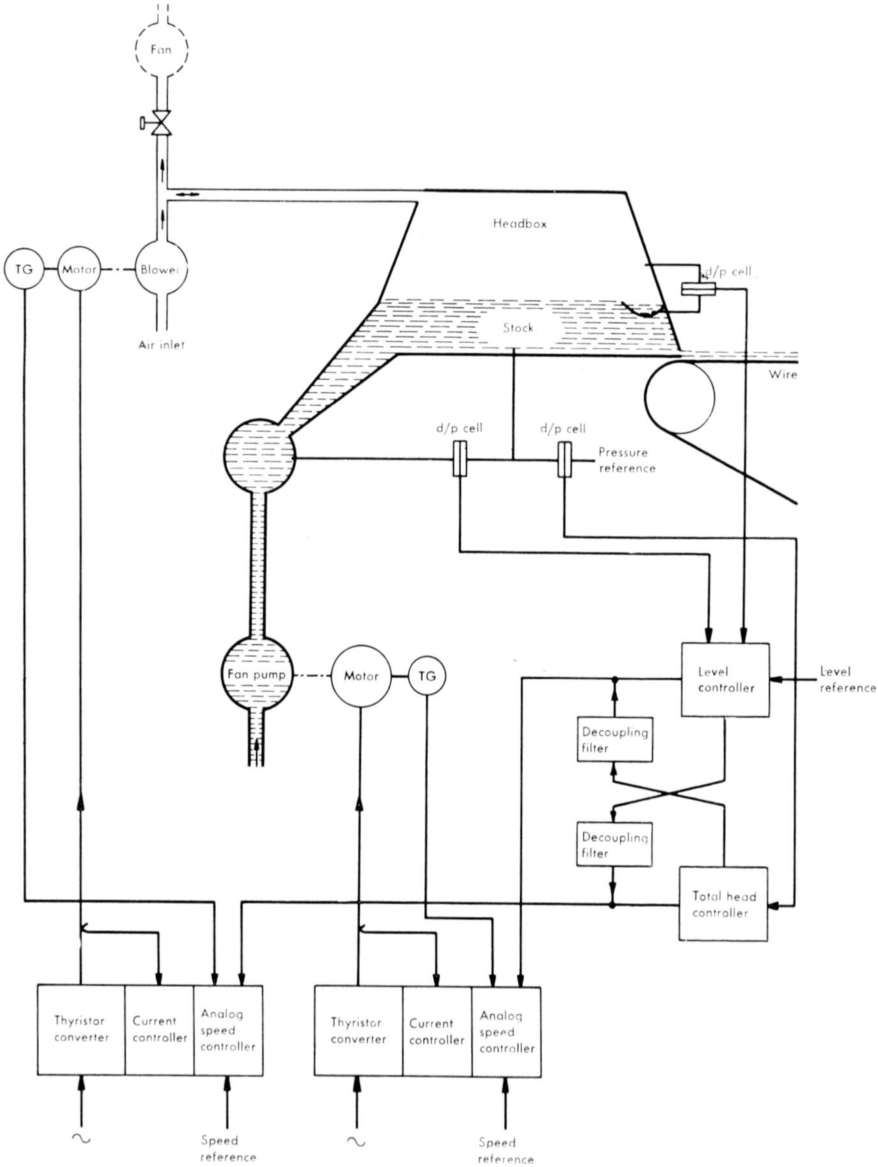

Fig. 1—ASEA head box control system: integrated total head level control in the overflow channel

instead in the pond. When the reference value for the speed is not set so that the flow of the fan pump results in a constant level, the level controller will yield an addition that rapidly eliminates the level error. To counteract as quickly as possible spontaneous flow variations in the inlet, another signal can be fed back to correct the speed of the fan pump, namely, the differential pressure across the manifold or distribution plate by the inlet to the head box. The reference value for the level is set in the level controller.

The control equipment for the total head comprises a total head controller with a speed-controlled blower as actuator. The total head reference is obtained from a column of water, whose height can be steplessly adjusted with the aid of a motor-operated actuator. A dp cell senses the differential pressure between the total head of the head box and the reference pressure. This differential pressure, which is thus the difference between the desired value and the actual value of the total head, is converted into an electrical signal and fed to the total head controller. This yields an addition to the speed reference for the speed controller of the Roots blower with static converter motor drive, according to the same principle as that used for the level control.

In head boxes for which the air cushion head is low, a fan is provided to maintain a constant air exhaust.

The integrated total head and level control is provided also with circuits for the mutual decoupling of the two controllers. If these decoupling circuits were not provided, any trimming measures in one of the systems (for example, the total head control) would influence the other system (the level control). This would mean that the level control would also have to be retrimmed. After this retrimming, the total head control would have to be finely trimmed and so on.When the decoupling circuits are trimmed, each control system can be individually taken into operation, which greatly facilitates the commissioning.

Thyristor converters are well suited for feeding the motors driving the fan pump and blower for the head box control. Very short settling times can be achieved with the current and speed controllers included in the converter equipment (50–100 ms for the elimination of minor disturbances). This ensures that the reaction time of the actuator itself can be nelected. By means of suitable designing of the fan pump, blower and associated drive and control equipment the head box control system will not therefore impose the limit for the resetting time when, for example, the machine speed or slice opening is altered. The resetting rate instead is limited by other factors in the process itself.

If this head box control system is to be utilised, it is essential (as already mentioned) that the fan pump, also the Roots blower should be provided with speed control. In most cases, speed control is also the most economic alternative for fan pump drives. An uncontrolled drive with, for example, induction motor may be justified from an economic point of view only if the pump is run without any significant throttling of the stock flow.

Conclusion

Main features for the head box control system described are—

1. Extremely fast total head control reduces surges previously intangible for control.
2. Total head and level control use as actuators speed-controlled thyristor drives.
3. Both controllers are suitably interconnected in order to eliminate cross-coupling between the two systems.
4. Level control for the overflow channel is adjusted for minimum consistency variations in the head box.

An integrated total head and level control system was very recently started up for a Wärtsilä head box on the papermachine at Lövholmens Bruk, Piteå, Sweden, making kraftliner.

As a result, very accurate total head control is achieved. Typical spontaneous surges with a rise time of less than 2 s have been reduced by a factor of 2·5.

Discussion

Dr L. A. Kirk As we at Manchester University have a similar kind of papermachine, I would be very interested to know the total cost of computerising your papermachine. In view of the papers we have heard on the exact extent to which computer control has been applied in industry, do you consider it worthwhile for our machine to be computer-controlled at this stage?

Mr S. Bauduin The total cost of the project, which extends over a three year period (until the end of 1970) is approximately £140 000, including—

Instrumentation and wiring	£30 000
Computer	£10 000
Ancillaries	£15 000
Personnel salaries, etc.	£85 000

Our machine is quite different from an industrial one, although time constants are nearly the same. We think that the approach to control problems will have to be the same. Besides, such a system is very convenient for training purposes.

Dr J. A. Robinson In the text, it is stated that the response of the model appears to be better than the theoretical response, but the theoretical response in Fig. 7 appears to be better.

Mr Bauduin Our paper states about Fig. 7 that 'an explanation has not yet been found for the fact that the response of the full model appears at first sight to be better than the theoretical response'. Thus, the 'theoretical' curve shows a permanent deviation, whereas the 'simulation' (full model) curve does come back to the original signal level.

We have now found the answer: it lies in an ill-adapted coefficient in the model used for the 'theoretical' curve (polynomial division program).

In spite of that, we have still shown the two graphs (Fig. 7 and 8) to illustrate the difference between the one time constant model and the three time constant model.

Dr Robinson The only point is that it appears to be in the reverse order on the graph.

Dr D. Wahren You said that the effects of recirculation of whitewater could be ignored. Was this because you had a very high retention on the machine?

Mr Bauduin Yes, I think so. Most of our tests at that moment were done with kraft pulp. In fact, the model showed, at least for high frequency fluctuations, that it is still true for low retention stocks for regulation purposes.

A Speaker In talking about head box control systems, we should remember that machines in operation today range 30–35 in to high speed newsprint machines with widths of over 400 in. A variation between 1 in and 35 in is quite different from 400 in. I think it would be helpful if any speaker who was talking of a specific machine would give the range in which he was working.

A comment on whether to install a computer on the University of Manchester papermachine. We should remember that the Fourdrinier machine is now in the state of obsolescence and, with the new formers available, before anyone can put a computer on such a machine, he should consider carefully whether it is adaptable to the high speed machines that are likely to be used in the near future.

The Chairman We are starting up a new Fourdrinier machine in two weeks' time; I hope such machines are not as obsolete as this speaker suggests.

Mr N. C. Underwood Will the authors and committee consider a uniform set of units for the measurement of gains, frequencies and response functions so that the results presented in the transactions may be more directly compared?

The Chairman These are very good points that the authors should attempt to take care of in editing.

Mr A. J. Ward Would Mr Gunnarson like to indicate the difference in cost between his rapid response electrical system and the more conventional pneumatic approach?

Mr K. G. A. Gunnarson The total cost for an ASEA system depends on delivery time. The estimated price for a complete system, including complete dc drive (500 kW) for the fan pump, Roots blower with complete dc drive, dp cells, pneumatic reference device for the total head is about 20 000 dollars.

Mr L. D. Edenborough We have now seen two examples in which Mr Jones uses the CONRAD system. It seems to me that perhaps the building block approach of which he has spoken is to some extent comparable with the small mill that cannot justify a large computer system, but wants to do something a little cleverer than is possible with conventional analog control systems, a comparison of these two approaches may be worthwhile. Could Mr Jones please attempt to contrast them from the technical and economic viewpoints?

Mr R. E. Jones Comparing and contrasting is rather a tall order, because it would take quite a long time to cover all the implications of the two techniques. What I will say is that the control strategies that we have used with a CONRAD package certainly could be implemented by analog modules. I think it would be up to the individual customer or systems engineer to decide on the relative costs of going to these two different methods of implementing control. I would point out that *behind* the CONRAD package is a central processor that can do much background work as well.

Dr N. K. Bridge We have heard a lot about head box control this morning and speakers have quoted in passing figures for the control that they can achieve. I wonder whether it might not be of use to people when they get back to their mills if they could have a consensus of opinion from the experts on the platform on how well a flow box might perform with a good controller. What sort of variations might be expected in total head on grade, also how fast should it be possible to change the head?

Dr D. B. Brewster I would like to comment on Dr Sanborn's paper. The head box is a good example of a process for which a mathematical model may be developed directly from physical laws. The advantage of this type of model over the black box empirical model is that there are generally fewer parameters that need to be determined experimentally by identification. Identification experiments tend to make production superintendents nervous, particularly on machines that are being pushed hard. One feature of the head box identification described by Dr Sanborn is that the pond level response is determined open loop—that is, with the analog level controller disconnected. Special care must therefore be taken to ensure that level control is not lost.

It is not necessary to perform the extensive identification described by Dr Sanborn if the valve behaviour is modelled. Dr Al-Shaikh at Westvaco did this and was able to identify the head box completely, using only one experiment with the head box pond level controller on control. A DDC algorithm was designed with the controller parameters being calculated through the model. The validity of this approach became evident when the machine speed was changed by a factor of 2 and the automatic tuning worked perfectly.

The major problem in the Sanborn paper seemed to be the sluggish valve action that resulted in dead time. It would seem to be preferable to solve that problem mechanically.

Dr I. B. Sanborn The way we set up our system is in some respects rather cunning. We always left the analog control on when there was a mechanical failure in hardware between the computer and the head box or, if we wished to open the loop between one to the other, we always fell back on the controller. The net effect was that, as far as the head box was concerned, it was no particular problem to open loops in identification. The magnitudes of the upsets necessary in identification were in the order of 1–2 in water and this was in total head operating ranges of 50 in down to 20 in. In many instances, unless we told the operator we were making an identification, he never knew it took place.

In this particular instance, identification was not as upsetting as it might otherwise have been. This is not the case with basis weight moisture control, however, which causes a decided upset in the process. I agree that a little cleverness can avoid a lot of identification. In fact, as I pointed out during the discussion after my paper, we have developed means now by which, having identified the parameters at one particular operating point, we have calculated at all other points as well. We probably identify initially when we throw the loop, then vary very infrequently, perhaps once every six months or a year.

Mr W. D. Hoath My first question is how representative is this machine of the process efficiency found on full-scale machines? Do Ramaz *et al.* expect to modify the basic process design on those machines to improve the end product—in other words, do they intend to make major process changes linked with the computers to exploit its power and versatility?

We have heard quite a lot from Dr Smith and Dr Sanborn on the need to improve the process to achieve even better control, but have we anywhere achieved major improvement design of computer control? One suspects that possibly we have gone a little way towards this at Grove Mill, but I would like to know whether any contributions could be made to highlight whether we are able to achieve it.

Mr Bauduin I am not quite sure I understand the question, but I take it that you wonder whether our experimental machine is quite representative of an industrial one, as far as our computer project is concerned. We have already said that it is rather different from an industrial machine, but may I remind you that our project deals mainly with studying the principles of computer control. We might undertake a similar study on an industrial machine as a future project; for the time being, we do not really care about production cost or things like that. Therefore, we are not concerned with 'major process changes linked with the computer'.

The Chairman I would like to ask a question of the audience. Is there anyone who has the experience of designing new systems or making major modifications to existing systems and who have made major departures in design as a result of their process control work? In other words, have they changed the size of chests or eliminated chests or other equipment primarily for control purposes?

Mr I. McKnight I should like to comment mainly on Dr Smith's paper with a more detailed question to Dr Sanborn. I think my remarks follow on from those of Mr Hoath and the Chairman.

I wonder, if the Fourdrinier papermachine had not been invented 150 years ago, but just a few years ago, whether the flow box would have been a rather different piece of equipment. I suspect the designer would have produced a device rather like an extruding machine under careful flow control. I suspect also that the measurement of total head would have been scarcely considered. My point is to stress the measurement of flow as the major control parameter and to ask how this conception would have modified the mathematics.

We discussed the Chartham flow box in an earlier paper and there are other machines for which flow control is the main control parameter of the flow box. If a lid was put on this box, I am certain it would continue to run for a reasonable length of time under control without even a blower or level control. If the air cushion was then pared away, you would end up with a solid flow box. Surely, accurate flow control should give the correct answer at the breast roll.

My particular question to Dr Sanborn concerns the flow loop of his flow box scheme. He has created a slightly long-term flow control, whereby the slice is used as a giant slide valve. Now, flow control is a fully established practical technology, with even controllers being manufactured with fixed terms. It is generally accepted that the equal percentage valve gives the correct characteristics. I would like to ask whether this was taken into account when programming his computer to characterise his linear slice valve.

Dr Sanborn No, we have not; we have accepted the valve as it is, but some of the gain calculations made simply take the change in the valve constant into account.

The Chairman I would like to draw the last questioner's attention to the fact that there has been quite a lot of experimental work of a rather substantial nature done with nozzle type head boxes without any air cushion in the system. In general, most people have found that, in order to eliminate pulsations causing high frequency disturbances to the flow arising from pumps, for example, it is necessary to have a dampening effect of the air chamber.

SUB-OPTIMUM GRADE CHANGE PROCEDURES

R. E. JOHNSTON, and L. A. KIRK, Paper Science Section, Polymer and Fibre Science Department, University of Manchester Institute of Science & Technology

Synopsis A numerical optimisation technique has been used to show, computationally, that simple control trajectories can give considerably reduced grade change times. The models studied include one-pump and two-pump systems with and without changing wire rejection factors. Consideration has been given to some simple constraints.

Introduction

THE problem of manually controlling production changes on a Fourdrinier papermachine are numerous. With the advent of computer control, more sophisticated grade changes are possible.

Computers are now used for grade change control[1–4] and Sullivan & Schoeffler have undertaken a theoretical study of the problem.[5] No single policy has been generally adopted on the type of trajectory for this control.

This study demonstrates, computationally, the possible benefits of using simple control trajectories to effect grade changes. A *grade change* in the context of this paper refers to the operation of making a change from one particular paper substance to another, possibly at a different machine speed and different flow box consistency.

Mathematical models

THE particular type of wet end studied is shown in Fig. 1. A model of this type of machine can be used to study basic one-pump and two-pump systems, the former being the special case when flow A is zero. The mathematical model derived relates the system variables to the control variables. The control variables are the slice opening, the machine speed and the thick stock flow; the system variables are the oven-dry basis weight and the web consistency (both ex the suction box section) and the flow box consistency.

The assumptions of the model and the calculations involved at each time

Under the chairmanship of P. E. Wrist

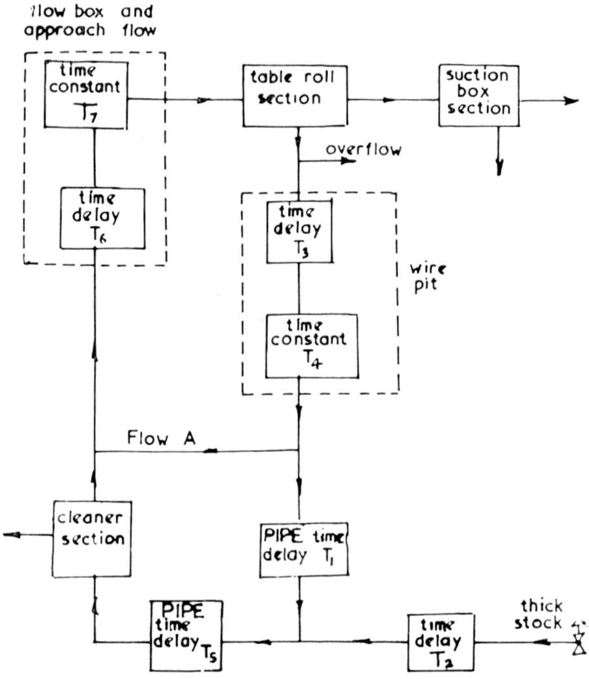

Fig. 1—Wet end system studied

interval for the digital simulation are included in the appendix. A calculation interval of 4 s was used after preliminary investigations indicated that this gave a good representation of the continuous model while keeping the total computation time to an acceptable level. The calculation procedure includes three basic steps at each calculation interval—

1. Choice of (new) values of control variables.
2. Determination of flows around the system by flow balances.
3. Determination of stock consistencies around the system by consideration of the time delays and first order lags.

Constraints of the process

THE constraints consist of both steady state desired values of basis weight, speed and flow box consistency and dynamic constraints dependent upon the particular machine.

Avoidance of web breaks is obviously very important, but it appears to be impossible to put reliable weightings on the relevant factors. In severe cases, it may be necessary to limit both steady state and dynamic states by the condition that the web consistency should remain constant. This restriction would make grade change trajectories very complicated.[5] In this study, the only restriction directly concerned with web breaks has been a minimum web consistency ex the suction box.

In one computer installation,[1] grade changes are controlled within the limit of a specified rate of change of basis weight that appeared to be limiting from previous grade changes made manually. It would appear that this form of grade constraint leads to a safe, if somewhat slower grade change than may otherwise be possible.

Types of trajectory studied

FIG. 2 shows the types of trajectory studied, together with the parameters (X values) describing them. Trajectory *d* was rejected after preliminary investigations indicated that it gave no improvement over trajectory type *a*.

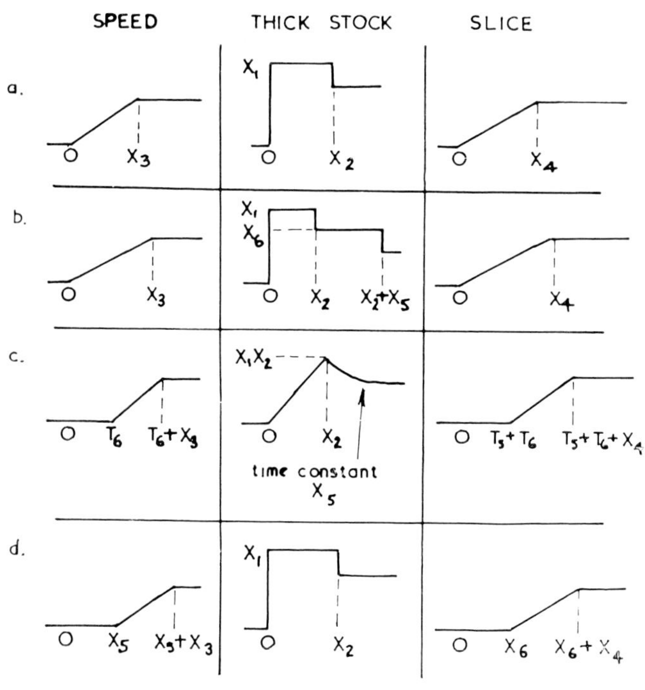

Fig. 2—Trajectory types (control variable plotted against time)

Trajectories *a* and *b* have been used in the simulation of sub-optimum grade changes and trajectory *c* has been used to study trajectories of the type previously referred to.[1]

Sub-optimum control

THE parameters of trajectories *a* and *b* have been varied in a search for a trajectory that minimises a performance index dependent upon the particular constraints assumed. When there is no constraint, the performance index is the time to settle within ± 1 per cent of the desired basis weight. When a constraint is applied to web consistency, the above performance index is modified by the addition of a penalty function proportional to the time integral of 'below limit' web consistency.

A performance index satisfactory for studying specified maximum rate basis weight grade changes, using trajectory type *c*, is the sum of the squares of the differences between actual and desired basis weights at each calculation interval.

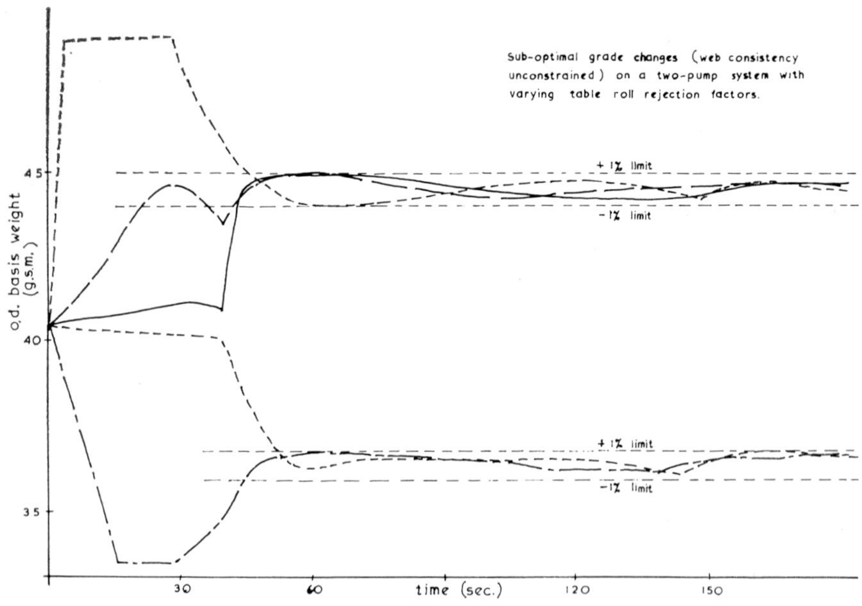

Fig. 3—Sub-optimum grade changes (web consistency unconstrained) on a two-pump system with constant table roll rejection factors

In searching for a minimum performance index, the numerical search technique of Nelder & Mead[6] has been used. This technique is based upon the simplex method, but has the added feature that the search simplex is expanded or contracted, depending on the local 'geometry' of the function space. Its application is highly non-linear problems is recommended by Box[7] for cases in which up to six search variables (parameters) are used. Each optimisation consists of approximately 250 search iterations, each extending over some 80 calculation intervals and requires approximately 30 million basic computer operations.

Results and discussion

SOME typical results are presented in Fig. 3–8 to show basis weight trajectories resulting from sub-optimum controls. The associated Tables 1–6 give details of the machine constants, initial conditions and particular grade changes. Settling times are shown, together with the type of control trajectory and, for comparison, settling times are also presented for control variables following single step trajectories to their final steady state values.

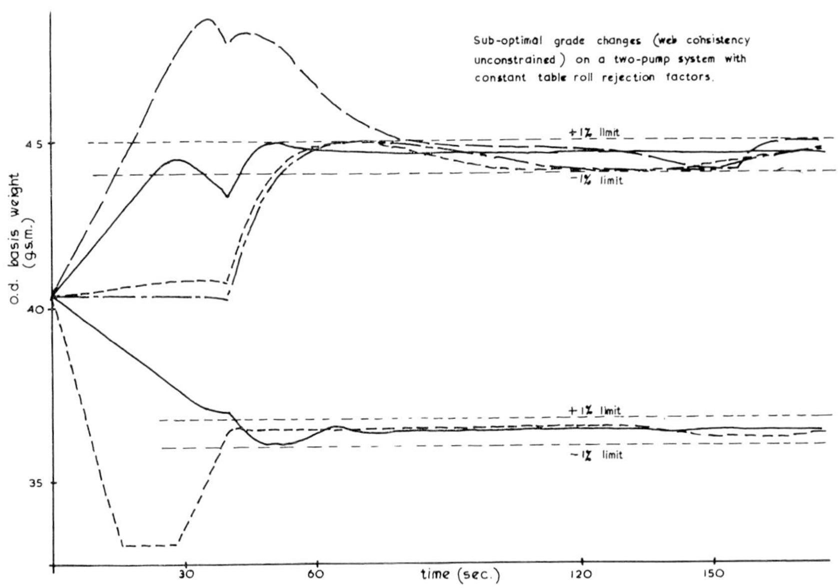

Fig. 4—Sub-optimum grade changes (web consistency unconstrained) on a two-pump system with varying table roll rejection factors

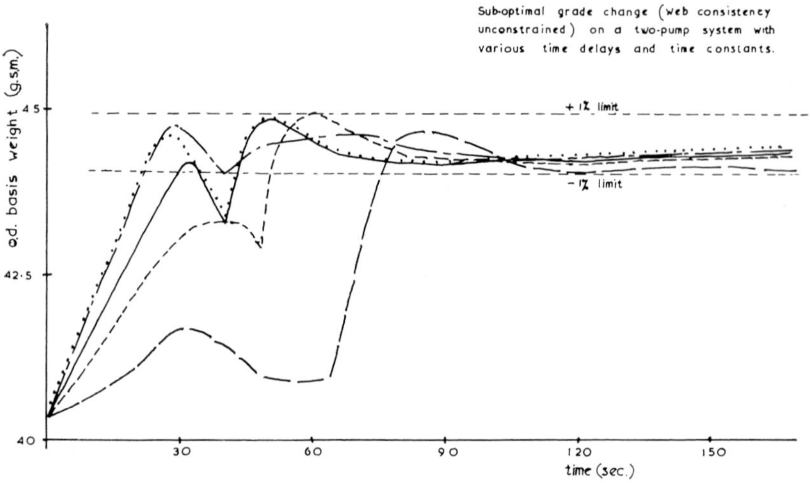

Fig. 5—Sub-optimum grade changes (web consistency unconstrained) with various time delays and time constants

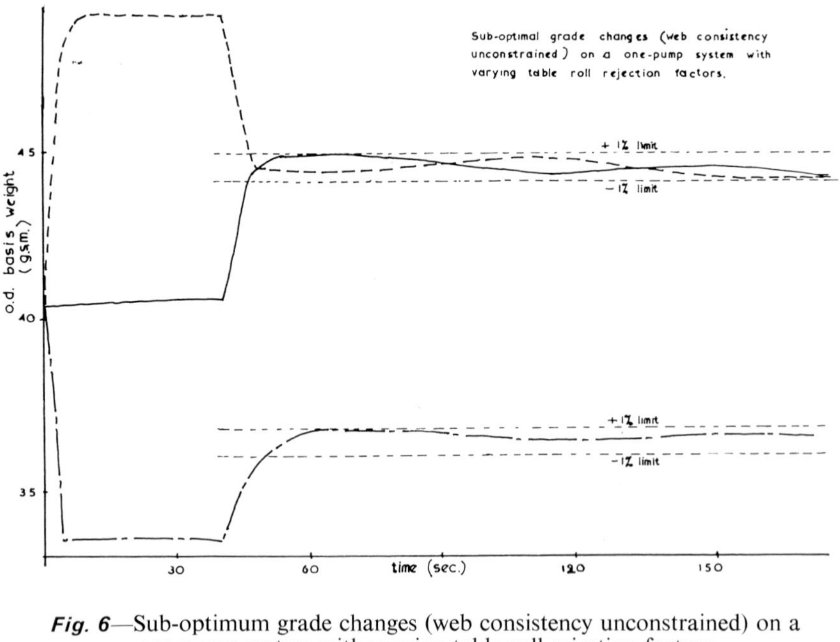

Fig. 6—Sub-optimum grade changes (web consistency unconstrained) on a one-pump system with varying table roll rejection factors

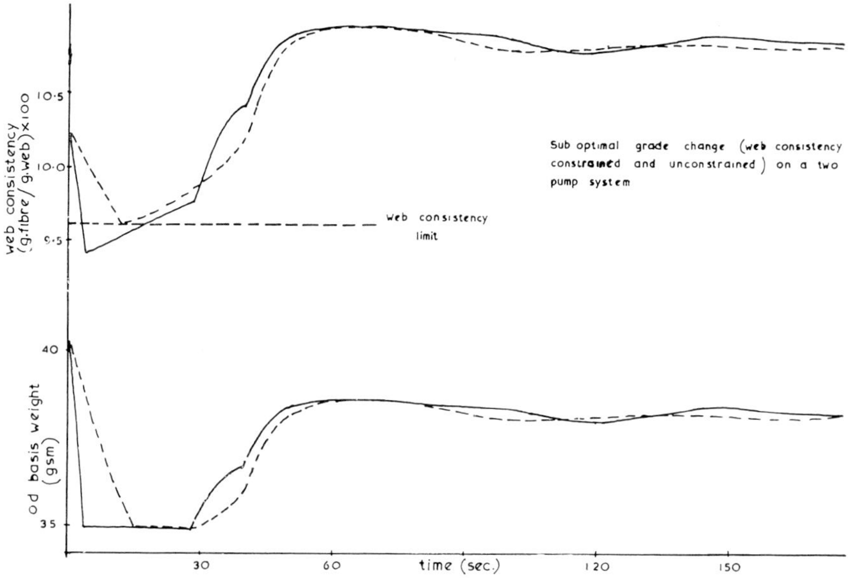

Fig. 7—Sub-optimum grade change (web consistency constrained and unconstrained) on a two-pump system

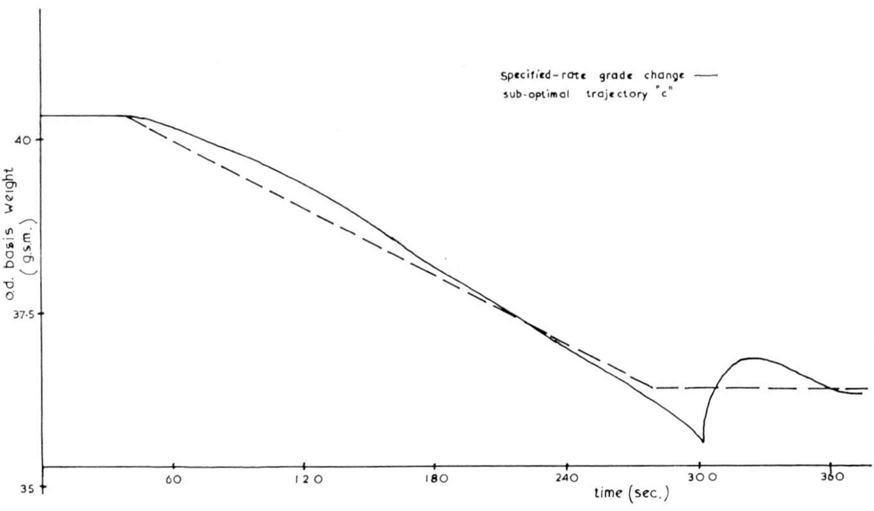

Fig. 8—Specified rate grade change—sub-optimum trajectory *c*

TABLE 1

Sub-optimum grade changes (web consistency unconstrained) on a two-pump system with constant table roll rejection factors (see Fig. 3)

Details
Two-pump system
Table roll rejection factors—constant

Initial conditions—

Machine speed	366 m/min
Basic weight	40·3 g/m^2
Flow box consistency	0·68 per cent

Time delays and time constants—

$T_1 = 12$ s	$T_2 = 16$ s	$T_3 = 44$ s
$T_4 = 16$ s	$T_5 = 12$ s	$T_6 = 28$ s
$T_7 = 8$ s		

Grade changes —unconstrained

Code	———	- - - -	— — —	— — —	———	- - - -
Changes— Basis weight, per cent	+10	+10	+10	+10	−10	−10
Speed, per cent	+10	+10	+10	+10	−10	−10
Flow box consistency, per cent	0	+5	+10	−10	0	+10
Trajectory type	*a*	*a*	*a*	*a*	*a*	*a*
Settling time for sub-optimum trajectory, s	44	52	52	84	40	40
Settling time for step trajectory, s	56	176	256	240	160	232

TABLE 2

Sub-optimum grade changes (web consistency unconstrained) on a two-pump system with varying table roll rejection factors (see Fig. 4)

Details
Two-pump system
Table roll rejection factors—
Solids—constant
Volume—affected by (*1*) machine speed −0·2 per cent/per cent
(*2*) flow box consistency −0·1 per cent/per cent

Initial conditions—

Machine speed	366 m/min
Basic weight	40·3 g/m^2
Flow box consistency	0·68 per cent

Time delays and time constants—

$T_1 = 12$ s	$T_2 = 16$ s	$T_3 = 44$ s
$T_4 = 16$ s	$T_5 = 12$ s	$T_6 = 28$ s
$T_7 = 8$ s		

TABLE 2—*continued*

Grade changes—unconstrained

Code	— — —	———	- - - - -	-- - - -	— - —
Changes— Basis weight, per cent	+10	+10	+10	−10	−10
Speed, per cent	+10	+10	−10	+10	+10
Flow box consistency, per cent	0	+10	−10	−10	+10
Trajectory type	*a*	*a*	*a*	*a*	*a*
Settling time for sub-optimum trajectory, s	44	44	44	52	44
Settling time for step trajectory, s	152	288	280	272	276

TABLE 3

Sub-optimum grade changes for a particular grade change on a two-pump system with various time constants and time delays (see Fig. 5)

Details

Two-pump system

Table roll rejection factors—

Solids—constant

Volume—affected by (*1*) machine speed −0·2 per cent/per cent

(*2*) flow box consistency −0·1 per cent/per cent

Initial conditions—

Machine speed	366 m/min
Basic weight	40·3 g/m²
Flow box consistency	0·68 per cent

Time delays and time constants, s

Code	T_1	T_2	T_3	T_4	T_5	T_6	T_7	*Settling time for sub-optimum trajectory, s*
———	12	16	16	36	12	28	8	44
- - - -	12	16	16	36	20	28	8	52
— — —	12	16	16	36	40	28	8	76
— - —	12	16	16	36	12	28	20	20
· · · ·	12	16	44	16	12	28	8	44

Grade changes—unconstrained

Basis weight	+10 per cent
Machine speed	+10 per cent
Flow box consistency	0 per cent

TABLE 4

Sub-optimum grade changes (web consistency unconstrained) on a one-pump system with varying table roll rejection factors (see Fig. 6)

Details

One-pump system

Table roll rejection factors—
Solids—constant
Volume—affected by (*1*) machine speed −0·2 per cent/per cent
(*2*) flow box consistency −0·1 per cent/per cent

Initial conditions—
Machine speed 366 m/min
Basic weight 40·3 g/m²
Flow box consistency 0·68 per cent

Time delays and time constants—
$T_1 = 12$ s $T_2 = 16$ s $T_3 = 16$ s
$T_4 = 44$ s $T_5 = 12$ s $T_6 = 28$ s
$T_7 = 8$ s

Grade changes—unconstrained

Code	———	- - -	— — —
Changes— Basis weight, per cent	+10	+10	−10
Speed, per cent	+10	−10	+10
Flow box consistency, per cent	+10	−10	+10
Trajectory type	*a*	*a*	*a*
Settling time for sub-optimum trajectory, s	48	48	48
Settling time for step trajectory, s	296	352	292

TABLE 5

Sub-optimum grade change (constrained and unconstrained) on a two-pump system (see Fig. 7)

Details

Two-pump system
Table roll rejection factors—constant

Initial conditions—
Machine speed 366 m/min
Basis weight 40·3 g/m²
Flow box consistency 0·68 per cent

Time delays and time constants—
$T_1 = 12$ s $T_2 = 16$ s $T_3 = 44$ s
$T_4 = 16$ s $T_5 = 12$ s $T_6 = 28$ s
$T_7 = 8$ s

Grade changes
Basic weight −5 per cent
Speed +10 per cent
Flow box consistency +10 per cent
Trajectory type *a*

TABLE 5—*continued*

Code	———	- - - - - - -
Limitation	unconstrained	constrained
Web consistency minimum	does not apply	0·94 of initial consistency (that is, 9·62 g fibre/100 g web)
Settling time for sub-optimum trajectory, s	44	48

TABLE 6

Specified rate grade change—sub-optimum trajectory *c* (see Fig. 8)

Details

Two-pump system

Initial conditions—

Machine speed	366 m/min
Basic weight	40·3 g/m^2
Flow box consistency	0·68 per cent

Table roll rejection factors —

Solids—constants

Volume—affected by (*1*) machine speed	−0·2 per cent/per cent
(*2*) Flow box consistency	−0·1 per cent/per cent

Time delays and time constants—

$T_1 = 12$ s	$T_2 = 16$ s	$T_3 - 44$ s
$T_4 = 16$ s	$T_5 = 12$ s	$T_6 = 28$ s
$T_7 = 8$ s		

Grade changes details

Basic weight	−10 per cent
Machine speed	+10 per cent
Flow box consistency	−10 per cent

Code	
- - -	Desired (best) trajectory
———	Sub-optimum trajectory

The numerical values of the time delays, time constants and initial conditions were chosen to be representative of actual practice. They are in line with those used by Smethurst.[9]

It is evident from the results obtained that these simple trajectories can give quite substantial reductions in grade change times when compared with those obtained by using step trajectories for the control variables. In most cases, it is possible to reduce the grade change time to a time only slightly greater than the dominant time delay from the primary dilution point to the wire, that is, T_5 plus T_6.

Fig. 3–5 indicate an apparent simple classification of grade changes according to whether the basis weight does or does not overshoot its final desired value. The former case is associated with a change in flow box consistency in the opposite direction to the basis weight change. The latter is associated with either no flow box consistency change or with a change of flow box consistency in the same direction as the basis weight change.

The sub-optimum control variable trajectories can also be classified. From the results obtained for constant and changing rejection factors and for the various time constants and delays, on both one-pump and two-pump systems, it appears that the following general rules are applicable.

1. Changes in which the fractional flow box consistency change is the same in magnitude and direction to the fractional basis weight change

This is an important class, because no significant movement of the slice is involved and the control trajectories should therefore be applicable to machines without slice control.

Direction of speed change	*Trajectory type required*	*Rate of speed change*	*Thick stock trajectory*
Same as basis weight	*a*	Slow $X_3 \simeq 4(T_5 + T_6)$	Slight overshoot ($X_2 \simeq T_6$)
Opposite to basis weight	*b*	Fast as possible ($X_3 \to 0$)	Temporary change in the direction of the basis weight change ($X_2 + X_5 \simeq 4T_6$)

2. Changes in which the flow box consistency is to remain the same

Direction of speed change	*Trajectory type required*	*Rate of speed change*	*Rate of slice change*	*Thick stock trajectory*
Same as basis weight	*a*	Quick ($X_3 \simeq T_5 + T_6$)	Quick ($X_4 \simeq T_5 + T_6$)	Step ($X_2 \to 0$)
Opposite to basis weight	*a*	Fast as possible ($X_3 \to 0$)	Fast as possible ($X_4 \to 0$)	Variable

3. Changes in which the flow box consistency change is in the opposite direction to the basis weight change

Direction of speed change	*Trajectory type required*	*Rate of speed change*	*Rate of slice change*	*Thick stock trajectory*
Same as basis weight	*a*	Fas as possible ($X_3 \to 0$)	Fast as possible ($X_4 \to 0$)	Undershoot $X_2 \simeq 4T_6$)
Opposite to basis weight	*a*	Slow ($X_3 \simeq 4(T_5 + T_6)$)	Fast as possible ($X_4 \to 0$)	Variable

There appear to be no general rules for carrying out grade changes when the fractional change of flow box consistency is the same in direction, but not in magnitude to the fractional change in basis weight. Trajectory type *a* is adequate for this type of grade change.

The results for case *2* (also see Tables 1 and 2) show that grade changes when the desired flow box consistency change is zero are the simplest to carry out—there is very little scope for improvement over step control trajectories.

Web consistency restriction

IN CONSIDERING the possible restricting effect of a minimum allowable web consistency ex the suction box, it is important to use an accurate representation of the action of the suction box. In the simulation, it has been assumed that the volume rejection factor of the suction box section is a linear function of the volumetric flow rate to it; more specificially, it has been assumed that a 1 per cent increase in flow rate gives a 1·4 per cent increase in the volume rejection factor. This implies that drainage is improved through the better sealing of the wire over the boxes obtained by having more stock over them. The value of 1·4 per cent was chosen as a reasonable value on the basis of steady state data obtained by Brauns & Oskarsson,[8] who carried out experiments on a pilot-plant machine at speeds approaching 400 m/min.

Under these particular conditions, the results obtained so far do not indicate any severe restrictions, owing to the application of a web consistency constraint. In some cases, the sub-optimum grade change time is increased if the web consistency is not allowed to drop below 0·92 of the lowest steady state value. (Fig. 3 E shows an example in which the allowable limit is 0·94 of the initial web consistency and no appreciable increase of the grade change time results.)

Specified maximum-rate basis weight grade changes

FIG. 7 shows the results obtained by 'optimum' selection of the parameters of trajectory type *c*. A reasonable approximation to the best possible basis weight trajectory (within the limits of the constraint) is obtained.[1]

Conclusions

THE application of a numerical optimisation technique to the search for sub-optimum grade change procedures on a simulated papermachine has been shown to be successful.

The results indicate that simple control trajectories are sufficient to obtain considerable reduction of the settling time involved in a grade change. This conclusion is shown to hold for both one-pump and two-pump systems with two different wire drainage effects.

The imposition of a minimum web consistency constraint does not markedly affect the above conclusions under the particular conditions assumed to describe the drainage effect of the table rolls and the suction box section.

Some general rules governing the sub-optimum procedures are evident from consideration of the results.

References

1. Bennett, J. A., *Paper Maker*, 1968, **155** (5), 39, 40, 42
2. Larking, J. H., Paper presented at Norwegian Pulp and Paper Research Institute, Oslo, February 1966 (Elliott Process Automation reprint)
3. Alsholm, O. and Sangregorio, G., Technical Paper TP 18.170 (IBM Systems Development Division Nordic Laboratory, Sweden, October 1966)
4. Jones, R. E. and Whight, W. T., *Measurement and Control* (G.B.), 1968, **1** (8), 287–291, 293 and 194
5. Sullivan, P. R. and Schoeffler, J. D., Paper 22B (IFAC Congress, London, June 1966)
6. Nelder, J. A. and Mead, R., *Computer J.*, 1965, **7**, 308
7. Box, M. J., *ibid.*, 1966, **1**, 67
8. Brauns, O., and Oskarsson, R., *Paper Maker*, 1954, **127** (3), 190–196
9. Smethurst, J. A., Paper 22C (IFAC Congress, London, June 1966)

Appendix—Assumptions in the mathematical model

1. Level control in the flow box and thin stock flow rate control are such that the efflux ratio can be maintained constant. (This eliminates the possibility of an independent thin stock flow control variable.)

2. Volume and solids rejection factors at the table rolls are either constant or linear functions of flow box consistency and machine speed.

3. Volume rejection factor of the suction box section is a linear function of the volumetric flow rate to the suction box section.[8]

4. Pipes can be characterised by constant time delays in transmitting stock consistencies.

5. Time constants of perfectly mixed tanks remain constant.

6. Mixing characteristics of both the flow box and the wire pit can be described by a combination of a time delay and a perfectly mixed tank.

7. Consistency of the suction box rejects remains constant.[8]

8. Flow rate to the cleaning section is constant in the two-pump system.

9. Volume and solids rejection factors of the cleaning section are constant.

10. There is zero time delay at the wire.

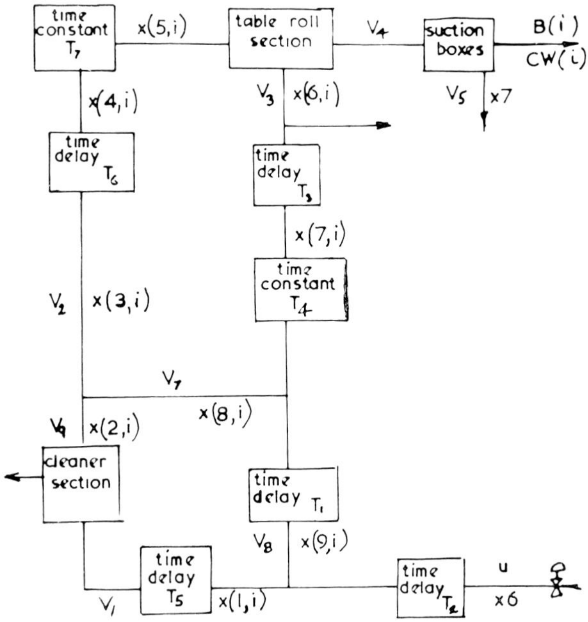

Fig. 9—Terminology for simulation calculation of interval *i*

Simulation—calculations after time interval number i

Terminology (see also Fig. 9)

Volume rejection factors	
Cleaner section	*k*1
Table roll section	*k*3
initially	*c*3
speed effect	*c*5
flow box consistency effect	*c*6
Suction box section	*SBR*
effect of flow	*C*10
Solids rejection factors	
Cleaner section	*k*2
Table roll section	*k*4
initially	*c*4
speed effect	*c*7
flow box consistency effect	*c*8
Basis weight after time interval *i*	*B*(*i*)
Web consistency ex the suction box section after time interval *i*	*CW*(*i*)

Slice opening	h
Machine speed	v
Thick stock flow rate	u
Efflux ratio	e
Flow box jet constant	FBK
Deckle	w

Primed conditions denote initial conditions
Volumetric flow rates denoted by V values
Stock consistencies denoted by x values

Calculations

1. Select control variables as functions of trajectory parameters (see Fig. 2)

$$u = f1(X \text{ values})$$
$$h = f2(X \text{ values})$$
$$v = f3(X \text{ values})$$

2. Calculate flows

$$V_2 = FBK*e*w*h*v$$

Calculate $x(4,i)$, $x(5,i)$ so that $k3$ can be found

$$x(4,i) = x(3,i-T_6)$$
$$x(5,i) = (x(4,i)+T_7*x(5,i-1))/(1+T_7)$$
$$k_3 = c3+c5*(v-v')/v'+c6*(x(5,i)-x5')/x5'$$
$$V_4 = (1-k3)*V_2$$
$$V_7 = V_2-V_9$$
$$V_8 = V_1-u$$
$$V_9 = (1-k1)*V_1$$
$$V_5 = V_4*SBR*(1+C10*(V_4-V_4')/V_4')$$

3. Calculate stock consistencies, basis weight and web consistency

$$x(9,i) = x(8,i-T_1)$$
$$x(7,i) = x(6,i-T_3)$$
$$x(8,i) = (x(7,i)+T_4*x(8,i-1))/(1+T_4)$$
$$x(2,i) = (1-k2)*x(1,i-T_5)/(1-k1)$$
$$x(1,i) = (V_8*x(9,i)+u*x6)/V_1$$
$$x(3,i) = (V_7*x(8,i)+V_9*x(2,i))/V_2$$
$$k4 = C4+C7*(v-v')/v'+C8*(x(5,i)-x5')/x5'$$
$$x(6,i) = k4*x(5,i)/k3$$
$$B(i) = (V_2*x(5,i)-V_3*x(6,i)-V_5*x7)/(v*w)$$
$$CW(i) = B(i)*v*w/(V_4-V_5)$$

Discussion

Dr I. D. McFarlane Are not our basis weight and other control strategies likely to become much simpler when we have better instruments for on-line measurement? For instance, there is no fundamental reason not to measure oven-dry basis weight profiles at the wet end instantaneously with electronic instead of mechanical scanning.

Mr R. E. Johnston Since my paper was concerned with feedforward control, feedback from any instrument will make no difference. I hope I made it clear that we were looking for simple trajectories.

Mr E. Justus I would like to ask about design for control or lack of need for control, perhaps. First of all, I agree with Dr McFarlane. One question to be considered in the design of head box control is whether we are trying to control a process or trying to eliminate noise—hydraulic noise—in the system. We should therefore design a flow system and the rest of the system to eliminate what are really noise problems so that we can then consider the basic variables of the system. It is most important to rule out noise, just as in audio amplifiers.

The Chairman Lest there be any confusion, the noise referred to was random flow variations that are uncontrolled rather than any audible sound.

Mr G. Gavelin Referring to an earlier question by the Chairman, I think there are two parameters we should definitely take into account in head box control—the volume of the air cushion and the area of the liquid surface. The effect of these two variables on head box control has been studied quite extensively at the Swedish Institute and the results have recently been published (in Swedish). I hope they will be made available to you later on.

The Chairman This is an important point. Control systems become effective only when noise has been reduced to a reasonably low level, otherwise any improvement will be lost in the uncontrollable variations that persist.

Mr J. S. Harris Dr Sanborn and Dr Brewster mentioned the problems associated with process identification—finding the value of constants for empirical equations. For dynamic studies, it is unnecessary to carry out changes in big machine equations to find these constants. Tracer studies at the wet end and sampling down the machine are effective methods of process identification.

Mr J. A. Bennett With regard to Mr Johnston's paper, the original white-water time constant at Grove Mill was of the order of 20 min. This was reduced to 3 min before computer control was applied.

Secondly, I am indebted to Mr Hoath this morning for highlighting one of the most important aspects of computer control. Our work on computer control has taught us to question accepted papermaking practices and this has led to a very substantial reduction in the complexity—hence, high capital cost—of our papermaking process.

COMPUTER CONTROL OF A PAPERMACHINE USING A LINEAR STATE SPACE MODEL

S. HEM, Control Group, University Engineering Department, Cambridge*

Synopsis Perturbation experiments on a Fourdrinier papermachine have confirmed that its dynamic behaviour can be represented adequately by a state space model in the form of a matrix difference equation. The basic equations involved have been treated in general terms, but the discussion on the model building and control system design is made more explicit by reference to a specific system.

Methods have been developed for investigating and describing the papermaking system as a process to which modern control theory can be applied. It has been shown how the model can be used to determine possible control strategies to change grade in such a way that the grade change time is at a minimum and certain papermaking criteria are obeyed. The control objectives have been stated by analytical performance criteria in the form of quadratic cost functions.

A simple grade change at constant machine speed was achieved by altering the thick stock flow according to a trajectory determined by the rate constraint of the flow valve. It was found necessary to manipulate and synchronise the thick stock and thin stock flows together with the machine speed in order to change grade at constant production rate.

Based on optimum control and filter theory, an on-line controller has been designed to manipulate the thick stock flow in order to minimise the variance of the measured basis weight. The developed formulation incorporates optimum estimation of inaccessible state variables as an implicit feature. The control action is given by a proportional term together with a memory term to account for past values of control and basis weight. The controller has been implemented on a machine and is shown to have stabilised the system considerably.

Introduction

A SERIOUS problem in the production of multi-grade paper is to make fast and smooth grade changes and to maintain uniform and reproducible properties of the paper.

* Present address—Reed Group Ltd., Research & Development Centre, Aylesford, Maidstone, Kent

Under the chairmanship of P. E. Wrist

The process is characterised by transport delays and long-term transient behaviour, owing to the inherent capacitance of hold-up vessels and pipe network, which can cause the paper to drift off specification, especially immediately after a grade change.

A number of variables such as certain consistencies and flows are inaccessible to measurement, consequently information about the state of the process is incomplete and it may not be possible to compensate for disturbances.

There is a major economic incentive to optimise grade changes and thereby minimise production of off-grade paper. Furthermore, the system is generally not in a steady state immediately after a grade change and it is most important to ensure that the paper is on specification while the system settles down and that it is maintained steady for as long as it is required. Paper breaks caused by undesirable transients during grade changes can result in a considerable loss of production.

A basic prerequisite for the design of a control system to perform the tasks indicated above is to establish an adequate mathematical model of the process.

The papermaking process comprises a certain number of basic operations such as the transport of fibre and additives between tanks and chests where mixing takes place. Another important feature is the coming together at a point of several streams of stock with different consistencies, then emerging as one stream. Thus, four basic concepts are fundamental to the process—

1. Transport delays.
2. Mixing in chests and tanks.
3. Mixing at a point.
4. Flow dynamics.

Some of these are essentially non-linear in character. It is assumed in the present investigation, however, that we are interested only in the dynamic behaviour of the system in the neighbourhood of some specified operating level; therefore, it will be sufficiently accurate to consider a linear model. Further direct digital control is to be used and the attention can be focused on a linearised, discrete time model as follows—

$$x(n+1) = \sum_{i=o}^{h} F_i x(n-i) + Eu(n) + d \quad . \quad . \quad . \quad (1)$$

where n is the independent, discrete time variable,
x is the m $\times$ 1 state vector of the papermachine,
u is the s $\times$ 1 control vector,
d is the s $\times$ 1 plant noise vector

and F_i and E are m $\times$ m and m $\times$ s transition and input matrices, respectively.

Present methods of automation in papermaking provide for some control of the basic process variables, but this control can be described as being partial.

Effective control, however, must provide automatic control of all the basic parts of the process and must take into account the interrelationships of the corresponding controlled quantities. If this is to be achieved, it is necessary to develop a method for investigating and describing the papermaking system as a process to which modern control theory can be applied.

This integrated approach of dealing with the interactions between variables and their effect on the finished product remain to be established.

The objectives of the present investigation were as follows—

1. To devise a manageable and realistic model of the process in the form of equation (*1*).
2. To use the model to determine possible control strategies to change grade in such a way that the grade change time is small and certain papermaking criteria are obeyed.
3. To design a computer controller that maintains constant grade with minimum deviation from the desired specifications.

System description and control objectives

THE actual process on which the controls is to be implemented will be used as an example to make the developed general methods of model building and control theory more explicit.

The papermaking system is shown in Fig. 1. From a constant head tank, thick stock of about 3·5 per cent consistency is continuously supplied at a rate of 600–800 gal/min to the system at the mixing pump in a suction pipe, which also supplies recycled backwater with a concentration of 0·4–0·5 per cent fibre.

Thin stock (0·9–1·0 per cent) is formed by mixing, then pumped by a constant head pump at a rate of 7 000 gal/min through the cleaning equipment.

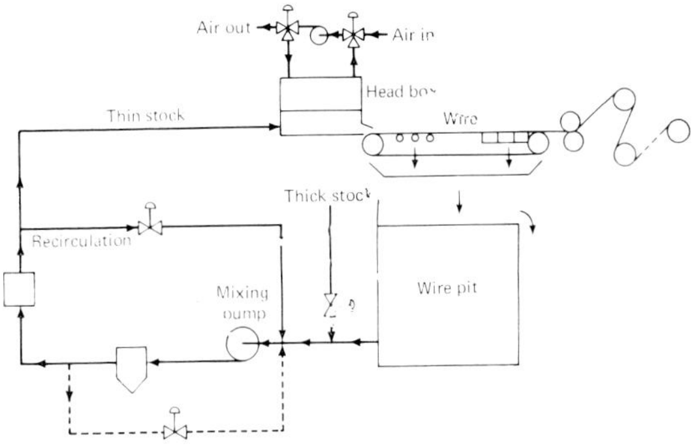

Fig. 1—Layout of process

Recirculation pipes are fitted to maintain a high flow through the cleaning equipment for efficient performance. The flow of thin stock into the head box is about 5 000 gal/min and is discharged through the slice opening on to the moving wire. The machine and wire speed ranges 650–950 ft/min.

A fraction of about 0·6 of fibre and additives filters through the wire together with about 98 per cent of the water into the wire pit. This is fitted with an overflow in order to supply the mixing pump from a constant head.

For papermaking reasons, it is necessary to maintain a constant low level of stock in the head box; the head box is therefore pressurised and air is continuously circulated through the head box airspace by a pump as indicated in Fig. 1.

The wire pit is the largest of the tanks and contains a constant volume of 2 300 ft^3 of backwater. This gives rise to a time constant of about 3 min to changes in consistency. The combined effect of consistency changes arising from recirculation dynamics associated with the cleaning, together with the head box and wire pit mixing dynamics, yields a system with a very long settling time of about 30 min. This makes manual control difficult, especially at grade changes, when it is necessary to manipulate several input variables simultaneously to force the system to go from one state to another.

The majority of grade changes for this particular machine are carried out at constant machine speed or constant production rate. The first method is mainly used for small changes in basis weight and when it is necessary to alter production rate for papermaking reasons. A major economic incentive is to maintain the highest possible production rate, usually constrained by drying capacity. Under these conditions, a grade change is best effected by keeping the production rate constant.

Sheet strength and formation depend to a great extent on the ratio of stock velocity in the slice to wire speed, called the efflux ratio. It is particularly important to keep the efflux ratio constant during a grade change.

A grade change subject to these papermaking criteria (constant production and efflux ratio) can be realised by simultaneously manipulating the three input variables— thick stock flow, thin stock flow and wire speed. If the papermaking criteria are violated, the result is that often the sheet breaks or the basis weight and moisture can be outside the tolerance limits for long periods of time. A general strategy for a constant production change can now be formulated. It is required to change the basis weight from one level to another as quickly as possible, subject to saturation and rate constraints of the input variables while production rate and efflux ratio remain constant.

Grade changes at constant speed are carried out by changing the thick stock flow, subject to constraints on the valve movements.

The third control objective is to design a direct digital computer controller that minimises the variance of basis weight by manipulating the thick stock flow based on measurements of the basis weight.

State space representation of the process

WATER is used as a vehicle to transport fibre in a network of pipes and tanks, hence lags will occur with respect to changes both in flow and in concentrations. It will be necessary to consider the past states of the system, as well as the present states, owing to transport delays that affect the concentration of fibre and additives. Finally, the interactions between flow and concentration dynamics must be considered, as these determine the distribution of fibre concentrations throughout the system.

The object of the modelling problem is to evaluate these functions and determine the resultant dynamic behaviour of the plant.

We are interested mainly in the dynamic behaviour in the neighbourhood of a fixed operating level and certain assumptions can therefore be made—

1. Simplified fluid flow equations in the form of a direct analog between fluid flow and electric current in networks is considered to be adequate.
2. Perfect mixing takes place in the head box and wire pit.
3. The delays are time invariant and independent of the state variables.
4. Changes in the fraction of fibre and additives that filter through the wire is taken to be proportional, but of opposite sign to the changes in the initial basis weight on the wire.

The last approximation has been verified and used by several investigators[1, 2] and provides the mathematical link between the flow and consistency dynamics of the head box and wire pit.

The total hydraulic pressure at the slice opening is of particular interest and this, together with the stock level, form the variables of the simultaneous differential equations that describe the head box flow dynamics—

$$C_1\frac{dP_1}{dt} = q_{ai}-q_{ao}+(C_1+C_2)\frac{dP_2}{dt} \qquad (2)$$

$$C_2\frac{dP_2}{dt} = q_i-q_o \qquad (3)$$

The input and output concentrations in the mixing tanks, head box and wire pit are assumed to be related by the first-order differential equations—

$$\frac{dV}{dt} = -q_o+\sum_{j=1}^{r}(q_i)_j \qquad (4)$$

$$\frac{d(c_oV)}{dt} -c_oq_o+\sum_{j=1}^{r}(c_i)_j(q_i)_j \qquad (5)$$

The thick stock flow, the recirculation flow and the wire pit flow converge immediately before the mixing pump. The output concentration from the pump can be derived as a special case of equations (*4*) and (*5*) by putting $V = 0$ and $r = 3$.

The head box and wire pit equations were derived with $r = 1$.

The basic equations (*2*)–(*5*) were linearised about a chosen operating level. In order to obtain a discrete model as indicated in the introduction, it was assumed that information of the state variables was available only at periodic intervals of time (every 10 s) and the forcing functions were to be held constant throughout the interval and changed in a step manner at the sampling instants.

In order to account for delays, a modified version of a method described by Tou[3], was used to form the difference equations of the linear continuous equations. This yields the complete deterministic model in the discrete form given by equation (*1*), where now h = 3, m = 5 and s = 4.

$$x(n+1) = \sum_{i=o}^{3} F_i x(n-i) + Eu(n) \quad . \quad . \quad . \quad (6)$$

The system is characterised by five state variables and four control variables, thus the transition matrices F_i and control matrix E are 5×5 and 5×4, respectively.

The dynamic behaviour of the system is the n given by the discrete time history of it state variables, which constitute—

x_1 = Hydraulic pressure at the slice,
x_2 = Stock level in the head box,
x_3 = Consistency immediately after the mixing pump,
x_4 = Head box consistency,
x_5 = Wire pit consistency.

The dynamic behaviour can be controlled by the four inputs—

u_1 = Head box air flow,
u_2 = Thin stock flow,
u_3 = Thick stock flow,
u_4 = Machine speed.

It is difficult to determine a rational criteria for the adequacy of the model; nevertheless, it is of paramount important to establish that the model represents the plant behaviour sufficiently well in order to devise practical control schemes.

The thick stock flow on the machine was changed according to a predetermined pseudo-random binary sequence of thick stock valve positions and the response of basis weight was observed from beta-gauge measurements.

The model was then perturbed by an identical sequence of assumed thick stock flows u_3. The response was observed by augmenting the state space of equation (*6*) by a linear discrete output equation to represent changes in basis weight—

$$x_6(n+1) = Mx(n+1) \quad . \quad . \quad . \quad . \qquad (7)$$

where M is a 1×5 matrix and x is the state vector of equation (*6*). The simulation thus represents a tie between analysis and experiment and, as can be seen in Fig. 2, the agreement between the predicted behaviour of basis weight and the actual behaviour seems quite satisfactory for our purpose.

It should be pointed out that the basis weight trace contained a small amount of noise and was smoothed by eye before it was plotted on Fig. 2.

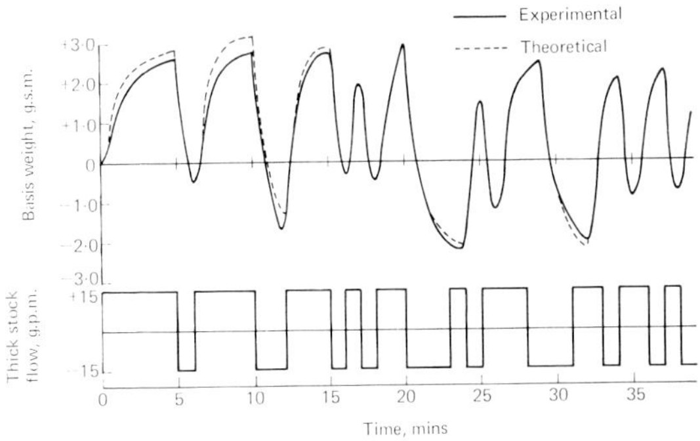

Fig. 2—Perturbation experiment (this refers to a plant and model with the additional recirculation indicated in Fig. 1)

Statement of control problems

WE WILL attempt to control a process that can be described by a linear, discrete matrix difference equation of the following form—

$$x(n+1) = \sum_{i=0}^{h} F_i x(n-i) + Eu(n) \quad . \quad . \quad . \qquad (8)$$

For grade changes, it is required to find the input vector—

$$u(n),\ n = 0,1,---,p-1,$$

which minimises a performance criterion of the form—

$$J = \sum_{n=0}^{p-1} \tfrac{1}{2}\|x_d(n+1) - x(n+1)\|^2 \quad . \quad . \quad . \qquad (9)$$

subject to equation (*8*) and the rate constraint—

$$|u(n) - u(n-1)| \leq v \quad . \quad . \quad . \quad . \qquad (10)$$

A quadratic non-linear programming method developed by Kishi,[4] suitable for on-line computation, was extended to multi-variable systems with delays in order to solve the above problem.[5] With some modifications, the same method was used to solve the regulator problem by minimising a performance criterion of the form—

$$J = \sum_{n=o}^{p-1} \tfrac{1}{2}\|x(n+1)\|_Q^2 + \tfrac{1}{2}\|u(n)\|_R^2 \quad . \quad . \quad . \quad . \quad (11)$$

subject to equation (*8*). This leads to a feedback control law—that is, the controls are given as a function of the state variables, thus—

$$u(n) = -\sum_{i=o}^{h} D_i x(n-1) \quad . \quad . \quad . \quad . \quad (12)$$

where D_i are $s \times m$ matrices.

The discrete form of dynamic programming[6] can also be used to derive the feedback law, but it is then necessary to augment the state space of equation (*8*) to account for the delays.

Control strategies

Grade change at constant machine speed—The object was to determine the input trajectory of thick stock flow u_3 in order to raise the basis weight 10 g/m² from the assumed operating level. At the same time, it was required to minimise the variance about the target subject to the rate constraint—

$$|u_3(n) - u_3(n-1)| \leq 0{\cdot}15 \text{ ft}^3\text{/s per 10 s}$$

Fig. 3 shows the determined optimum trajectories. It can be seen that it is required to increase u_3 as fast as possible for about 60 s, to slightly more than twice the change in flow necessary to achieve the new grade. The thick stock is then gradually brought down towards the demanded change in flow of 0·37 ft³/sec. As a comparison, the response of basis weight to a hypothetical step change of 0·37 ft³/sec in u_3 is also shown in Fig. 3.

The minimisation of quadratic performance criteria often results in oscillatory input trajectories being determined, as that exhibited by the thick stock flow. In this case, a less oscillatory grade change could be obtained by imposing a more severe rate constraint—for example, 0·05 ft³/sec per sampling interval (10 s). The time taken to achieve the new grade is then obviously longer, but compares very favourably with the faster oscillatory response.

The method used above to determine thick stock trajectories can be implemented on-line as a feedback control system, especially with a single input/single output system like the one considered here. A large grade change at constant machine speed is likely to cause the dryline to change position (the

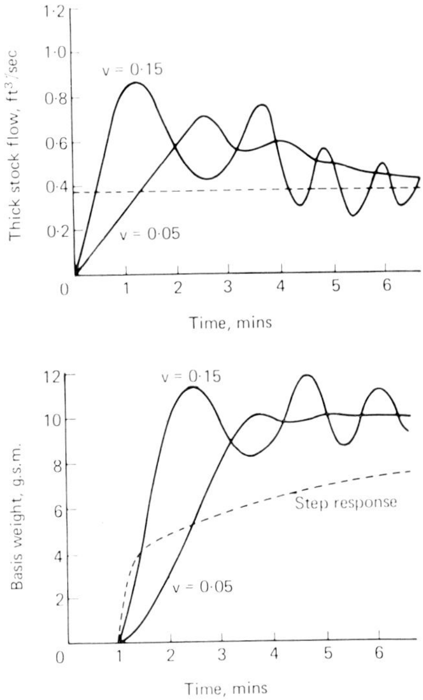

Fig. 3—Grade change trajectories at constant machine speed rate constraints of thick stock flow $V = 0{\cdot}05, 0{\cdot}15^3$/s/10 s

position on the wire at which no stock remains above the sheet) with the result that the sheet is either too dry or too moist to be transferred from the wire without breaking. In the present investigation, the dryline position was assumed to be adequately represented by—

$$s_d = \frac{kWv_m}{\Delta P} \quad \quad (13)$$

which was derived from general filtration theory, as stated by Meadley.[7] A grade change at constant machine speed is therefore likely to require adjustment of ΔP, the suction box vacuum.

A more elaborate and complicated equation for the dryline position has been used by Sullivan & Schoeffler,[8] who considered grade changes subject to the constraint that the dryline position be invariant.

Grade change at constant production rate and efflux ratio—For the sake of ease in presentation and discussion of this topic, it is considered necessary first to deal briefly with the problem of head box control.

The formation of the sheet depends to a great extent on a constant stock level being maintained in the head box. This problem has previously been studied using conventional means.[9] In the present investigation, a time optimum control law was determined and used both on-line and in simulations to regulate the stock level by the air-flow valves, thus—

$$u_1(n) = a(x_1(n)_d - x_1(n)) + bx_2(n) \qquad (14)$$

where a and b are constants computed from the head box model and $x_{1,d}$ is the desired value of the total pressure at the slice opening. Equation (*14*) was then substituted for u_1 in equation (*6*) to reduce the number of control inputs to those of particular interest in the present grade change problem—namely, thin stock flow u_2, thick stock flow u_3 and machine speed u_4.

Before the grade change trajectories are determined for these three control inputs, it is necessary to examine the specific performance requirements of the present problem. It is assumed that the following linear output equation relates changes in total head, head box consistency and machine speed to basis weight, production rate and efflux ratio—

$$\begin{bmatrix} \Delta W \\ \Delta M \\ \Delta \phi \end{bmatrix} = M \begin{bmatrix} x_1 \\ x_4 \\ u_4 \end{bmatrix} \qquad (15)$$

For the grade change problem in question, $\Delta M = 0$ and $\Delta\varphi = 0$. In addition, the change in machine speed is severely limited by the rate constraint given by—

$$|u_4(n) - u_4(n-1)| \leq 0{\cdot}278 \text{ ft/ s/ 10 s}$$

This simplifies the problem, as the desired trajectories of x_1 and x_4 can now be determined explicitly assuming that it is required to change u_4 as fast as possible to the new value, governed by the choice of ΔW.

We then envisage using the available control inputs in an optimum sense such that the process outputs (x_1 and x_4, in particular) are kept close to the desired trajectories. This operation has then to be maintained throughout a predetermined optimisation interval so that the process is left in a state ready for the on-line regulator to maintain constant conditions.

The transport delay from the mixing pump to the head box makes it impossible to alter the head box consistency x_4 immediately by the thick stock flow u_3. No changes should therefore be made to either thin stock flow u_2 or machine speed u_4 until this delay has elapsed.

All the available control inputs affect the consistency distribution in the system and this is, of course, taken into account by the model when the input trajectories are determined. The control inputs must be synchronised to produce the desired trajectories of total head and head box consistency. Failing to do so often results in paper breaks, owing to insufficient strength, excessively high moisture content or low basis weight. Experience on the plant has shown that the dryline position remains constant for grade changes carried out at constant production rate; this can also be inferred from equation (*13*) as $M = Wv_m$.

In order to raise the basis weight by 12 g/m² at constant production rate and efflux ratio, it was required to manipulate the control variables as shown in

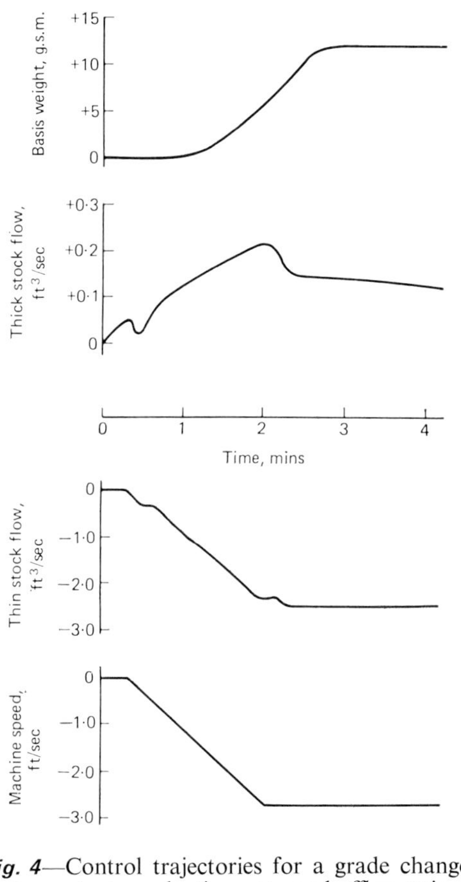

Fig. 4—Control trajectories for a grade change at constant production rate and efflux ratio

Fig. 4. The demand was to increase the head box consistency from 0·88 per cent to 0·99 per cent and decrease the total head by 42·8 lb/ft², according to the predetermined trajectories. The machine speed was lowered as fast as possible by 2·78 ft/s.

The control strategy can be resolved into three distinct phases—

1. Initial recirculation dynamics and head box level control demand an abrupt change in thick stock flow and a slight relaxation in the thin stock flow in order to realise the initial path of the desired trajectories of x_1 and x_4.
2. The system is quickly stabilised and a period of almost uniform change in the input controls follows until just before the desired targets are reached.
3. During the grade change, the system has gained a considerable consistency 'momentum', which is dissipated by the gradual decrease of thick stock flow in the last phase. In fact, the thick stock flow will change until it is slightly below the reference value before the system has reached the new steady state.

The control strategies described above are at present being implemented on-line.

Basis weight control

THE specific task of maintaining the basis weight constant is particularly important, not only immediately after a grade change, but also in face of disturbances such as drifts and fluctuations in thick stock consistency and changes in the drainage properties of the fibre.

In order to realise a feedback control law as the one suggested in equation (*12*), it is required that all the papermachine state variables are known or can be measured. The object of the regulator problem must therefore be to design a controller that makes optimum use of information about the state variables contained in the basis weight measurement.

Kalman[10] has derived a linear estimator for the state variables that has the property that it minimises the discrepancy between observation and the state variable estimate according to a criterion based on the noise characteristics of the process and the measurements.

The present problem requires estimates of the state variables at the sampling instants as well as estimates of the past state to account for the transport delays in the system. In order to do this, it is best to think of the system in an augmented state space, thus—

$$x^*(n+1) = \begin{bmatrix} F_0 & F_1 & . & . & F_h \\ I & O & . & . & O \\ O & I & . & . & O \\ . & . & . & . & . \\ O & O & . & I & O \end{bmatrix} x^*(n) + \begin{bmatrix} E \\ O \\ O \\ . \\ O \end{bmatrix} u(n) \quad . \quad . \quad . \quad (16)$$

where the augmented state vector $x^*(n)$ includes all the vectors—

$$x(n), x(n-1), \ldots, x(n-h); \text{ as follows—}$$
$$x^*(n)^T = [x(n)^T x(n-1)^T \ldots x(n-h)^T]$$

The augmented plant state $x^*(n)$ and the measurement $y(n)$ can then be described in a stochastic environment by the following matrix equations—

$$x^*(n+1) = Ax^*(n)+Bu(n)+z \qquad (17)$$
$$y(n) = Mx^*(n)+w \qquad (18)$$

where M is an $1\times mh$ output matrix.

It is assumed that z and w are Gaussian random independent variables representing the plant and measurement noise, respectively.

The linear estimator has the following structure—

$$\hat{x}^*(n,n) = \hat{x}^*(n,n-1)+K[y(n)-M\hat{x}^*(n,b-1)] \qquad (19)$$
$$\hat{x}^*(n+1,n) = A\hat{x}^*(n,n)+Bu(n) \qquad (20)$$

The initial estimate $\hat{x}^*(0,0) = 0$, that is, the best estimate of the state at time $n = 0$ given 0 measurements (the intial state has zero mean). After taking the measurements $y(n)$, the best estimate of the state at time n given all n measurements is equal to the best estimate before the nth measurement plus a weighting matrix times the difference between the actual measurement and the expected measurement.

The control law is given by—

$$u(n) = -Cx^*(n,n) \qquad (21)$$

The gain matrix K and control matrix C can both be computed by various iterative methods.[11]

The thick stock flow was chosen as the control variable in the present investigation and, for a particular set of machine conditions, it was found to be given by—

$$u_3(n) = 0{\cdot}544[3{\cdot}79x_3(n)+0{\cdot}02x_4(n)+4{\cdot}34x_5(n)+1{\cdot}17x_3(n-1)+1{\cdot}07x_3(n-2) +0{\cdot}08x_3(n-3)] \qquad (22)$$

with 10 s sampling interval. It can be seen that this is not immediately realisable, as u_3 is a function of inaccessible state variables. If equation (*22*) is to be used as a control law, it is first necessary to estimate the state variables involved from the measurement of basis weight.

A considerable computational load is placed on the computer control system if equations (*19*), (*20*) and (*21*) have to be evaluated on-line at every sampling instant to produce a control action. To meet this problem, the basis weight controller was here based on a canonical form of minimum arithmetic for the given state space representation, as developed by Lee.[12] For ease of on-line implementation, the control action was formulated as a z-transform of the basis weight. A general computational procedure was developed to derive the controller directly in the required z-transform—

$$u(z) = c\left[-d\times\frac{\sum\limits_{n=1}^{mh} a_n z^{-n}}{1-\sum\limits_{n=1}^{mh} b_n z^{-n}}\right]e(z). \qquad (23)$$

which is based on the developed state space model of the process together with plant and measurement noise characteristics.

Åström[13] has derived similar control expressions based on experimentally identified processes and noise models. Equation (*23*) represents an optimum control transform relating the change in thick stock flow necessary to compensate for an observed deviation in basis weight. The control and estimation are both implicit features of this formulation, which consists of a proportional term and a memory term to account for past values of both control action and basis weight.

In place of equation (*22*), the new controller, which accounts for inaccessible state variables and noise, is given by—

$$u_3(n) = 0{\cdot}544\left(-0{\cdot}681\times\frac{0{\cdot}144z^{-1}-0{\cdot}294z^{-2}-0{\cdot}126z^{-3}-0{\cdot}06z^{-4}}{1+0{\cdot}058z^{-1}-0{\cdot}522z^{-2}-0{\cdot}322z^{-3}-0{\cdot}07z^{-4}}\right)x_6(n) \qquad (24)$$

with a sampling interval of 10 s.

This controller has been successfully implemented and tried on the process itself, but has not been in operation long enough to have been fully evaluated quantitatively. There is, however, ample indication that the system has been considerably stabilised as can be seen from the plant records shown in Fig. 5.

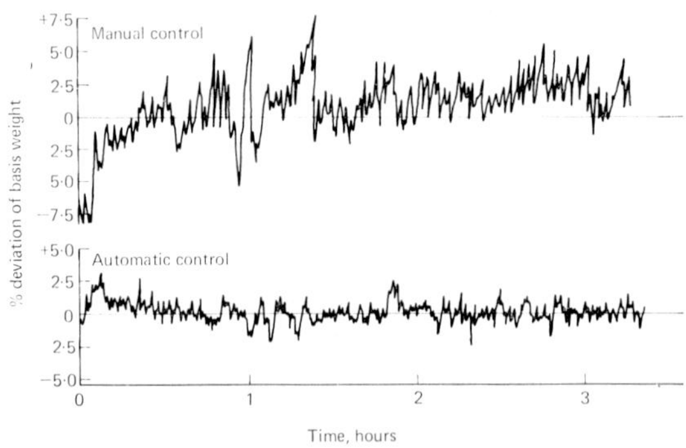

Fig. 5—Comparison between manual control and on-line computer control of basis weight

The above controller was derived under the assumption that the thick stock consistency could take on values with equal probability between certain limits. The developed methods are, however, not limited to problems with random noise disturbances. The following controller was derived by assuming the thick stock consistency to behave in a random walk manner—

$$u_3(n) = 0{\cdot}544\left(-2{\cdot}71 + \frac{0{\cdot}58z^{-1} - 1{\cdot}65z^{-2} + 0{\cdot}182z^{-3} + 0{\cdot}451z^{-4} + 0{\cdot}36z^{-5}}{1 - 0{\cdot}87z^{-1} - 0{\cdot}564z^{-2} + 0{\cdot}01z^{-3} + 0{\cdot}3z^{-4} + 0{\cdot}12z^{-5}}\right)x_6(n) \quad . \; . \; . \; (25)$$

This has not yet been implemented, but simulation tests indicate that more integral type action has been introduced, which will tend to eliminate offsets to a greater extent than with the first controller.

Conclusions

Consistent with certain assumptions, a linear, discrete time, state space model was derived for the papermachine in the form of a matrix difference equation. Perturbation experiments on the plant were used to confirm the validity of the model. The system dynamics was found to be adequately predicted by the model. It comprises an independent flow model (which represents the flow of water as a vehicle for the transportation of fibre and additives) and a fibre model (which depends on the flow model and represents the concentration and distribution of fibre throughout the system).

Having formulated and established a mathematical model, the control objectives were stated by analytical performance criteria in the form of quadratic cost functionals, which were to be minimised, subject to the model equations and constraints on the control inputs.

The intent of this work has been to exploit the state space model and to develop a number of direct digital control systems compatible with available instrumentation and on-line digital computers, in order to improve the performance of a papermachine.

The optimum grade change strategies were formulated, one for grade changes at constant machine speed and the other for grade changes at constant production rate.

It has been shown that a systematic treatment of the problem of inaccessible variables is possible provided the state equations are known. This approach led to the synthesis of an on-line basis weight controller that was found to stabilise the system considerably when implemented on the papermachine.

It is thought that further studies of the treatment of inaccessible state variables would reveal the importance of being able to measure other state variables such as thin stock, head box and wire pit consistencies in order to approach an optimum feedback control system where all the state variables are used.

References

1. Meyer, H., 'A filtration theory of compressible fibrous beds from dilute suspensions': *Tappi*, 1962, **45** (4), 296–310
2. Han, S. T., 'Retention of small particles in fibre mats': *Tappi*, 1964, **47** (12), 782–791
3. Tou, J., *Modern control theory* (McGraw-Hill, New York, 1963)
4. Kishi, F. H., *Advances in control systems theory and applications*, Vol. 1 (Academic Press, New York, 1964), 245–357
5. Hem, S., *Computer control of a papermachine*: Thesis to be submitted for the degree of Doctor of Philosophy at Cambridge University, 1969
6. Bellman, R., *Dynamic programming* (Princeton University Press, Princeton, N.J., 1957)
7. Meadley, C. K., 'High speed filtration of compressible fibrous media': *Consolidation of the paper web* (Technical Section, British Paper and Board Makers' Association, London, 1965)
8. Sullivan, P. R., Schoeffler, J. D., 'Dynamic simulation and control of the Fourdrinier papermaking process': IFAC Conference, London, 1966, Paper 22A
9. Hem, S., *The dynamic behaviour of a pressurised papermachine head box as determined by computers* (M.Sc.Tech. Thesis, University of Manchester Institute of Science and Technology, 1963)
10. Kalman, R. E., 'A new approach to linear filtering and prediction problems': *J. Basic Eng.*, 1960, **82,** 35–45
11. Noton, A. R. M., *Variational methods in control engineering* (Pergamon Press, Oxford, 1965)
12. Lee, R. C. K., *Optimum estimation, identification and control* (M.I.T. Press, Cambridge, Mass., U.S.A.)
13. Åström, K. J., 'Computer control of a papermachine—an application of linear stochastic control theory': *IBM J. Res. & Dev.*, 1967, **11** (4), 389–405

Nomenclature

Physical representation

C_1, C_2 = Hydraulic capacitance, of head box air space and stock volume, respectively.

c_i, c_o = Consistencies before and after mixing.

P_1, P_2 = Hydraulic pressure at slice and pressure attributable to head box stock level alone.

M = Production rate, defined as $M = Wv_m$.

ΔP = Applied differential pressure at suction boxes.

q_{ai}, q_{ao} = Airflows in and out of the head box.

q_i, q_o = Stock flows into and out of mixing tanks.

r = Number of converging streams before mixing.

s_d = Dryline position as measured from the point where suction is first applied.

t = Time, independent variable.
V = Volume of stock in mixing tanks.
v_m = Machine speed.
W = Basis weight.
φ = Efflux ratio.

State space representation

A = Augmented transition matrix (mh × mh).
B = Augmented input matrix (mh × s).
C = Augmented control matrix (s × mh).
D_i = Control matrices (s × m).
E = Input matrix (m × s).
$e(z)$ = Measured deviation of basis weight.
I = Identity matrix.
F_i = Transition matrices (m × m).
h = Number of delay matrices.
K = Filter gain matrix (mh × 1).
l = Number of accessible state variables.
m = Number of state variables.
n = Discrete time.
Q = Positive semi-definite weighting matrix (m × m).
R = Positive definite weighting matrix (s × s).
s = Number of input variables.
u = Input vector (s × 1).
v = Vector of input rate constraints (s × 1).
x = State vector (m × 1).
x_d = Desired state vector (m × 1).
$\hat{x}^*(i,j)$ = Best estimate of augmented state vector at time i, given j measurements.
y = Measurement vector (1 × 1).
z = Delay operator.
$\|x\|^2$ = Quadratic form (for example, $\|x\|_Q^2 = x^T Q x$).
T = Transpose of a vector or matrix.

Prepared discussion contribution

Incomplete Control Systems

D. B. Brewster, Stapleton Dowdeswell Kelly & Hunt Ltd., Canada

---

ENTIRE papermachine control implies at least the regulation of all the product quality variables such as basis weight, moisture content, strength, opacity and porosity. Implementation of such a control scheme is generally done in a piecemeal fashion. The usual starting point is automatic control of basis weight by manipulation of stock flow. The remaining variables are on manual control.

This partial automatic control system in general causes a change in the response behaviour of the remaining product variables to adjustments in the manipulated input variables, which are not being used for automatic control.

This sudden change in the effective process response can be quite disconcerting to the operator and has contributed to the failure of automatic control systems in a number of cases. A proper understanding of the phenomenon, imparted through operator training, can usually overcome the problem.

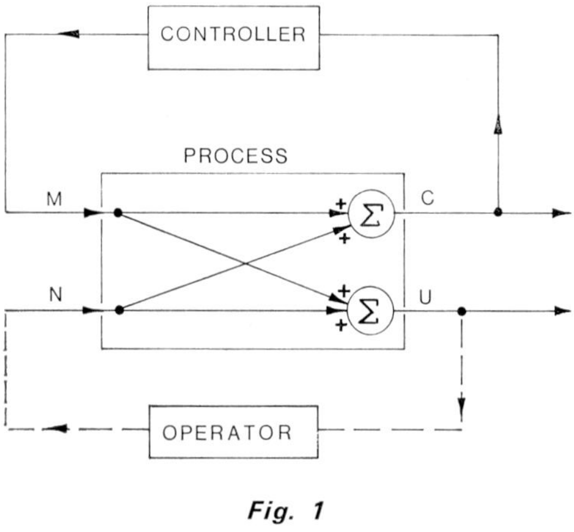

Fig. 1

Under the chairmanship of P. E. Wrist

Fig. 1 shows a general diagram of the situation with a set of quality variables C being automatically regulated by manipulation of a set of variables M. The remaining quality variables U are adjusted by manipulating the set of variables N.

As an example of the phenomenon, consider a two input/two output system for automatic control of basis weight (C) and manual control of moisture content (U), using stock flow (M) and steam pressure (N) as the manipulated variables.

If the moisture content (U) is running high, the operator will make an increase in the steam pressure (N), which, before automatic basis weight control, would have put the moisture content on target. This change will result in a drop in basis weight (C) arising from the interaction (nc). The drop in basis weight will cause the automatic controller to increase the stock flow (M). This stock flow increase will prevent the moisture content from dropping to the target because of the interaction (mu). This phenomenon in effect reduces the steady state response or 'gain' between steam pressure and moisture content. As mentioned earlier, unless the operator is warned of this occurrence, he is likely to be less than pleased with the automatic basis weight control system.

The extent of this problem can be analysed as follows. The stead state gain between N and U is defined as $[K(nu)]$ in matrix notation. The other three gains are similarly defined. The open loop steady state response is then simply—

$$[C] = [K(mc)][M]+[K(nc)][N] \quad . \quad . \quad . \quad (1)$$
$$[U] = [K(mu)][M]+[K(nu)][N] \quad . \quad . \quad . \quad (2)$$

where $[C]$ and $[U]$ are deviations from target and $[N]$ and $[M]$ are deviations from steady state.

With effective automatic control, variations in the controlled variable are minimised so that effectively $[C] \approx 0$.

Equation (*1*) then becomes—

$$[M] = -[K(mc)]^{-1}[K(nc)][N]$$

Substitution in equation (*2*) gives—

$$[U] = \{[K(nu)]-[K(mu)][K(mc)]^{-1}[K(nc)]\}[N] \quad . \quad . \quad (3)$$

It can be seen that the second term inside the curly brackets is the modification of the original open loop gain $[K(nu)]$. It is also clear that this modified term is zero if either of the interactions is zero. The problem then ceases to exist. Inspection of equation (*3*) shows that the problem is minimised whenever—

$$[K(nu)]^{-1}[K(mu)][K(mc)]^{-1}[K(nc)] \ll 1$$

For a two input/two output system, this can be simplified to—

$$K(mu).K(nc) \ll K(nu).K(mc)$$

which says that the product of the 'straight-through' gains should be much greater than the product of the interactions to eliminate the problem.

For pulp and papermaking processes, these interactions are usually quite strong so that the problem is normally present.

Discussion

Mr A. J. Ward Could you please indicate whether you tried this automatic grade change on the actual mill. If so, did you find that the constant values of the parameter in the equations remained valid over the whole range tried?

Mr S. Hem No. Unfortunately, I have been unable to implement these control strategies in full, but the grade change at constant production rate and efflux ratio have been implemented practically by synchronously changing the thin stock flow and wire speed. There is no doubt that, by manipulating the thick stock flow as well, an even better response could be achieved.

Mr H. B. Carter Can you give some figures on the improvement made in basis weight variation with the implementation of this control?

Mr Hem This is an experimental project, not long enough in operation to have reliable figures, but the sort of *variance* that we hope for on basis weight is 0·3 per cent.

Mr W. T. Whight A certain amount of investigation on the results has been done by Mr Burrows of our research and development department. We have at the moment another controller of our own that gives the basis weight in the machine-direction to within ±1·25 per cent. The information on Mr Hem's controller (in the experimental stage) is that it is as good as this and there is evidence to believe that it is suitable for our purposes. If more tuning effort were put into it, it would perform better.

Mr R. E. Johnston Would you like to guess at the major contributing effect to the difference between your controller and any other controllers that might be used? Is it the fact that a more complicated state space model than a linear estimator was used or is it that the amount of stock valve movement was included in the cost function?

Mr Hem There is a considerable difference between this controller and the one used by Åström, for instance. With this controller, we have the possibility of weighting the inputs in order to tune the controller. We can then prevent wear on valves, especially when, in this case, we were taking corrective action on the thick stock valve every 10 s. I think that the controller used here contains a larger memory than the Åström controller; in addition, it is a multi-variable approach applicable in a general sense.

Mr O. Alsholm I would add a few words that may be of interest. What I have encountered during this session is very much the same as Mr Johnston said earlier. It seems to me that everybody is trying to do the same thing, but using more and more complicated mathematics. I do not intend to discuss the differences between the Åström controller and other controllers presented here today or explain how much more efficient you could work with our DDC package than in CONRAD, but I would like to ask the authors to translate their nice mathematics into somewhat simpler terms. I enjoy listening to these excellent mathematicians and I really believe we need them for the future, but the majority of the problems that we implement today could be presented in a much simpler manner. If, instead of using the term Åström controller, for instance, one explains that there is a digital controller corresponding to PI plus dead time correction, people would not be so confused that they do not dare implement the strategy in practice. On the other hand, we should give credit to the mathematicians, because, if they do not continue with their advanced work, we will be left stranded.

Mr Hem May I say that, although the mathematics may sound awfully complicated, the actual process, once it has been done, can be performed on a fast computer in about 30 s and the implementation takes no longer than to implement the Åström controller.

Mr J. A. S. Newman Is it possible to use such mathematical models not only to predict how they can be controlled by the application of, say, DDC, but also how they can be made more inherently stable or controllable by modifications to their structural parameters such as pipework and tank sizes.

Mr Hem Yes, the structure of the plant is easily recognised in the formulation of the model. The parameters are very quickly changed if you want to investigate their effect on the plant behaviour.

Mr T. J. Boyle In 1959, dynamic programming was in vogue. I was in graduate school and applied this method on a chemical reactor problem similar to a grade change. I was successful in finding an optimum change on a simulated basis, but was very disappointed to find that some very simple strategies did quite as well as the optimum strategy. In this paper and that by Johnston & Kirk, we seem to have a similar situation with one being an optimum change, the other selecting the best version of a heuristically developed strategy. Has either author applied his technique to the other's model and thus developed a comparison?

Mr Hem Well, the grade change strategy at constant production rate and influx ratio was obtained purely and simply by solving the equations. There are no special optimisation procedures involved at all. This is mainly because the wire speed can be changed only at a certain rate. When one computes all the other manipulated variables, we find their trajectories never violate their constraints. It is therefore very easily obtained.

COMPUTERISED PAPER WEB PROFILE CONTROL

U. J. G. STRÖM, A. Ahlström Oy., Varkaus, Finland

Synopsis This paper presents the Warkaus Mill computer system and is focused particularly on the paper web moisture profile control developed in our company. The aim is to make use of the possibilities given by the second and the third presses with controlled crown rolls. Special attention is directed to profile measurement data analysis techniques and an on-line continuous statistical variation partitioning method is developed. The information is then used for operator guidance control, based on a derived control law. The profile control is not yet integrated in the system, but the control parameters are being established.

General

THE basic structure of the Warkaus Mill process computer system is represented in Fig. 1.

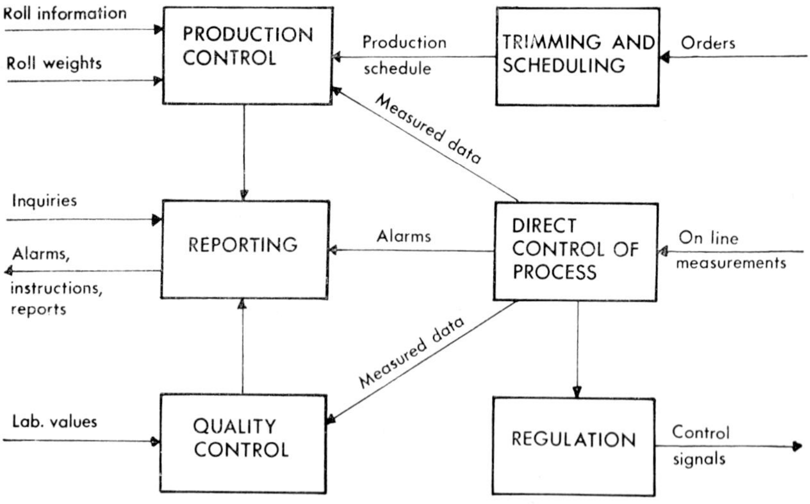

Fig. 1—General system's layout

Under the chairmanship of P. E. Wrist

The orders are allocated manually to three papermachines, then fed to the computer where the trimming and sequencing is done automatically. The possibility of criticising and making changes manually in the production schedule thus produced is an integral part of the production planning.

The automatically read scale readings are the main input to the production follow-up system, which has to work according to the approved production schedule. In this system, single-reel quality ratings are also accumulated from on-line and laboratory measurements and compared with specifications. The good reels are accepted and the rejected reels separated into two classes, those that are unconditionally broke and those that are subject to quality inspection. The broke is then directed to the best suitable use.

The on-line measurements are processed in the process control system. The basic functions of the system are filtering, limit checking conversion, averaging, control action calculations, logging and the control of scanning instruments. One part of the process control system is paper web profile control discussed below.

The quality control system is a data processing and cumulating system. Its basic functions are filtering, limit checking, averaging, statistical analysis, prediction and cumulation per order, per grade and per time period. The inputs are both manual and automatic.

All the written reports are designed to form the basis of an effective information system serving the managerial, superintendent and machine personnel level. The reporting can be made periodically, at request or automatically as alarms.

As a typical example of our process computer system, we will discuss the paper web profile control, which affects the greater part of the whole system. For example, the profiles are important factors when judging reel quality, thus important information inputs to production control and quality control.

Paper web profiles

THE paper web cross-direction profiles of main interest are basis weight, moisture content and caliper. The first two can be reduced to the independent oven-dry basis weight and moisture content profiles. The main variables affecting the oven-dry basis weight profile are thin stock consistency distribution over the slice, thin stock flow on to the wire, hydrodynamic conditions on the wire and retention on the wire. After the Fourdrinier section, the oven-dry basis weight profile is not noticeably altered.

All these factors affect the moisture content profile. In addition, the moisture content profile is modified in the press section and the dryers. The caliper depends on the basis weight and moisture and is mainly formed in the press section, breaker stack and calender.

Before we take control of the profiles into consideration, any major deficiencies in the normal working conditions should have been taken care of, say, for moisture, the consistency distribution over the slice, bad ventilation conditions, plugged press felts, etc.

The control policies for these profiles are the following. The oven-dry basis weight profile gross level is determined by the thin stock flow and average consistency by dry stock flow at the thick stock valve, head box pressures, temperatures and gross slice opening. The variations in gross level—that is, the machine-direction basis weight variations—are controlled by the computer through feedback from the dry end basis weight measurement to the thick stock valve. The other adjustments mentioned are determined and made by the machine personnel. The control of the oven-dry basis weight profile is made through slice screw settings suggested by an analysis of the dry end profile measurements. These setting deviations from the gross level are restricted by certain limits with which we may control, for example, the maximum deviation on the moisture profile caused by the slice screw settings.

The Fourdrinier section will cause deviations from the ideal moisture profile. To correct this and to compensate for uneven drying in the dryer section, we intend to manipulate the pressure settings on the press section, thus to control the moisture profile.

There are three presses in the press section of our No. 1 papermachine under consideration. Press No. 1 is a rubber-covered couch roll pressing against a stone roll. Press No. 2 is a controlled crown roll with a grooved stainless steel cover pressing the paper and the press felt against a stone roll from above. Press No. 3 is similar to press No. 2, but the paper is pressed against the stone roll from below. The average line pressures in the presses are determined and set by the machine personnel. As there are two similar presses in the grouping mentioned above, the personnel can control such quality variables as two-sidedness and oil absorption. The moisture profile is controlled by the ilne pressure profiles at the nips of the second and third presses. This can be done by changing the balance of the back and front loads of the press rolls and by changing the crown pressures.

Then changes to be made in the pressure settings are suggested by the computer after analysing the dry end profile measurements. The control policy of making changes that correct the dry end profile without caring about the profile anywhere else (such as before the dryer section) gives us the chance to work only with dry end measurements. A wet end scanning moisture gauge is, however, foreseen.

The restrictions imposed on the pressure sections by other quality factors—such as caliper, smoothness and running conditions, fluttering edges in the dryer section or similar conditions—are inserted in the control by limits on the pressure setting deviations from the normal level. The caliper is controlled in the calender, where there is also a controlled crown roll. When judging the evenness of the paper web, we have given the profiles mentioned different priorities. This has been done more from the customers' viewpoint. The classification in decreasing importance is thus caliper, moisture content and basis weight.

The control problem can be divided in three parts—

1. Defining the process.
2. Measurements.
3. Design of the control law.

Case study: moisture profile

Definition of the process

Considering only the third press, we want the dry end moisture profile expressed as a function of the pressure and load settings. These settings are—

F_A = front load,
F_B = back load,
K_{pc} = relative excess crown line pressure ratio.

We have the following equalities—

$$q_o = \frac{F_A + F_B}{l}$$

$$q_c = D.P_c$$

$$q_{co} = q_o$$

$$K_{pc} = \frac{q_c - q_{co}}{q_{co}}$$

where P_c = crown pressure,
q_o = average line pressure,
q_c = crown line pressure,
q_{co} = crown equilibrium pressure,
l = length of press roll,
D = diameter of press roll.

The dry end moisture k_d may be represented as a rough approximation by the expression—

$$k_d = L(x)M(k_o, q_o, v, w, c_f(t))G(t).y(x, F_A, F_B, K_{pc})$$

The first term $L(x)$ takes into account non-linear drying characteristics in the dryer section, which are reflected in the profile variation sensitivity at the dry end. The second term $M(k_o, q_o, v, w, c_f(t))$ takes into account the press section working level characteristics and is primarily a function of ingoing moisture ratio k_o, nip line pressure q_o, machine speed v, basis weight w (mainly through rewetting) and felt quality factors $c_f(t)$, which vary slowly with time.

The dynamics of the press section drying process is mostly determined by the moisture equilibrium between the paper web and the press felt. Any small change will thus be followed by a certain settling time. The press dynamics are taken into account in the third term $G(t)$.

The last term $y(x, F_A, F_B, K_{pc})$, where x denotes the cross-direction, takes into account profile variations around the working level. This term depends on the settings.

If we assume that to water removal depends linearly on the paper web and press felt compression, we may calculate the compression profile and thus the water removal profile as follows.

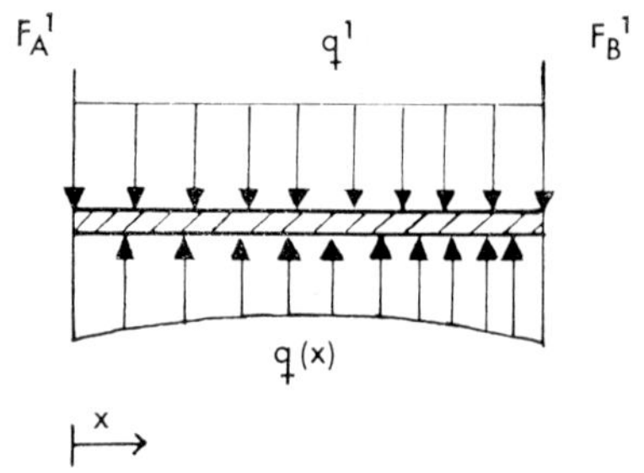

Fig. 2—Force diagram for the press roll crown considered as a bar on an elastic mat

We equalise the swimming roll shell with a long bar that lies on an elastic mat (Fig. 2). The pressure inside the shell is even so that the force per length unit is a constant q_1. In addition, to this force is added on the front end a force through the bearing F'_A that equals $F'_A = F_A - \frac{1}{2}q_1 l$ and, analogously, we have a force $F'_B = F_B - \frac{1}{2}q_1 . l$ on the rear end. The reaction is a force per length unit $q(x) = cy$. This leads to a fourth degree differential equation—

$$EI\frac{d^4y}{dx^4} = q_1 - cy$$

which has the following solution—

$$y = \frac{q_1}{EI} \cdot \frac{1}{4\beta^4} + e^{\beta x}(A \cos \beta x + B \sin \beta x) + e^{\beta x}(C \cos \beta x + D \sin \beta x)$$

where $\beta = \sqrt[4]{\frac{C}{4EI}}$

E = elasticity constant,
I = stiffness.

The parameters A, B, C, and D in the equation are evaluated by applying the boundary condition—

$$\frac{d^2y(o)}{dx^2} = 0 \qquad \frac{d^2y(l)}{dx^2} = 0$$

$$\frac{d^3y(o)}{dx^3} = -\frac{F'_A}{EI} \qquad \frac{d^3y(l)}{dx^3} = -\frac{F'_B}{EI}$$

The parameters listed above are all functions of F_A, F_B and q_1—that is, the load and pressure settings. Because of the complexity of the equation and the approximations made in deriving it, we have decided to take another approach to the problem.

Measurements

The moisture is measured with an on-line moisture gauge traversing in one direction. The gauge is read periodically and the start of the traverse is in phase with the gauge reading. Thus, we in effect take readings at N equispaced (Δx) machine-direction strings on the paper web. The readings of one scan lie on a diagonal over the paper. The data are exponentially filtered in the machine-direction along the mentioned sample strings. The filtering formula is—

$$X_F = X_F(t-\Gamma) + K[x(t) - X_F(t-\Gamma)]$$

where X_F = the filtered value,
Γ = sample interval,
K = filtering constant.

This filtering method is mainly used because of its compatibility with the computer. It requires little computer calculating time and little computer storage, because we only have to store the last filtered value.

Effect of sampling—The process of converting continuous data through sampling and quantification to discrete numbers we call digitising.

If the sampling frequency is f_s the cut-off frequency of the attainable information is $f_c = \frac{1}{2}f_s$. For any frequency of f in the range $0 \leq f \leq f_c$, the higher frequencies that are aliased with f are defined by—

$$(2f_c \pm f), (4f_c \pm f), \ldots, (2nf_c \pm f), \ldots$$

To avoid this aliasing problem, we could either choose f_s sufficiently high so that it is physically unreasonable for data to exist above f_c or filter the data before sampling so the information above a preselected maximum cut-off frequency is no longer included in the filtered data.

With a scanning measuring unit, the aliasing will be substantial. The practical scanning velocity is limited by the gauge response time and mechanical rigidity, etc., which will make the sampling time between adjacent samples on the same string relatively large. Prefiltering of the data before sampling is not permitted, because this will bring unwanted information from nearby sampling points.

We have run several autocorrelation studies and have found that, owing to the long sampling time and the aliasing effect, the aliased power spectrum on the frequency area ($0-f_c$) is essentially constant (white). The band width of the band width limited Gaussian white noise is thus $B_s = f_c$.

We consider the case where $x(t)$ has a mean value. We assume that the power spectral density function is—

$$G_x(f) = \frac{1}{B} + \mu_x{}^2\delta(f) \qquad 0 \leqq f \leqq B$$
$$= 0 \qquad f > B$$

where $\delta(f)$ is the Dirac delta function. The associated autocovariance function is—

$$C_x(\Gamma) = \frac{\sin 2\pi B\tau}{2\pi B\tau}$$

The variance of the mean value estimate is then—

$$\text{Var}\,[\mu_x] \approx \frac{1}{2B\Gamma}$$

If $C_x(0) \neq 1$, we get—

$$\text{Var}\,[\hat{\mu}_x] \approx \frac{1}{2B\Gamma}\,\delta x^2$$

From this, we get the mean value estimate variance for an N-sample moving average—

$$\text{Var}\,[\hat{\mu}_x] \approx \frac{\delta x^2}{N},$$

which is familiar from statistical analysis.

We are interested in how good a mean square value estimate we get with the data filtered using the above approximate exponential formula.

$$\text{Var}\,[\hat{\mu}(t)] = E\{(x_F(t) - \mu(t))^2\}$$

where

$$x_F = \sum_{m=0}^{\infty} K(1-k)^m x(t-m\Gamma)$$

The autocorrelation function values $R_x(mT)$ are zero for the band width limited Gaussian white noise and we thus get—

$$\begin{aligned}\text{Var}\,[\hat{\mu}(t)] &= E\{(X_F(t)-\mu(t)-\mu(t))^2\}\\ &= R_x(o)\sum_{m=o}^{\infty} K(1-K)^{2m}-\mu(t)^2\\ &= \sigma x^2\,\frac{K}{1-K}\end{aligned}$$

If we set this equal to the moving average estimate variance, we get an expression relating N and K for equally good estimates—

$$N = \frac{2-K}{K};\; K = \frac{2}{N+1}$$

These conditions apply when the sample number is very large. If the sample number M is finite, we get—

$$\text{Var}\,[\hat{\mu}(t)] = \sigma x^2\left[\frac{K}{2-K}+(1-K)^{2(M-1)}\left(1-\frac{K}{2-K}\right)\right]$$

Problems of a diagonal scan—The aim is to get a true picture of the cross-direction stable profile with a continuously scanning measuring unit. The measuring path will be a diagonal across the paper web. The measuring data will thus be contaminated with machine-direction variations. It is of interest to know how well a single scan represents the true momentary cross-direction profile. The error introduced into the measured data for two adjacent measuring points by the machine-direction variations may be estimated with the one-step Markov process prediction method. If the measured point represents the wanted momentary profile, the mean square error of the next point on the diagonal is—

$$\sigma^2 = R_x(o)-R_x(h) = C_k(o)-C_x(h)$$

The R values are the autocorrelation functions and the C values the autocovariance functions of the machine-direction data. In our measurements, the time interval h was 2 s and this gives a mean square error approximately to 10 per cent of the machine-direction mean square error.

Partitioning of the data—During one scan, we calculate the following values—

One-scan measurement average.
The stable profile by filtering.
One-scan mean square error (see below).
Total variance estimate by filtering.

In connection with special profile inquiries, the total average estimate and the stable profile fixed effects variance are also evaluated.

Thus, we have always available in the computer the profile measurement data approximately partitioned into—

Machine-direction variation.
Cross-direction profile variation.
Stable profile evennness figure.

We define the following values—

1. The cross-direction profile fixed effects variance estimate—

$$\hat{s}^2{}_{CD} = \frac{1}{M-1}\sum_{j=1}^{M}(x_{Fj}-\bar{x}_F)^2$$

2. The total random variation variance—

$$\sigma^2{}_T = E\left\{\frac{1}{M}\sum_{1}^{M}(x_j(t)-\bar{x}_j(t))^2\right\}$$

3. The machine-direction variation variance—

$$\sigma^2{}_{MD} = E\{\bar{x}((t)-\hat{\mu}x)^2\}$$

4. The cross-direction profile random effects variance—

$$\sigma^2{}_{XD} = E\left\{\frac{1}{M}\sum_{1}^{M}(x_j(t)-\bar{x}_j(t)+\bar{x}(t)-\mu_x)^2\right\}$$

$$= \sigma^2{}_T - \sigma^2{}_{MD}$$

If we introduce the above filtering formula and assume that all the samples are statistically independent (this is not true, but will give a rough picture), we may calculate the variances—

$$\sigma^2{}_{FT}(t) = E\left\{\frac{1}{M}\sum_{1}^{M}(x_j(t)-x_{Fj}(t))^2\right\} = \frac{(1-K)^2}{1-K/2}\sigma^2{}_T(t)$$

$$\sigma^2{}_{FMD}(t) = E\{(\bar{x}(t)-\bar{x}_F(t))^2\{ = \frac{(1-K)^2}{1-K/2}\sigma^2{}_{MD}(t)$$

$$\sigma^2{}_{FMD}(t) = \sigma^2{}_{FT}(t)-\sigma^2{}_{FMD} = \frac{(1-K)^2}{1-K/2}\sigma^2{}_{XD}(t)$$

In the above expressions, $x(t)$ is the measured datum and μ_x the entire average. The subscripts *j* denote the different cross-direction measuring points and a bar above a symbol denotes an average.

$\hat{\sigma}^2(t)$ is the time-varying estimate of the expected values σ^2. If we approximate these estimates with the same filtering formula, we get—

$$\hat{\sigma}_F{}^2(t) = \sum_{i=0}^{\infty} L(1-L)^i \hat{\sigma}^2(t-i\Gamma)$$

The expected value is—

$$E\{\hat{\sigma}_F{}^2\} = E\{\hat{\sigma}^2\}$$

and the variance is—

$$E\{(\hat{\sigma}^2{}_F - \hat{\sigma}^2)^2\} = \sigma^2{}_\sigma \frac{L}{2-L}$$

where σ^2 denotes the estimate of the variance and $\sigma^2{}_\sigma$ as its expected value. The filtered variance estimation is thus consistent. For convenience, we may choose $L = K$. We thus have continuously available in the computer the following standard deviations—

$S_{CD} = \sqrt{S^2{}_{CD}}$ which is a measure of profile evenness

$\sigma_{MD} = \sqrt{\sigma^2{}_{MD}} = \sqrt{a\sigma^2{}_{F(FMD)}}$ which is a measure of machine-direction stability

$\sigma_{XD} = \sqrt{\sigma^2{}_{XD}} = \sqrt{a\sigma^2{}_{F(FXD)}}$ which is a measure of profile stability

where $a = \dfrac{1-K/2}{(1-K)^2}$

and $\sigma^2{}_{F(F..)}(t) = L\sigma^2{}_{F..} + (1-L)\sigma^2{}_{F(F..)}(t-\Gamma)$

We have thus an effective tool for judging papermachine operations.

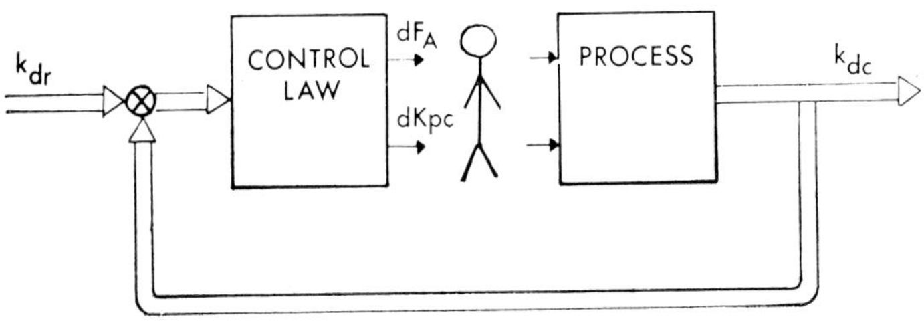

Fig. 3—The control flow diagram

Design of the control law

The control loop under consideration is illustrated in Fig. 3. The target moisture profile is denoted k_{dr} and the actual profile k_{dc}. The input to the control system consists of the measured and filtered dry end moisture profile. The profile k_{do} is subtracted from the measured data, then the term $L(x)$ is removed from the process model by multiplying the profile with a first order polynomial in k_d—

$$dk'_w(x) = (k_d'(x) - k_{do})(E + Fk_d(x))$$

In effect, we are reflecting the dry end profile to the wet end. We then fit the data to a second-order profile model—

$$dk'_w(x) = \simeq A'x^2 + B'x + c'$$

The wet end moisture after the presses is written—

$$k_w(x) = MGy$$

where the term y is the profile model—

$$y = Ax^2 + Bx + C$$

If we differentiate the wet end moisture model thus defined, we get—

$$dk_w(x) = M.G.Z.y(x)$$

where Z denotes the differential operator.

$$Z = dF_A \frac{\partial}{\partial F_A} + dF_B \frac{\partial}{\partial F_B} + dK_{pc} \frac{\partial}{\partial K_{pc}}$$

We may write the wet end profile—

$$dk_w(x) = M.G.x'DU$$

$$\text{where } x = \begin{bmatrix} x^2 \\ x \\ 1 \end{bmatrix}$$

$$D = \begin{bmatrix} \frac{\partial A}{\partial F_A} & \frac{\partial A}{\partial F_B} & \frac{\partial A}{\partial K_{pc}} \\ \frac{\partial B}{\partial F_A} & \frac{\partial B}{\partial F_B} & \frac{\partial B}{\partial K_{pc}} \\ \frac{\partial C}{\partial F_A} & \frac{\partial C}{\partial F_B} & \frac{\partial C}{\partial K_{pc}} \end{bmatrix}$$

$$U = \begin{bmatrix} F_A - \frac{1}{2}q_o l \\ F_B - \frac{1}{2}q_o l \\ k_{pc} \end{bmatrix}$$

In the matrix U, we have linearised the process around the normal operating conditions. If we require that—

$$q_o = \frac{F_A + F_B}{l}$$

and disregard the constant term in the profile model, we may rewrite the above expression for changes in the profile arising from changes in the setting values in the form $dk_{we}(x)$,

$$\text{where } x = \begin{bmatrix} x^2 \\ x \end{bmatrix}$$

$$D = \begin{bmatrix} \dfrac{\partial A}{\partial F_A} - \dfrac{\partial A}{\partial F_B} & \dfrac{\partial A}{\partial K_{pc}} \\ \dfrac{\partial B}{\partial F_A} - \dfrac{\partial B}{\partial F_B} & \dfrac{\partial B}{\partial K_{pc}} \end{bmatrix}$$

$$U = \begin{bmatrix} dF_A \\ dK_{pc} \end{bmatrix}$$

If

$$G(s) = \frac{1}{s+a}$$

we may write the proportional control law—

$$dU(t) = \frac{1}{M} D^{-1}[-A_1(t) - aA_1(t) - aA_1(t-\Gamma)]$$

where

$$A_1(t) = \begin{bmatrix} A'(t) \\ B'(t) \end{bmatrix}$$

The terms $A'(t)$ and $B'(t)$ are actually the differences between the second-order and first-order terms in the target profile and the actual profile.

Implementation of the control

The control interval is chosen rather long, say, 30 min, to be convenient for the machine personnel, thus to be accepted. The control program prints out the filtered profile and the following values defined above—

The filtered profile average.
The given moisture upper limit.
The profile evenness figure.
The machine variation standard deviation.
The profile random variation standard deviation.

The computer furthermore suggests changes in the third press loading and pressure settings, according to the control law derived above. The operator then responds to this by making the suggested adjustments and notifies the computer that the control action has been performed. If he does not notify the computer, the program will not consider the suggested adjustment an effected control action.

Any change made in the settings without notifying the computer will be

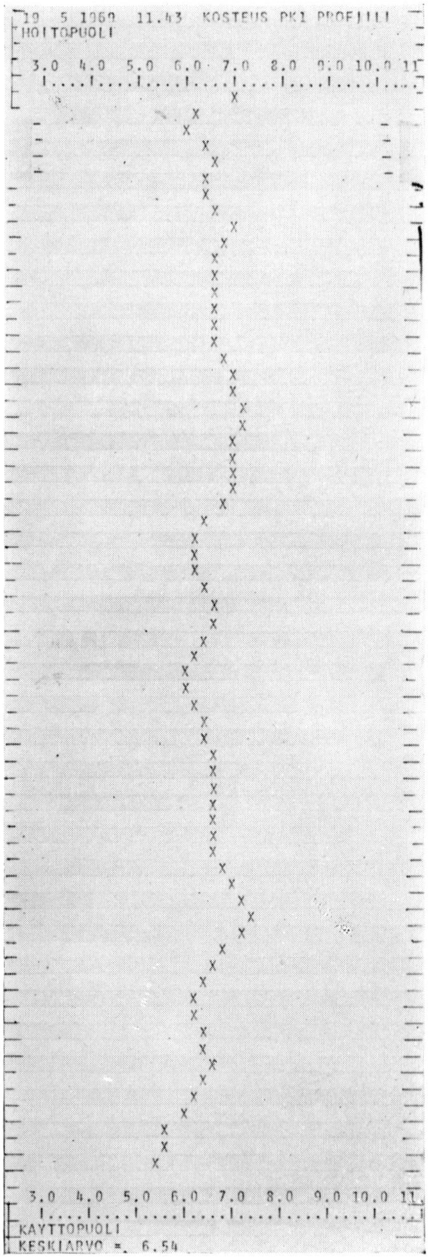

Fig. 4—The complete output of a filtered moisture profile

The text reads—
19/5/69 11.43 h Moisture content profile No. 1 machine
Control side
Drive side
Mean value = 6.54 per cent

considered as process noise. A feature of the program is that it can continuously change the set point of the automatic machine-direction moisture control so that it is 2 σ below the moisture upper limit. The term σ is the total moisture variation standard deviation. The profile fixed effects must be taken into account when this deviation is calculated.

State of the study: future objectives

THIS preliminary study forms the basis of the research work on profile management that is being done at our mill today. The control is not yet wholly integrated in the system. Work is being done to establish the values of the parameters in the derived control law. In Fig. 4, we have a typical output of the profile control program as it is today. The filtered profile was printed on an IBM 1053 printer. The part of the profile exceeding the prescribed upper limit is printed in red.

The press felt factors vary slowly with time, and thus also the gain of the control law. In order to avoid this, the total gain should be made adaptive. Another possibility is to make the parameters in the matrix D adaptive. This would make it possible to follow the evolution of the cross-direction felt characteristics, thus to detect faults in the felts automatically.

One major object of this study has been to develop a method by which the evenness of the paper web can be judged by quality control. This may be accomplished by presenting at request, the paper web profiles and standard deviations in a clear common summary report.

References

1. Wahlström, B., 'Our present understanding of the fundamentals of pressing' (Second International Symposium on Water Removal at the Presses and Dryers, Mont Gabriel, October 1968)
2. Bendat, J. S. and Piersol, A. G., *Measurement and Analysis of Random Data* (John Wiley & Sons, New York, 1967)
3. Ylinen, A., *Theory of Elasticity and Strength* (Vol. 1, Porvoo, Finland, 1948)

Discussion

Mr B. Radvan May I be the first to congratulate the author that he has tackled something that we have not heard enough about up to now—the control of cross-direction profile. With all the instrumentation and its ability to take profiles continuously and to analyse them so fast, would it be not worthwhile to derive the autocorrelogram, as well as the standard deviations? It is surely almost as informative as some of the deviations, particularly the standard deviation of the permanent profile. Would it be very difficult to alter the programs to produce the autocorrelograms as well as standard deviations?

The Chairman Unless I am confused, you can only do an autocorrelation within a series, which in this case would be a single cross-direction profile.

Mr Radvan I stand to be corrected, but I understand that a set of graphs such as were shown would yield an autocorrelogram.

Mr D. L. Cooper There is no reason for one not to carry out autocorrelation for cross-direction variation, but one problem is the degree of accuracy of the result. One assumes that one is not going to have very many values from a single cross-directional scan. I imagine you could do something by averaging the coefficients as different lags, using the results from each scan. If you are taking a very short sample, you might find yourself in some difficulty.

Mr U. J. G. Ström Just a small comment—these were in a way average profiles. We have made autocorrelation studies on the machine-direction and found that the correlation between two samples at 1·5 min interval is very small in our machine.

Dr I. B. Sanborn A recent paper* recommended that the first derivative of the profile be taken, then plotted and the number of oscillations that show a greater rate of change than some specified value be determined. Most of those

* Wingrove, H. G., Madeley, G. D. and Shabi, F. A., *Paper Tech.*, 1969, **10** (1), T1–T7

who have worked with profiles, whether they be basis weight, caliper or whatever, have found that the total change in profile is not as important as the rate at which it changes.

Mr M. M. Kaila If I understood correctly, Mr Ström did not actually try to find out whether this is a statistical set of values, then define a standard deviation of these set of values, but instead he tried apparently just to characterise how much these values vary so as to give the amplitude of the variations by condensing all this information into one number, exactly in the same way as one does when giving a standard deviation.

Mr Cooper We have done many such analyses on behalf of our members. On very rare occasions, do we find the autocorrelation function tends to zero in such a short time. I did in fact show a slide in my talk earlier this week—taken from actual results—where we saw variations occurring with wavelengths in the 10–15 min range. I would not like it to be thought natural for all machines to show zero relationship between consecutive points when those points reach as little as a few minutes separation.

Mr H. B. Carter Our experience might be helpful. We installed a moisture profiler on the same traversing gear as the basis weight unit. The average of each profile (about every 4 min) was recorded. For four months, we were recording only these averages until we put basis weight on stuff valve control, then we noticed an immediate improvement in basis weight variation. We therefore did a distribution of basis weight averages before and after control: a 50 per cent reduction in the standard deviation was found. We then repeated this with moisture content and found that it had practically an identical improvement. From this, I infer that, if basis weight is perfectly controlled, absolutely no trouble with moisture content will be experienced. The control method may be of interest. We computed the average of the profile and compared it with the standard: 70 per cent of this error is fed as a timed pulse to reposition the stuff valve. This is the elementary system that gave us this kind of improvement and seems to be better than the very complicated one referred to a few minutes ago.

The Chairman As we are leading into the topic of our last paper, namely, the interaction between moisture content and basis weight in the dry section, perhaps we could leave that comment until after the session's last paper.

Mr G. Gavelin It might be relevant to point out that the standard deviation obtained with moisture and basis weight gauges depends entirely on the time constants of the instruments. One cannot therefore compare one installation with another for improvements achieved.

Dr Sanborn Might I ask Mr Ström to repeat his figures with respect to the correction that he was able to achieve with his crown roll and whether the standard deviation was the term used to measure the variation of moisture across the profile.

Mr Ström With good basis weight profile management, we could reach about 65 per cent reduction in the sample standard deviation. With normal basis weight profile management, this reduction still reached 40 per cent. A typical value of the sample standard deviation, considered normal before control at an average moisture content of 7 per cent, is 0·30 per cent moisture (peak-to-peak value 1·50 per cent).

Mr L. D. Edenborough Although not pretending to understand all Mr Ström's mathematics, his techniques have obviously led to some useful results. I would raise some queries on the basic measurements and consider the factors that affect the profile obtained from a traversing instrument. If the instrument has a finite head width, a finite time constant and it traverses the sheet at a finite speed, the read-out is not the same as the actual variability of the paper over which the head has travelled. How does one establish the criteria by which these three variables are chosen, also has any work been done to correlate the profile results to gravimetric determinations on the paper scanned?

Mr Ström The time constant of the instrument was 1s, the distance in time between two measuring points was 2 s, so the shift to nearby points is not very large. The measuring window diameter is 60 mm. What we actually want is not a diagonal profile, but a measure of the stable cross-machine variation with one single profile.

The Chairman We are moving into an area thoroughly debated in the literature over the past 12 years. Whereas we must readily admit that scanning gauges with their finite time constants and scanning areas introduce complexities into the measurements, they are the only means on the machines for our purpose and they have a value that overrides their limitations.

Dr M. Judt Allow me to revert to Mr Hoath's question—have papermaking systems been changed under the impact of computer process control?

At Feldmuehle AG., a process control group was founded eight years ago and, together with the central engineering department and some mill managers, a number of rules were laid down that were adhered to when rebuilding and modernising existing papermachines, especially when designing completely new systems.

Some of the rules were—

1. Never return whitewater from the papermill to the groundwood mill.
2. When using sedimentation savealls, return the reclaimed stock continuously to pulpers for broke and pulp.
3. Create independent broke systems—
   (*a*) Treat coated and uncoated broke separately.
   (*b*) Thicken broke before returning to storage chest.
   (*c*) Never put broke into, say, groundwood or pulp chests.
4. Eliminate machine chests by installing continuous stock metering systems, paddle or magnetic flow meters and feed the furnish straight into the inlet of the fan pumps.
5. Build small wire pits and de-aerate the stock.
6. Try continuous mixing of coating colours when using all synthetic binder combinations.

Such rules helped tremendously to reduce dead times and interdependencies between, say, papermill and groundwood mill or broke department. I want to stress that you have to start rather early modifying your machines in order to make them easier to be put on computer control.

Feldmuehle's approach to computer control of a papermachine was quite different from what we have heard so far. I want to give Mr Mardon the assurance that he is not alone in his thinking. For three years, we have a 'hot' process control computer system working at our Reisholz mill. There, we try to achieve perfect production, low waste, excellent quality, which now and then occurs for a few days. We want to have this state of the machine for the total month, giving operators and foreman the information back from the ideal state all the time. I think you call this supervisory control. This approach was very successful. The operators and production management fixed their own targets and the variance per target. Deviations are printed out in red and action is demanded. There were no psychological problems in this mill; on the contrary, there is a waiting list now of Feldmuehle mills that would like to go into computer process control with their machines on similar lines.

INTERACTIONS IN A MULTI-VARIABLE DRYER CONTROL SYSTEM

R. A. HOLM and J. F. PERRY, Department of Engineering, The Institute of Paper Chemistry, Appleton, Wis. 54911, U.S.A.

Synopsis The design of more responsive control systems for papermachine dryer sections requires a knowledge of the coupling between web moisture content and changes in basis weight and machine speed. The interactive effect of changes in basis weight and machine speed on the measured moisture content must be known to allow proper compensation.

An estimate of the steady state coupling coefficients was made by forming an approximate mathematical model and calculating the steady state response surfaces for average web moisture content versus machine speed and basis weight. The model was a realistic representation of the partial differential equations relating heat conduction and vapour diffusion within the web. Approximate, constant values of web properties and transport coefficients and the alternating boundary conditions characteristic of conventional cylinder drying were used. Basis weight of 26–90 lb/1 000 ft^2 and speeds of 800–1 800 ft/min were studied.

The calculated coefficients showed that, for typical production conditions of 42 lb board at 1 400 ft/min, a 1 lb change in basis weight would be expected to produce a 2 per cent change in web moisture content; a speed change of 100 ft/min would produce a 5 per cent change in moisture content. The dependence of the coupling coefficients on the level of basis weight and speed was also demonstrated. The coupling coefficients for basis weight and for speed both increased approximately linearly with the operating speed, whereas the increases in both coefficients through basis weight changes were less significant. These results are of use in the design of non-interactive control systems for papermachine dryer sections.

Although the calculated results were restricted to the steady state gains for this system, an indication is given of the possible use of a more complex model in studying the dynamic behaviour of the drying system.

Introduction

THE control of web moisture content in a conventional cylinder dryer is complicated by the simultaneous changes in moisture that can occur because

Under the chairmanship of P. E. Wrist

of changes in basis weight and machine speed.[1, 2] Fig. 1 gives a schematic diagram of a simple linearised model of moisture content changes that can be used in a control system that compensates for the effect of changes in basis weight and speed on moisture level. Equations (*1*)–(*3*) show the definitions of the steady state coupling coefficients for the three-variable response surface.

$$M = M(P,B,S)\ldots\text{steady state}\ldots \qquad (1)$$

$$M - M_o = \left(\frac{\partial M}{\partial P}\right)_{B,S} \Delta P + \left(\frac{\partial M}{\partial B}\right)_{P,S} \Delta B + \left(\frac{\partial M}{\partial S}\right)_{P,B} \Delta S \qquad (2)$$

$$\Delta M = KP\Delta P + KB\Delta B + KS\Delta S \qquad (3)$$

If the conventional dryer system were a device that filtered out the effects of machine speed and basis weight, then the coupling coefficients KB and KS would be zero. This can occur in practice if the web is overdried to the point that the web moisture is only slightly affected by changes in speed or basis weight. Such a situation, though offering fewer difficulties in control, is inefficient, uneconomical and may be detrimental to the physical properties of the product. In most situations, then, the changes in basis weight and speed will be reflected in concomitant changes in moisture content. Because drying a heavier sheet is more difficult, an increase in basis weight will increase the final moisture content. Likewise, since the moisture content is a function of several complicated rate processes in the dryer section, an increase in machine speed will decrease the time available and increase the final moisture content of the paper. To design a control system, however, one must progress beyond these elementary qualitative ideas to quantitative descriptions of the system and the actual values of these coupling coefficients. To do this requires either a detailed model of the drying process itself or extensive experimentation on a mill basis.

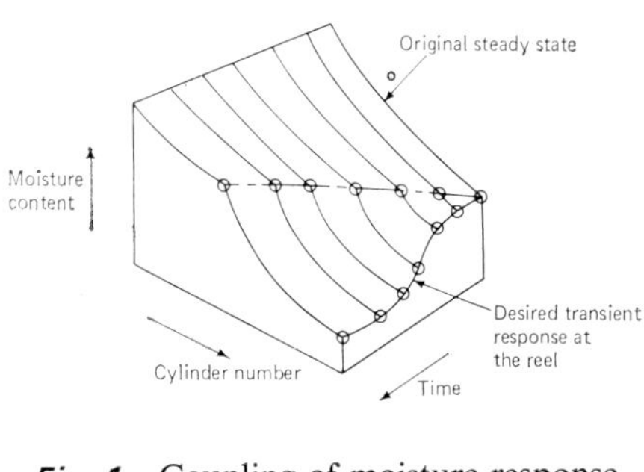

Fig. 1—Coupling of moisture response

Distributed parameter, constant coefficient model of cylinder drying

THE phenomena occurring during drying have been studied on the fundamental level by several investigators.[3–7] The primary possible mechanisms in this system are vapour diffusion, heat conduction and capillary transport. The form of the strongly coupled equations is as follows—

Energy balance

$$\frac{\partial}{\partial t}((T-T_d)C_w) = -\text{div}\ Q-\lambda\ vap-U\ \text{grad}\ T \quad (4)$$

Mass balances

$$\frac{\partial}{\partial e}(\rho_v E(1-S)) = -\text{dib}\ N_v+vap \quad \ldots\text{vapour}\ldots \quad (5)$$

$$\frac{\partial}{\partial t}(\rho_w ES) = -\text{div}\ (\rho_w U)-vap \quad \ldots\text{liquid}\ldots \quad (6)$$

$$\frac{\partial}{\partial t}(\rho_f(1-E)) = 0 \quad \ldots\text{fibre}\ldots \quad (7)$$

Rate equations

$$Q = -k_e\ \text{grad}\ T \quad (8)$$

$$N_v = V\rho_v+\frac{M_v}{(1+y_v)}\left(y_v\left(\frac{-N_a V\rho_a}{M_a}\right)-D_e\ \text{grad}\ y_v\right) \quad (9)$$

$$V = -\frac{k_v}{\mu_v}\text{grad}\ P \quad (10)$$

$$U = -\frac{k_c}{\mu}\text{grad}\ P_c \quad (11)$$

On the basis of knowledge of commercial behaviour of the system, local vapour accumulation and the capillary terms were neglected, so that the following system of approximate equations was used.

$$C_w\partial T/\partial t = k_e\ \text{div grad}\ T-\lambda\ vap \quad (12)$$

$$\rho_w E\partial S/\partial t = -vap \quad (13)$$

$$vap = -D^*\ \text{div grad}\ y_v \quad (14)$$

$$y_v = f(T, S) \quad (15)$$

The last equation (*15*) is the experimental isotherm for water vapour desorption.

A digital finite difference solution was developed under the reasonable assumption that transport at rightangles to the plane of the web was much more significant than that parallel to the web. The boundary conditions corresponded to the use of Schmidt's revised method for the steady state approximation[8] at the surfaces of the web with appropriate resistances of the cylinder and the ambient air. Since the surface of a board dries out quite rapidly, the approximation was made that the degree of saturation at the surface was zero, thus decoupling the heat and mass transfer equations for the surface increments only. This greatly simplified the calculations and is a realistic assumption for most of a dryer with well-ventilated pockets. This approximation would not give realistic results in a dryer with poor control of pocket conditions or very high partial pressures of water in the pockets.

The surface temperature and saturations were adjusted in the space-time sequence corresponding to 240 angular degrees of contact with 5 ft diameter cylinders with 3 ft draws between cylinders. A machine of 80 cylinders was used as an example. The integrations were performed using the simplest integration routine and a heuristic stability monitor. The average moisture was recorded at the moment of contact with each cylinder.

Selection of operating conditions

AN ARRAY of six basis weights covering 26–90 lb/1 000 ft^2 and six machine speeds ranging 800–1 800 ft/min were used. A constant steam temperature of 159°C corresponding to 70 lb/in^2 (gauge) was used. The calculation was continued until either the average degree of saturation dropped below 0·005 (roughly 2 per cent moisture content) or the 80th cylinder was reached. The initial time step in all cases was 0·005 s real time and was decreased as required by the stability monitor. The calculations for the 36 combinations of speed and basis weight were performed in 2 h 20 min on the IBM system 360, model 44 digital computer at The Institute of Paper Chemistry's computing centre. This initial calculation resulted in the response surfaces of average sheet moisture content versus cylinder number and machine speed for each basis weight. The data were then further processed to obtain the desired coupling coefficients. A typical response surface is shown in Table 1.

TABLE 1—RESPONSE SURFACE OF MOISTURE FOR 42 lb WEB
Average moisture content, g/g oven-dry fibre

Cylinder	*Machine speed, ft/min*					
	800	*1 000*	*1 200*	*1 400*	*1 600*	*1 800*
1	1·933	1·939	1·942	1·945	1·946	1·947
10	1·488	1·566	1·632	1·681	1·718	1·747
20	1·056	1·214	1·320	1·398	1·454	1·496
30	0·677	0·886	1·023	1·133	1·220	1·292
40	0·323	0·574	0·731	0·871	0·974	1·060
50	0·030	0·291	0·500	0·662	0·787	0·894
60	0·0	0·041	0·259	0·444	0·581	0·707
70	0·0	0·0	0·045	0·235	0·393	0·525
80	0·0	0·0	0·0	0·048	0·206	0·360

Coupling coefficients derived from the response surfaces

THE coupling coefficients defined earlier also depend very importantly on an implicit definition of the position in the machine where the measurement is taken. For example, the numerical values of gain may be different for moisture measurements at cylinder 60 and the values at the reel.

To illustrate this point, the cross-section of average moisture content at cylinders 60 and 80 are given in Tables 2 and 3. Successive column and row differences were used to estimate the coupling factors defined earlier and these these are given in Table 4. The very high moisture contents for the heavier basis weights at higher speeds do not represent usual operating ranges, but are given to indicate the shape of the response surfaces over a larger range. The gains were taken as the average of the slope at each side of the given data point. For example—

$$KB\,(80,42,1400) = \tfrac{1}{2}\left\{\frac{0{\cdot}206-0{\cdot}040}{42{\cdot}0-38{\cdot}0}+\frac{0{\cdot}924-0{\cdot}206}{69{\cdot}0-42{\cdot}0}\right\} = 0{\cdot}034\ \mathrm{lb}^{-1}\,. \qquad (16)$$

These tabulations show, as expected, that the steady state coupling coefficients are dependent on the speed and basis weight. If one concentrates on the region surrounding 42 lb basis weight at 1 400 ft/min, the following pattern of coupling coefficients emerges—

Pattern of coupling coefficients at the 80th cylinder

——+——→ B		KB (lb^{-1})			KS (100 ft/min)$^{-1}$	
↓		1·2			(1·2)	
S	(0·6)	2·0	2·1	(1·0)	5·2	6·5
		3·4			7·8	

Thus, at 42 lb and 1 400 ft/min, the calculations indicate that an increase in basis weight of 1 lb would increase the moisture by approximately 2 per cent and a speed increase of 100 ft/min would increase the moisture by about 5 per cent. This kind of information is useful in designing non-interactive control schemes for papermachine dryer sections.

TABLE 2—AVERAGE MOISTURE AT CYLINDER 60
Moisture content, g/g oven-dry fibre

Speed, ft/min	*Basis weight, lb*					
	26	*33*	*38*	*42*	*69*	*90*
800	0·0	0·0	0·0	0·0	0·588	0·928
1 000	0·0	0·0	0·0	0·041	0·801	1·087
1 200	0·0	0·0	0·073	0·259	0·948	1·212
1 400	0·0	0·038	0·279	0·444	1·059	1·306
1 600	0·0	0·193	0·435	0·581	1·155	1·375
1 800	0·0	0·350	0·565	0·707	1·236	1·433

TABLE 3—AVERAGE MOISTURE AT CYLINDER 80
Moisture content, g/g oven-dry fibre

Speed, ft/min	*Basis weight, lb*					
	26	*33*	*38*	*42*	*69*	*90*
800	0·0	0·0	0·0	0·0	0·230	0·647
1 000	0·0	0·0	0·0	0·0	0·482	0·851
1 200	0·0	0·0	0·0	0·0	0·667	0·989
1 400	0·0	0·0	0·0	0·048	0·815	1·105
1 600	0·0	0·0	0·040	0·206	0·924	1·197
1 800	0·0	0·0	0·184	0·360	1·013	1·274

TABLE 4—SUMMARY TABLE OF STEADY STATE COUPLING FACTORS

KB (lb)$^{-1}\times 100$
$N = 60$

Speed, ft/min	*Basis weight, lb*			
	33	*38*	*42*	*69*
1 000	0·0	(0·5)	(19)	2·0
1 200	(0·7)	3·1	5·5	1·9
1 400	2·4	4·5	3·5	2·0
1 600	2·7	4·3	2·9	1·6

KB (lb)$^{-1}\times 100$
$N = 80$

Speed, ft/min	*Basis weight, lb*			
	33	*38*	*42*	*69*
1 000	0·0	0·0	(0·9)	1·8
1 200	0·0	0·0	1·2	2·0
1 400	0·0	(0·6)	2·0	2·1
1 600	(0·4)	3·9	3·4	2·0

KS (100 ft/min)$^{-1}+100$
$N = 60$

Speed, ft/min	*Basis weight, lb*			
	33	*38*	*42*	*69*
1 000	0·0	(1·9)	6·5	9·1
1 200	(0·9)	7·0	10·6	6·5
1 400	4·8	9·6	8·1	5·2
1 600	7·9	7·7	6·6	4·4

KS (100 ft/min)$^{-1}+100$
$N = 80$

Speed, ft/min	*Basis weight, lb*			
	33	*38*	*42*	*69*
1 000	0·0	0·0	0·0	11·0
1 200	0·0	0·0	(1·2)	8·4
1 400	0·0	(1·0)	5·2	6·5
1 600	0·0	4·6	7·8	5·0

The results also show the way these proportionality factors change with machine speed and basis weight. Terms KS and KB both increase rapidly with machine speed and less strongly with basis weight. Over a small region, a reasonable representation of the results would be that the change of either coupling coefficient with speed is linear and the change with basis weight is negligible. There exists the possibility of using this information in more complex control schemes than that shown in Fig. 1.

Critique of the model

ALTHOUGH the results of the model calculations are useful to indicate relative trends, several shortcomings should be pointed out. The calculations in this model assume constant, average values of the transport coefficients. The data of Matters & Han[2] and McMaster[3] show the actual dependence of these factors on moisture content and web porosity. In addition the contribution of the caliper shrinkage of the web to the thermal and mass-diffusion dynamics is excluded. Of great importance is the simple boundary conditions that were used. Increases in pocket humidity or changes in cylinder or pocket temperature will certainly have an effect on the actual drying rate. The model is thus not suitable for dryer design calculations, but does show the relative interactions on a realistic basis without the use of equivalent web temperatures[9] or estimation of steady state contact factors.[10 11] The model should therefore be more dependable over a range of conditions for process control development than previous semi-empirical models.

Dynamics of dryer sections

A SIMILAR simplified mathematical model could be used to study the dynamic response of the dryer. The method of using such a model is described below, but the results are still incomplete. The present model could be used to study the time response of the system if one could assume a quasi-steady state boundary condition during the change. Such an assumption is not realistic enough, however, to obtain useful results. If the assumption were true, then on no account could the equilibrium time for the dryer exceed the transport delay of the system (approximately 1 min). Observed time lags are significantly larger than this. Therefore, the dynamic interaction of the system outside of the web must be imposed on the present model. A major contribution to the dynamic response is the thermal capacity of the cylinders and a compatible dynamic model of the cylinder boundary condition is in the process of being developed.

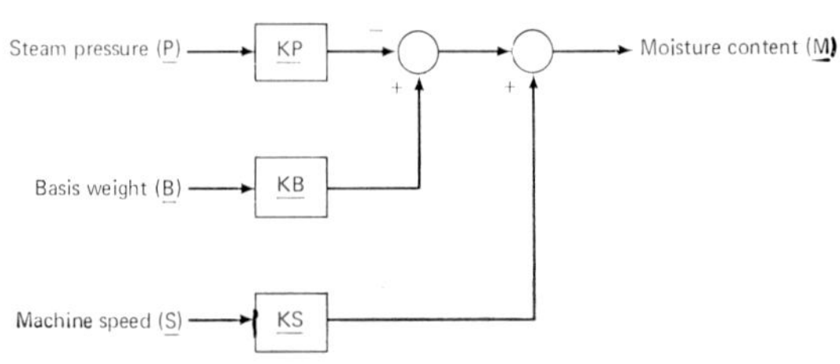

Fig. 2—Development of transient response from model calculations

Assuming that such a more complicated model were already available, however, how would one use it to study the dynamics of the dryer under various upsets? The proposed method involves the segmentation of the dryer and the imposition of a step function in the boundary conditions, for example, in the steam temperature. From a series of calculations at increasing distances from the reel, one could build up the step function response of the reel moisture as a function of time as shown schematically in Fig. 2. Unfortunately, the strongly coupled, non-linear characteristics of the drying process seem to preclude any less tedious method.

The resulting curves could then be compared with other empirical or semi-empirical response functions for less complex dynamic models in practical control systems.

References

1. Dahlin, E. B., Ziemer, R. L., Wickstrom, W. A. and Horner, M. G., 'Designing and timing digital controllers': *Instr. and Control Systems*, 1968, **41** (7), 87–91
2. Rounsley, R. R., 'Predictive control applied to a dryer section': *Pulp & Paper Mag. Can.*, 1969, **70** (2), 59–62
3. Han, S. T. and Matters, J. F., 'Vapour transport in fibre mats during drying': *Tappi*, 1966, **49** (1), 1–4
4. McMaster, D. G., 'Thermal conductivity of compressible porous materials': *Tappi*, 1964, **47** (12), 796–801
5. Han, S. T., 'Heat and mass transfer in hot surface drying of fibre mats': *Pulp & Paper Mag. Can.*, 1964, **65** (12), T537–T549
6. Parker, J. D., 'Permeability to water of partially saturated glass fibre beds': *Ind. Eng. Chem.*, 1960, **52** (3), 247–250
7. Dreshfield, A. C., 'The mechanism of hot surface drying of fibrous sheets': *Chem. Eng. Progr.*, 1957, **53** (4), 174–180
8. MacAdams, W. H., *Heat Transmission* (McGraw-Hill, New York, third edition) 1954), 48
9. Baines, W. D., 'Analysis of transient effects in drying of paper', Second International Symposium on Water Removal at the Presses and Dryers (Technical Section CPPA, October 1968, Mont Gabriel, P. Q., Canada)
10. Montgomery, A. E., 'Variation of drying tates of individual dryers through the dryer section': *Tappi*, 1954, **37** (1), 1–13
11. Norman, N. E., 'Drying on boardmachines': *Appita*, 1956, **10**, 132–149

Discussion

Dr L. A. Kirk There was a gentleman who wrote to a rather eminent professor, querying a particular point that he did not understand and he received the answer back, 'If you had read my earlier papers, you would have had no need to ask this question.' I hope I am not going to place myself in that position, but I think that the symbols are unexplained. Nevertheless, I am going to hazard a guess what they mean. If equation (*15*) refers to the rate of evaporation from a free water surface, an experimental relationship (of which there are many) would be required to evaluate it. Are the actual results sensitive to the particular form of rate of evaporation relationship that is adopted or have I misinterpreted equation (*15*) altogether?

Dr R. A. Holm Equation (*15*) is the equivalent equilibrium isotherm—that is, the amount of water vapour content in a web at the given temperature and saturation. This is not a rate equation, it is simply the normal equilibrium desorption curve. While the web is very wet, we have essentially the vapour pressure of water. Many of the approximate models that are used assume that you have pure water. If you get towards the dry end, it becomes more and more difficult to extract that water, the vapour pressure becomes lower and this has a significant effect on the final result.

Dr J. A. Robinson Could I ask for clarification about what you showed on the last slide? You said that the speed gain was calculated at constant stuff gate position. Does this mean that the gain between the speed and the moisture content is calculated at constant fibre flow to the machine and therefore under conditions of varying basis weight?

Dr Holm I think not. In this case, the condition of, say, 42 lb nominal basis weight was set at the wet end, then the drying was calculated, so the fibre weight is actually constant.

Dr Robinson The data then is very interesting in connection with the problem of controlling moisture content by speed on a dryer-limited machine. If one makes a speed change at constant fibre flow, there will be a change in basis rate as well. Using your data, at 42 lb and 1 400 ft/min, a 100 ft/min increase in speed would result in a drop in basis weight of approximately 3 lb. The net effect of summing the changes in moisture content due to *KS* and *KB* would be −6 per cent +5·2 per cent = −0·8 per cent. That is, speeding the machine up has caused a decrease in moisture content. This is, in fact, in line with what we have observed on a linerboard machine: the gain between speed and moisture content at constant fibre flow is close to zero, sometimes positive sometimes negative. This presents severe problems in implementing moisture/speed control, unless it is dynamically decoupled from basis weight control.

Dr Holm Thank you very much. We would welcome more detailed data on this point from other sources.

Dr B. W. Smith Your equation (*3*) has three constants in it and, as I understand it, you have shown how these constants varied with the basis weight and speed. I would have expected that they would vary also with moisture content of the web into and coming out of the dryers. Have you any evidence of this?

Dr Holm In this particular case, we assume conditions of constant pressing. I believe the moisture content was about 65 per cent and we did not vary this. It would be possible to insert a different level of moisture and recalculate to determine the response at different input moistures.

Mr M. I. MacLaurin I am becoming very confused by the basis weight units used by Kodak. These are neither lb per 3 000 ft², nor one of the metric forms, nor even a British or American lb per some peculiar size of ream. I would like to make a plea that, when these proceedings are published, the equivalent in g/m² be given when other units for basis weights are quoted.

Dr I. B. Sanborn I presume this is strictly a simulation result, but I wonder if part of the project that you are currently involved in would include correlation of your models with an actual machine.

Dr Holm I mentioned that we have two parallel programs under way. The process control work has been sponsored by the Institute of Paper Chemistry, therefore this is what I am discussing here. The design studies are sponsored by a private group and this includes such comparison, but I am not free to discuss these results now.

Dr D. Wahren For a very crude estimate of the interaction between rapid basis weight and moisture content variations in the machine-direction, it is assumed that the drying capacity is constant, owing to the high thermal inertia of the dryer section. This means that the amount of water evaporated per unit time and unit area is assumed to be constant for short intervals of time. The drying capacity per unit width is thus given by—

$$D = (W_{wo} - W_{wp}) . u$$

Furthermore—

$$W = W_f + W_w$$
$$m = W_w / W_f$$

Combining and differentiating—

$$\partial m_p = (m_o - m_p)\left[\frac{\partial W_f}{W_f} + \frac{\partial u}{u}\right]$$

Since $m_o \gg m_p$, one obtains approximately for constant speeds—

$$\partial m_p \approx m_o \frac{\partial W_f}{W_f},$$

which is not a quantitatively correct expression, of course, but serves to show the importance of the moisture ratio entering the drying section. At constant thick stock flow rate and varying speed, $W_f . =$ const. or

$$\frac{\partial W_f}{W_f} + \frac{\partial u}{u} = 0$$

which yields $\partial m_p = 0$. This checks well with Dr Sanborn's remark.

W = Basis weight
m = Moisture ratio
u = Machine speed

Subscripts
f = fibre
w = water
o = after presses
p = at reel-up

Dr A. Kohl Has your Institute made similar studies to find the influence of pH changes and freeness changes on the moisture content?

Dr Holm The parallel design studies are more exact. Freeness changes have been investigated and found to have an effect in the expected direction. We have had nothing to do with pH changes so far.

A Speaker Did you investigate the dynamics of the drying cylinder? Secondly, has the work of the Institute of Technology and Design Study been incorporated or used as a base for this additional design study work?

Dr Holm I will answer your second question first. The work done by Dr Snow at the Institute of Technology was done independently and parallel to this work. This is an independent study and it is not built up upon that work. We both use finite difference techniques, as did Nissan and many other researchers in the past.

Dr Sanborn The Snow model was not a dynamic model, it was to determine the steady state of the operation of the drying section for various operating conditions. Is your detailed model also a static design model?

Dr Holm As you can see from the results we presented today, the work so far concerns the steady state. It is a dynamic calculation, but for a particular fixed condition on the machine.

Mr I. Tanner Assumptions and approximations have to be made for achieving a solution to these equations. Various assumptions have been made by researchers in this field in the past and they have reached different solutions for the steady state operation of the dryer system. They appear to choose assumptions that give solutions that best match the physical system against which they wish to test their theory? Thus, no general and all-embracing solution exists. Could you comment on the accuracy of your model? You say that you still wait for someone to test your model against a physical system. Could you therefore comment on the accuracy of your method, compared perhaps with the crudest mathematical model one could envisage, which would be to assess the amount of water entering the dryer section and to consider a heat energy balance?

Dr Holm This model was developed to be used in such studies. As reported, it is at the stage that it is about to be used. It has not yet been compared with all the various approximate empirical models.

Mr E. Justus I thought that your simulation was most ingenious and I tried to follow it with interest, but there is one point that I did not quite understand about the curve in the film that represented dryness. I believe that the 'legs' of the diagram keep going up and down. What is the meaning of this?

Dr Holm The jumping 'legs' show the fluctuations in the calculated moisture content right at the interface.

Mr Justus Then according to this calculation, this moisture returns to that interface, to that surface of the sheet, immediately on leaving the dryer?

Dr Holm No, rather immediately on contacting the dryer. The majority of the drying, at least for a heavy board like this, occurs as the moisture is flashed off when the hot face of the board is exposed to the pocket.

Mr J. D. Maloney A point that may be of interest to this symposium is what happens to a computer process control installation on a papermachine after it has been in operation for a number of years—that is, after the initial training programs are over, after the installing team of system engineers, programmers and process and control engineers have long left the scene and when a mill process engineer is left to maintain and develop the system.

The Mead Corporation has such a system, one of the oldest in the industry, installed late in 1962. This application is still in operation after two changes in basic hardware. Although the system has continued to evolve by modifications to accommodate machine changes by adding new control features, progress has been slow. This has been primarily due to the fact that development has been limited essentially to the efforts of one man.

In retrospect, it is quite evident that the use of a digital computer in a papermill is much more than adding a piece of sophisticated control equipment to the operation. In many respects, it requires a whole new philosophy of mill operation in order to obtain full benefit. In addition to understanding and co-operation from all levels of mill operation and management, there has to be a strong and continuing desire to supplement the art of papermaking with technology. This is not easily accomplished, at least in the Unted States; there are a number of mills where a generation of management will have to go by before this new philosophy can be fully implemented.

The addition of a computer system may be likened to a heart transplant. There are many factors that tend to reject it. Similarly, in a papermill, there are factors that cause a computer system to deteriorate or be rejected, such as lack of a champion for the cause, lack of enthusiasm from mill management, fear on the part of the operators and foremen of what is new and reluctance to give up the art, lack of adequate instrument maintenance support and lack of process and control engineering talent to name the most important.

Many operators say that they like computer control and would not want to run the machine without it. In the most part, this is because it makes their job easier. It might be compared to power steering on a motor car; it is nice to have, because it makes driving easier, but it gets you there no faster. In terms of a computer installation, just because the operators like it, does not necessarily mean that it is a profitable venture.

Over the years, I have developed a rule stating that, in operating and line management, enthusiasm is inversely proportional to the levels above the operators. Operators like it, but management easily sees the costs, not the benefits that justify the installation. This is not to say that computer systems do not contribute to profit. It is to say that to sort out the benefits attributable to the systems approach using a computer from many other changes made to improve productivity and quality is next to impossible. Management judgment must be used in making such a partition.

In summary, a distinct and continuing effort must be made in the training of operators, foremen, superintendents and managers, as well as in the upgrading of instrument maintenance if a process control computer installation is going to be a continuing success.

Mr R. A. C. Stables I would like to comment further on some of the considerations in computer systems architecture.

As users, you can choose between two alternative approaches when planning your process computer installation—and most computer suppliers will be prepared to go either route with you—

1. Install a single, large central processor with a very complete and sophisticated high-level software system (and the latter is important to minimise system costs and the need for computer expertise).

2. Install a hierarchy of several smaller computers each dedicated to one portion of the process and receiving general objectives only from higher-level computers (as described in Mr Keyes' presentation).

If you choose the single, central processor route, do not overlook the need for the following—

1. Convenient access to a large programming facility for program generation.

2. A good means of preventing program errors on one section of the system from destroying good on-line programs operating elsewhere.

3. A method to ensure that low priority programs will in fact operate when the computer becomes really busy.

I should mention that these problems have been solved by hardware in the new Bailey 855 machine, but it is not my intention to commercialise. Instead, I would recommend to your attention the following considerations on the hierarchy system of several smaller dedicated machines—

1. With the availability of low cost hardware, such an organisation is now very competitive with the single computer approach on a totally installed and operating basis (software and start-up costs included).

2. The software is broken down into well-defined, manageable packages, for which it is much easier to predict time and memory requirements, therefore total costs (this has probably been the most important single problem in planning and evaluating computer installations).

3. Multiple computers provide greater reliability in that it is far less likely that a single hardware or software fault can have a catastrophic effect on the entire process.

4. This approach permits the user to proceed one step at a time as psychological, technological and economic limitations (in that order of importance) permit. (This is the most important argument for the multi-computer approach —as others have stressed this week, the major limitations to success are psychological.)

Although not a hierarchy of *general* purpose mini-computers, these were some of the reasons for the multi-computer system organisation chosen for the Powell River system, which Mr Mardon described on Wednesday. This was an early approach to what many believe will be a trend of the future in computer system architecture.

Mr W. T. Whight I feel that we have had today too much imbalance between theoretical background or learned dissertations on what should be done and what actually *has* been done and the results obtained or expected. Mr Madeley's contribution was a refreshing contrast to today's heavy going.

I was surprised that no comment was made on the computer control of flow box contribution from Mr Jones and myself. We stated that we regarded flow box head as a dependent variable—this comment was in no way disputed. We quoted *results*—an improvement in total head variation by a factor of 5 over an analog system, a final variability of 0·2 per cent or 0·1 in w.g.

Perhaps, I may say a few words with reference to Mr Johnston's paper on sub-optimum grade change procedures. We have intuitively developed a strategy similar to what we described. We calculate by simple flow ratios the new grade desired values, by taking the current running conditions and new required substance and (if required) wire speed. We modify these values to allow for expected retention changes, then ramp the DDC controller settings to the final values. We have obtained results that we feel cannot be much improved by any more sophisticated method. Wire speed changes at a rate of 100 ft/min per min and basis weight changes are complete in 5–6 min or less.

With modelling, we have obtained basis weight variability to within $\pm 1{\cdot}25$ per cent of desired value without the benefit of erudite modelling techniques. We estimate we have obtained 90 per cent of possible improvement without going to these costly and time-consuming preliminaries. These may be done later (in our case by Mr Hem), although benefits are already being felt.

Dr Smith We have had complete accounts of the savings that were achieved with analog control computers and we have had complete agreement among all digital computer users that it is impossible to discover what the precise savings are. I suggest that this is inherent in the difference between the two approaches—but what precisely is the inherent difference?

Mr J. Mardon I regard the current existence of *read only memory* as a very important development that relates to the hierarchal system described by Mr Staples.

Mr H. B. Carter I would like to raise a question of using head as a dependent variable. At the speeds reported, I have no experience; but, in high-speed newsprint machines, where the position of the jet landing on the forming board is extremely critical, I think the use of head is utterly taboo.

The Chairman This comment reinforces once again the dangerous nature of generalisations in papermachine operation and in papermaking.

Dr O. L. Forgacs In listening to these proceedings, one gets the impression that the sensing and measuring devices at our disposal in the pulp and paper industry are generally satisfactory. Yet, all of us still have great difficulty in obtaining reliable on-line measurements for quantities as basic as consistency. I feel that particularly the research institutions, well represented at this meeting, can make an important contribution by working towards better and more reliable instrumentation.

The Chairman Although supporting your position for improved instrumentation, we have nevertheless expected papermakers to make paper for the last 100 years with the present lack of instrumentation. Perhaps our first objective in process control is to achieve consistently what they can do on their best day with present tools and not to wait for perfect instrumentation before making any moves at all to improve our operations.

Dr Sanborn Could Mr Maloney say, in the handling of process engineers out in the mill, how they fit into the mill organisation after a considerable time following the initial installation?

Mr Maloney There is process engineer associated directly with the computer and working with the operators. He is part of the mill technical staff, not a part of operations.

Mr G. D. Madeley I would like to take up some points made by the last two or three speakers. At a fundamental research symposium, one expects a lot of theory, but this theory is of no use unless it is related to practice.

I am a little disappointed that so few people have given the results of what they are doing. In terms of the cash returns, our computer is giving a return that balances the cash outflow, so we have passed the point when the installation changes from loss to profit. We wish we could capitalise on our improved substance control.

Our market has changed in the last few months so that we have been unable to reduce our substance level. Without our computer control, however, we would have to sell 2 per cent extra fibre on area-based orders. Now that we have moisture content under control, we know that we can increase it at the reel by 1 point. Anyone can calculate the savings in these areas.

The problem of instrumentation is absolutely vital to a computer installation. We have had many months of trouble with some instruments. One that I would particularly like developed is for moisture measurement.

CUSTOMER FEEDBACK AND FEEDFORWARD

W. STOPPER, IBM Deutschland Gmbh., Sindelfingen, W. Germany

Synopsis Control systems, whereby a final customer may obtain a satisfactory product after it has passed through several stages of manufacturing, are characterised by considerable time delays in both feedforward and feedback paths. These systems are examples of the industrial dynamics problems investigated by Forrester and others. The matter under discussion in this paper is concerned with the establishment of sensible control systems and the maintenance of their reliability and accuracy. It specifically considers paper products used in data processing as punched cards and forms for optical character recognition.

Generalities

SIMILAR control engineering techniques as described in this paper are those used for dealing with hardware systems. In Fig. 1, an ideal case for a supplier-converter-user control system is compared with a moisture control circle of a papermachine.

Some of the principal weapons that can be used in a sensible control system are—

1. Use of cascade systems.
2. Local loops for accuracy checking.
3. Development of refined control algorithms.
4. Improvement of measurement accuracy.
5. Updating systems for control of drifts.
6. Reduction of time delays.

Before going into the specific subject of this paper, it is necessary to define the position of the parties concerned.

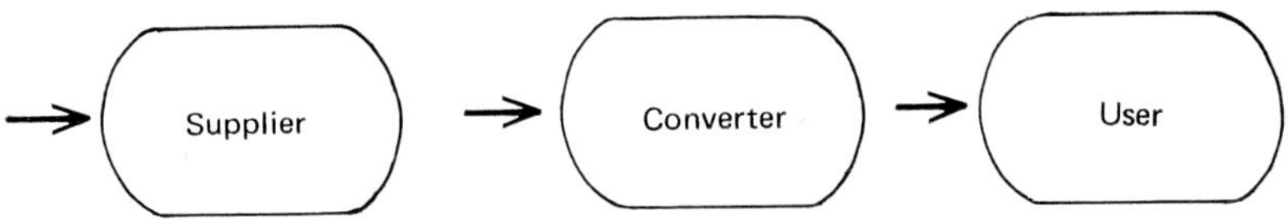

Supplier	Papermaker
Converter	Data processing industry and/or printing industry
User	User of data processing equipment

Under the chairmanship of M. I. MacLaurin

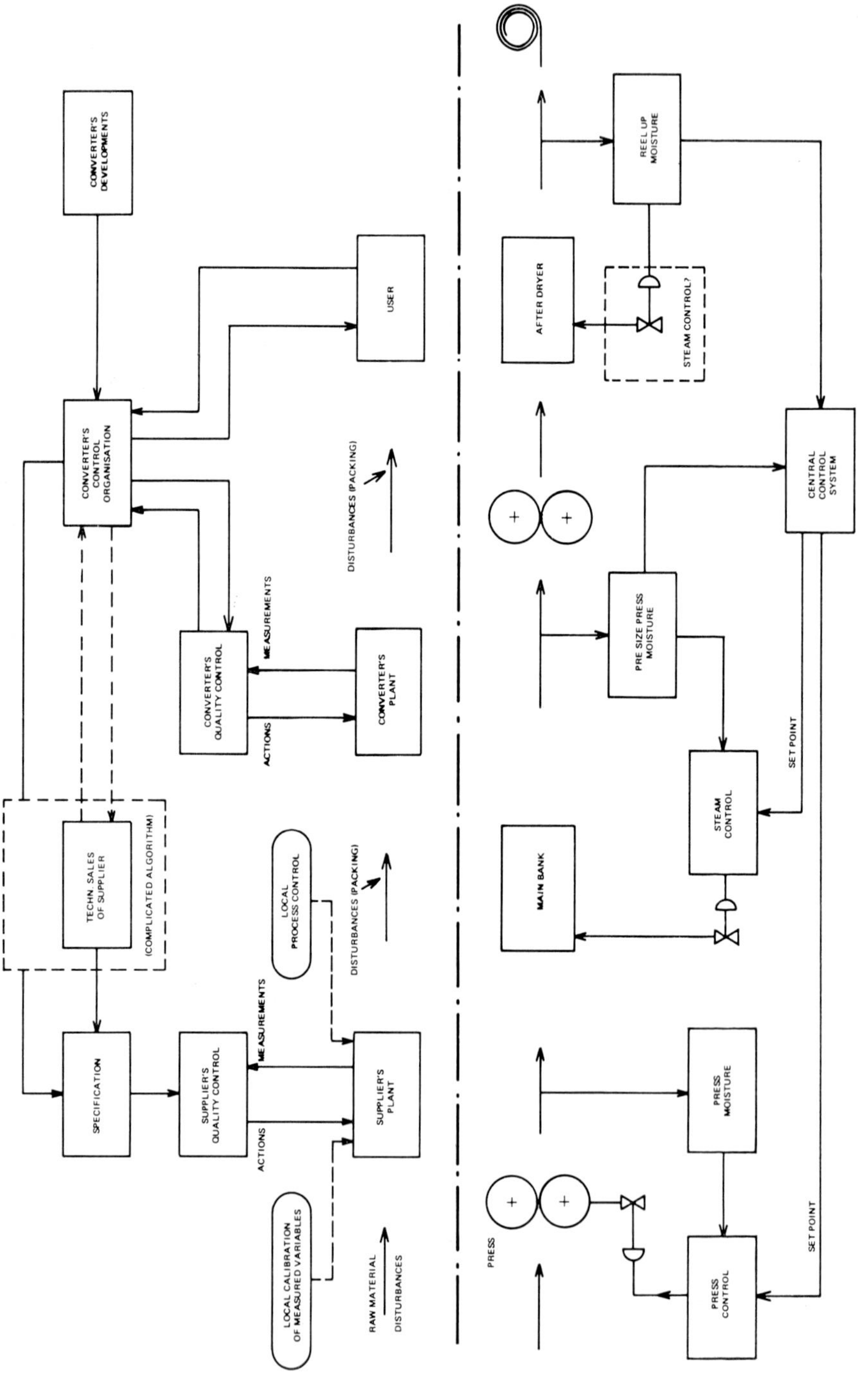
CONVERTER'S DEVELOPMENTS
USER
CONVERTER'S CONTROL ORGANISATION
DISTURBANCES (PACKING)
MEASUREMENTS
CONVERTER'S QUALITY CONTROL
CONVERTER'S PLANT
ACTIONS
TECHN. SALES OF SUPPLIER
(COMPLICATED ALGORITHM)
LOCAL PROCESS CONTROL
DISTURBANCES (PACKING)
MEASUREMENTS
SPECIFICATION
SUPPLIER'S QUALITY CONTROL
SUPPLIER'S PLANT
ACTIONS
LOCAL CALIBRATION OF MEASURED VARIABLES
RAW MATERIAL
DISTURBANCES
REEL UP MOISTURE
AFTER DRYER
STEAM CONTROL?
CENTRAL CONTROL SYSTEM
PRE SIZE PRESS MOISTURE
SET POINT
STEAM CONTROL
MAIN BANK
PRESS MOISTURE
PRESS
PRESS CONTROL
SET POINT

Fig. 1

Establishing the control system

In general, most of the activities start with a feedback from the converter, the papermill receives a specification, which should be based on existing standards, as established by a national or an international standardisation organisation. In the case of a new product or new specification, it is recommended that co-operation between converter and papermill should prove that it can be realised.

Based on the specification, paper samples will be handed over to the converter together with a test report of the product. At this point, it must be stressed that among reliable partners the customer cannot be the king as far as justification of test results is concerned. In other words, test results from a supplier cannot be ignored even if they are not in correlation with the customer's report. Too often it happens, especially in paper testing, that, because of specific conditions such as changes in relative humidity, wrongly adjusted test equipment or unskilled test personnel, deviations of results occur that can cause drastic and costly reactions to nobody's benefit.

We all should stick to the basic test philosophy that the test results of the supplier are as good as those obtained in our own laboratory. The only way in cases of disagreement is the performance of a correlation test.

A correlation test between converter's and supplier's equipment should be executed under all circumstances, after having decided to step into closer co-operation and before any test runs of paper. After completion of a successful correlation test, both partners should agree to a small test run of paper. The initial small test run of paper is mainly destined for—

1. Determination of the possibility of making paper according to given specification.

2. Detection of difficulties during converting on printing presses or other machines.

3. Performance of product tests of the finished product on processing units such as computer input and output equipment or business machines.

These test activities should finally result in a justification of the developed specifications.

It is depressing that so often negligence of these rules wastes money and serious psychological and legal problems occur between the partners involved.

From Fig. 2, one can see the time consumption that, in my experience, is rather optimistic. Time delays are certain if one or more points of the program are not in line.

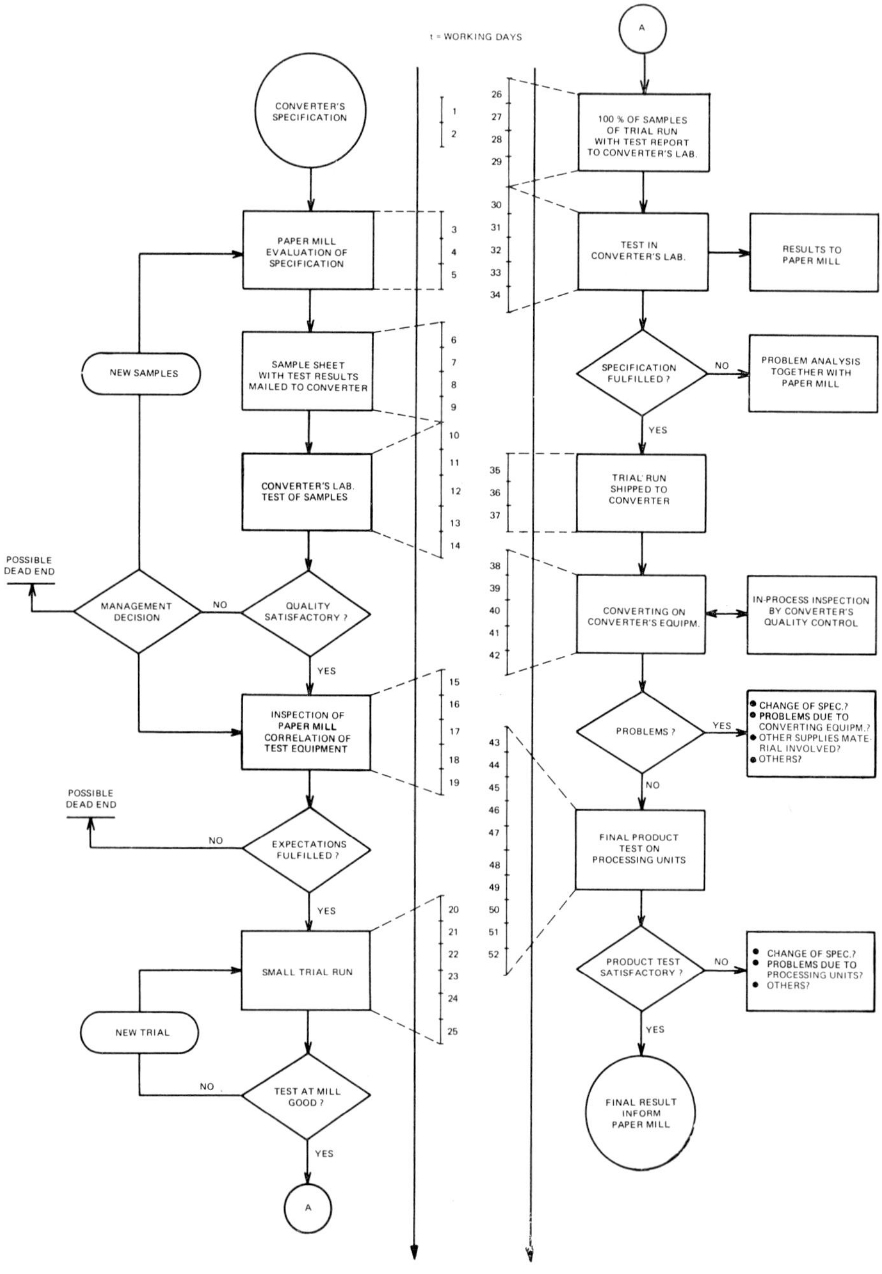
t = WORKING DAYS
CONVERTER'S SPECIFICATION
PAPER MILL EVALUATION OF SPECIFICATION
SAMPLE SHEET WITH TEST RESULTS MAILED TO CONVERTER
CONVERTER'S LAB. TEST OF SAMPLES
QUALITY SATISFACTORY ?
NO
YES
MANAGEMENT DECISION
POSSIBLE DEAD END
NEW SAMPLES
INSPECTION OF PAPER MILL CORRELATION OF TEST EQUIPMENT
EXPECTATIONS FULFILLED ?
NO
YES
POSSIBLE DEAD END
SMALL TRIAL RUN
NEW TRIAL
TEST AT MILL GOOD ?
NO
YES
A
1 2 3 4 5 6 7 8 9 10 11 12 13 14 15 16 17 18 19 20 21 22 23 24 25
A
100 % OF SAMPLES OF TRIAL RUN WITH TEST REPORT TO CONVERTER'S LAB.
TEST IN CONVERTER'S LAB.
RESULTS TO PAPER MILL
SPECIFICATION FULFILLED ?
NO
PROBLEM ANALYSIS TOGETHER WITH PAPER MILL
YES
TRIAL RUN SHIPPED TO CONVERTER
CONVERTING ON CONVERTER'S EQUIPM.
IN-PROCESS INSPECTION BY CONVERTER'S QUALITY CONTROL
PROBLEMS ?
YES
• CHANGE OF SPEC.?
• PROBLEMS DUE TO CONVERTING EQUIPM.?
• OTHER SUPPLIES MATERIAL INVOLVED?
• OTHERS?
NO
FINAL PRODUCT TEST ON PROCESSING UNITS
PRODUCT TEST SATISFACTORY ?
NO
• CHANGE OF SPEC.?
• PROBLEMS DUE TO PROCESSING UNITS?
• OTHERS?
YES
FINAL RESULT INFORM PAPER MILL
26 27 28 29 30 31 32 33 34 35 36 37 38 39 40 41 42 43 44 45 46 47 48 49 50 51 52

Fig. 2

Papermill qualifications

IN FURTHER test runs with increasing quantities, the papermill should be qualified as a good source of supply. The main reason for qualifying a papermill is to reduce one's own test activities and to procure material without time delays. One should finally aim at having the supplier make his own release without approval of the shipment by the converter's quality control.

Product acceptance

PURCHASED material is normally inspected by sampling, conducted on a statistical basis. This entails a factor of risk for both customer and supplier. The element of risk, however, is maintained at a known level. All characteristics should be classified and, as a result, inspection effort is concentrated on those characteristics that are considered most important for the satisfactory function and appearance of the material.

In Fig. 3, an acceptance method is described that is used in my company for paper from unqualified vendors. The acceptance test is performed on sample strips, not on paper reels. Our experience with this procedure is extremely good, especially if shipments to foreign countries are involved, when duties have to be paid and transport charges are high. Depending on the paper grade, the time from making up to release of shipment by customer's quality control can vary considerably. In this example, it takes approximately 10 days.

A qualified vendor should be allowed to release shipments by himself, although he should continue to send sample strips for correlation purposes. For products released by the supplier's quality control, considerable time can be saved.

At the beginning of the flow chart in Fig. 3, it is recommended to elaborate mill limits. The main reason for this suggestion is to avoid discrepancies between converter and supplier, caused especially by variation of test results through the effects of test equipment, handling and environmental conditions. In quality control at the papermill, it is advisable to plead for a separation of responsibility between production people and those carrying out the final test. It is actually the man on the papermachine who makes the quality—it is obviously his responsibility. In modern papermills, he has so many means of controlling the process—nowadays, even by computers—that he needs simply good co-ordination with the mill's quality control department to obtain optimum quality.

Once the paper has arrived at the customer's location, it is most important to check its arrival condition. It is well known that the method and kind of transportation can have significant effect on the quality of the raw material and can give rise to difficulties during converting processes. Since it is often

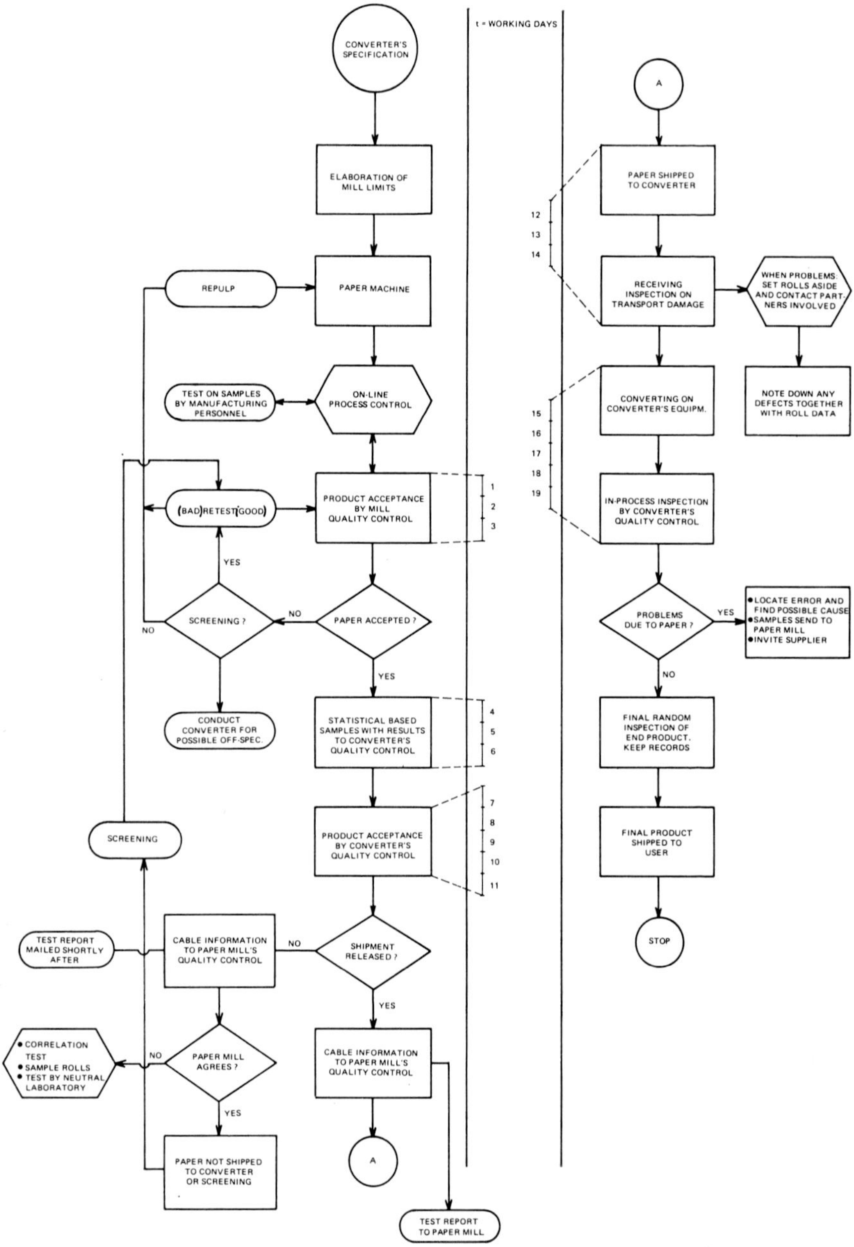
t = WORKING DAYS
CONVERTER'S SPECIFICATION
ELABORATION OF MILL LIMITS
REPULP
PAPER MACHINE
TEST ON SAMPLES BY MANUFACTURING PERSONNEL
ON-LINE PROCESS CONTROL
(BAD) RETEST (GOOD)
PRODUCT ACCEPTANCE BY MILL QUALITY CONTROL
1
2
3
YES
NO
SCREENING ?
NO
PAPER ACCEPTED ?
YES
CONDUCT CONVERTER FOR POSSIBLE OFF-SPEC.
STATISTICAL BASED SAMPLES WITH RESULTS TO CONVERTER'S QUALITY CONTROL
4
5
6
SCREENING
PRODUCT ACCEPTANCE BY CONVERTER'S QUALITY CONTROL
7
8
9
10
11
TEST REPORT MAILED SHORTLY AFTER
CABLE INFORMATION TO PAPER MILL'S QUALITY CONTROL
NO
SHIPMENT RELEASED ?
YES
• CORRELATION TEST
• SAMPLE ROLLS
• TEST BY NEUTRAL LABORATORY
NO
PAPER MILL AGREES ?
YES
CABLE INFORMATION TO PAPER MILL'S QUALITY CONTROL
PAPER NOT SHIPPED TO CONVERTER OR SCREENING
A
TEST REPORT TO PAPER MILL
A
PAPER SHIPPED TO CONVERTER
12
13
14
RECEIVING INSPECTION ON TRANSPORT DAMAGE
WHEN PROBLEMS: SET ROLLS ASIDE AND CONTACT PARTNERS INVOLVED
CONVERTING ON CONVERTER'S EQUIPM.
NOTE DOWN ANY DEFECTS TOGETHER WITH ROLL DATA
15
16
17
18
19
IN-PROCESS INSPECTION BY CONVERTER'S QUALITY CONTROL
PROBLEMS DUE TO PAPER ?
YES
• LOCATE ERROR AND FIND POSSIBLE CAUSE
• SAMPLES SEND TO PAPER MILL
• INVITE SUPPLIER
NO
FINAL RANDOM INSPECTION OF END PRODUCT. KEEP RECORDS
FINAL PRODUCT SHIPPED TO USER
STOP

Fig. 3

difficult to state where these problems have been derived from it, goes without saying that records must be kept with all the necessary data and photographs being taken to demonstrate the defects.

Being familiar with the converter's side of the business, I must admit that not all of our people have enough insight into paper technology to be able to give definitive statements whether or not the paper was the most likely cause of the problem with which they have been faced. It happens very often that they ignore converting machine problems and trace everything back to the raw material. It is essential therefore to co-operate very closely with the papermaker and to give him the best information feedback. For all steps of the converting process, reel data must be kept in order to simplify problem analysis. In complicated situations or in applications, joint studies are often the only way to achieve the best solution.

Inspection of product

THE final inspection of the final product is interesting not only for product quality, but also for possible changes in the raw material in the course of the converting process. This happens more often than one believes. Surface properties such as friction and smoothness also such other physical characteristics as thickness and stiffness, are subject to mechanical and/or atmospheric influences. Finally, it pays to be aware of these facts and to submit these findings to the papermaker. Both can take steps to prevent or at least to simplify potential problems.

Complaints from user

DEPENDING on existing contracts or legal obligations, the converter as well as the papermaker may be responsible for complaints arising from the final product at the user's location. In the first place, it is the converter's duty to contact the user and to evaluate the complaint. Fig. 4 describes the steps that are suggested for the converter to follow. The example is taken from the business machines industry, in which paper is used as input and output media. Especially in this field of business, it is sometimes extremely difficult to locate the possible cause. Quite frequently, we have a multiplicity of parameters causing the complaint. Only sophisticated methods of problem analysis, complete collection of all available data and sound record keeping lead us to an optimum solution.

As the majority of the sales people and field engineers of the converter industry lack a sound background of paper technology, it is absolutely necessary to establish questionnaires to be used in complaint handling. Data obtained by

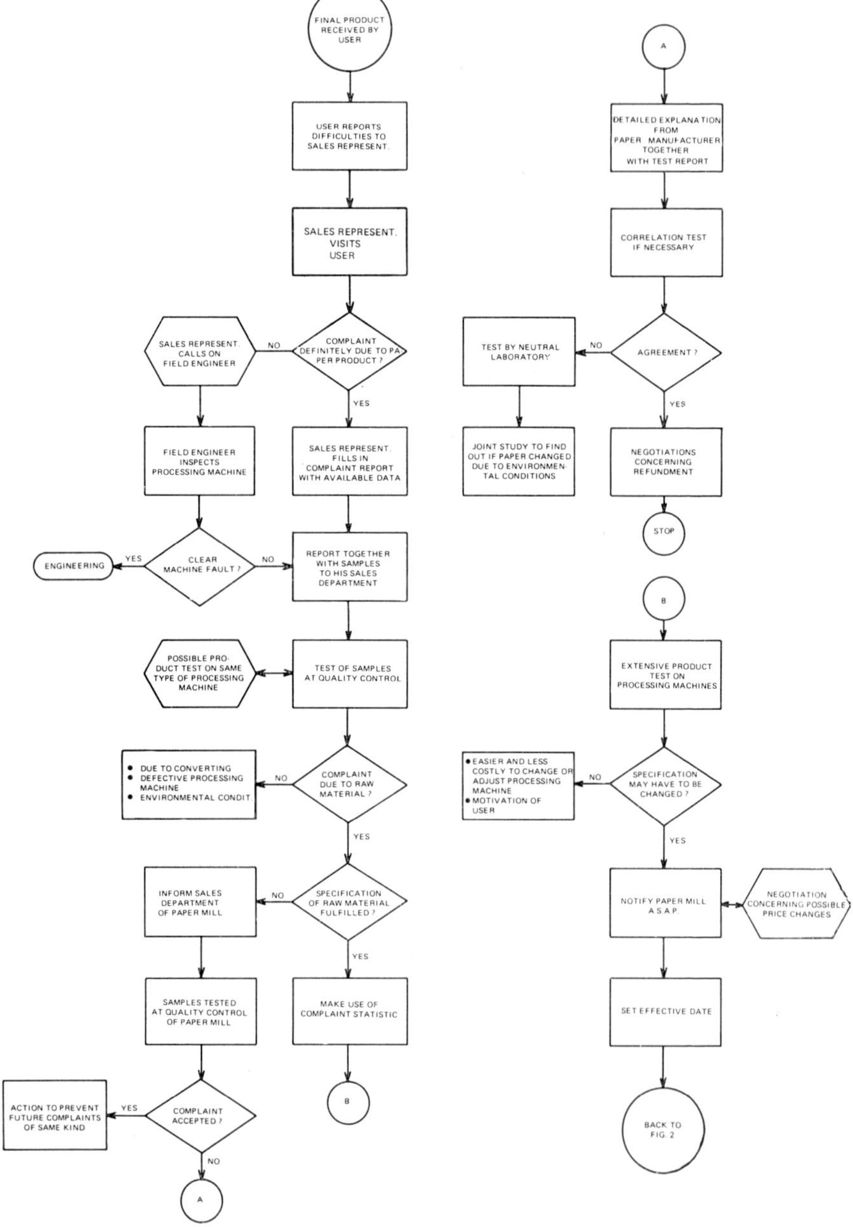
FINAL PRODUCT RECEIVED BY USER
USER REPORTS DIFFICULTIES TO SALES REPRESENT.
SALES REPRESENT. VISITS USER
COMPLAINT DEFINITELY DUE TO PAPER PRODUCT ?
NO
SALES REPRESENT. CALLS ON FIELD ENGINEER
FIELD ENGINEER INSPECTS PROCESSING MACHINE
CLEAR MACHINE FAULT ?
YES
ENGINEERING
NO
YES
SALES REPRESENT. FILLS IN COMPLAINT REPORT WITH AVAILABLE DATA
REPORT TOGETHER WITH SAMPLES TO HIS SALES DEPARTMENT
POSSIBLE PRODUCT TEST ON SAME TYPE OF PROCESSING MACHINE
TEST OF SAMPLES AT QUALITY CONTROL
COMPLAINT DUE TO RAW MATERIAL ?
NO
• DUE TO CONVERTING
• DEFECTIVE PROCESSING MACHINE
• ENVIRONMENTAL CONDIT.
YES
SPECIFICATION OF RAW MATERIAL FULFILLED ?
NO
INFORM SALES DEPARTMENT OF PAPER MILL
SAMPLES TESTED AT QUALITY CONTROL OF PAPER MILL
COMPLAINT ACCEPTED ?
YES
ACTION TO PREVENT FUTURE COMPLAINTS OF SAME KIND
NO
A
YES
MAKE USE OF COMPLAINT STATISTIC
B
A
DETAILED EXPLANATION FROM PAPER MANUFACTURER TOGETHER WITH TEST REPORT
CORRELATION TEST IF NECESSARY
AGREEMENT ?
NO
TEST BY NEUTRAL LABORATORY
JOINT STUDY TO FIND OUT IF PAPER CHANGED DUE TO ENVIRONMENTAL CONDITIONS
YES
NEGOTIATIONS CONCERNING REFUNDMENT
STOP
B
EXTENSIVE PRODUCT TEST ON PROCESSING MACHINES
SPECIFICATION MAY HAVE TO BE CHANGED ?
NO
• EASIER AND LESS COSTLY TO CHANGE OR ADJUST PROCESSING MACHINE
• MOTIVATION OF USER
YES
NOTIFY PAPER MILL A.S.A.P.
NEGOTIATION CONCERNING POSSIBLE PRICE CHANGES
SET EFFECTIVE DATE
BACK TO FIG. 2

Fig. 4

the questionnaires can be used for statistical evaluation, which most probably are the best tools for complaint analysis and for decision making.

Because of possible variables, it is impossible to give any time schedule for complaint handling. It goes without saying, however, that one must give priority to all complaint situations.

Discussion

Dr F. L. Hudson Mr Stopper, like Mr Mardon earlier, has referred to psychological problems. My recent experience with paper users has led me to believe that the delays and confusion caused by misleading specifications are still serious and perhaps the weakest part of the whole paper system. The example I quote is very old, old enough to be safely discussed, yet the worst I have seen and quite the most apposite to Mr Stopper's paper.

On joining a quality control laboratory about 35 years ago, where tabulating card was tested, I found the number of cards per lb, according to the customer's specification, was supposed to increase by $2\frac{1}{2}$ per cent only when oven-dried. The moisture content was therefore specified at $2\frac{1}{2}$ per cent, but as any technologist would know this is quite unworkable and the card was actually sold at 7 per cent. No one would have the specification altered, however, for it was psychologically inadvisable to annoy the customer by mentioning it and more convenient to apologise to periodic complaints. In the event, the specification remained intact for the ten years for which I held that position, quite possibly for a long time after I left it. Surely a time delay of 0·1 century for feeding information forward to a customer must be unusual even for psychological reasons. This would not happen now for this particular material, but I am afraid there are still serious time delays with other papers due to stupid specifications.

The answer to this kind of thing is more education and I would refer here to Mr Ball's contribution to the discussion following McFarlane & Hoath's paper (page 321). His 1956 publication* did not pass unnoticed in Manchester. One student memorised (I will not say learned) it so thoroughly that he reproduced an obvious misprint in an equation for an examination script!

In 1957, Dr Kirk was appointed at Manchester University to lecture on automatic control to paper science students and he has done so for 12 years. Suitable laboratory apparatus was obtained in 1958. A number of Manchester graduates who have taken this course have attended this symposium; one of them has read a paper. In this kind of education process, too, the time delay before one gets obvious results is about a decade. I trust that Mr Balls will get some satisfaction from what he initiated. Perhaps, on Fig. 3 of

* *Proc. Tech. Sect. B.P. & B.M.A.*, 1956, **37** (2), 201–210

Mr Stopper's paper, there should be an outer loop with a long time delay and with a paper science school somewhere in it.

Mr W. Stopper I am very glad not to have been in the company at that time, because, coming to that particular department, those specifications had been changed. I believe they are very definite now and very strict, but I think everyone can understand them. On the other hand, we shall be connected more and more with standardisation organisations—for example, the paper for OCR forms are already being discussed by national standardisation bodies. Thus, by the time the volume of such forms is increasing, there should already be existing specifications.

The Chairman Thank you. I must say, when I heard you use the words 'definite' and 'strict', you describe Mr Stopper's point of view fairly well in most cases.

Mr H. B. Carter I must congratulate the speaker for a magnificent display of a system: it clearly indicates that, in order to reduce the time delay, some rather obvious changes in the system are required. This system essentially is to produce cards for the user from paper made in a papermill and there seems to be at least one completely irrelevant stage in the process—perhaps two.

It seems unnecessary to have the intermediate user (the equipment supplier) monitor the quality of the paper being shipped to the converter and monitoring the quality of the converter's output. Surely, he can do this for himself. This may be only the development process, but, if it is the system used in routine business, the time delay is quite easily reduced. It is not only the speaker's company that does business in this way. This is a clear case for the system to be re-organised.

Mr Stopper Well, we cannot, firstly, for psychological reasons and, secondly, for cost reasons, have a man sitting next to the papermill or supplying directly to the mill. Companies might have ten or even more paper suppliers having a man sitting directly in the mill. We cannot pay for that.

Mr R. H. McClaren In everything that Mr Stopper has described, I find complete opposites. Possibly this is why Xerox competes so well with other large companies like IBM. We frankly do not believe that anyone can describe paper or make paper for our machines as well as we can. From our viewpoint, only God can write specifications by looking at a product and we do not propose to be God. In our relationships with our papermills, the mills themselves write their specifications based on a given papermachine, using a given raw material supply. Each specification defines a product from a machine in a mill. Mr Hopkins and myself control somewhere between 10–20 machine/mill outputs and, although we have provided them with guidelines that are general to our Xerox machine needs, the paper that is made by these mills based on these guidelines and paper that controls our specification writing is a product of a given mill using a given papermachine.

When the process of using these guidelines, a manufacturer makes a paper that is suitable for our machines and can triplicate this within narrow limits of control, we then derive a specification from the analysis of these productions. In purchasing paper from the mill, we require a 95–99 per cent statistical conformance with the specification. If this performance is not met, we reject that paper. We are fully aware that our Xerox machines may meet that product later, but we have no control over it, nor do we care what the manufacturer does with the rejected paper.

From my personal viewpoint, the major problem in any papermill is personnel. If we could remove people from the papermachine and provide statistical control derived from a computer, we would be able to make uniform paper within a run and from run to run. In other words, the people running the papermachines that make the paper we buy are ordered not to vary the machines once they are running properly. In fact, if they need something to adjust, I suggest that we provide them with computers, because the operator of a papermachine, in his manipulations, can do us far more harm over performance goals in meeting specifications than any inherent machine variation could provide. I hope in the near future to find a papermill where we would find no people at all, except those taking the product away from the reels.

I suggest that Mr Keyes had some of the priorities for the computer's potential misplaced. He identified quality as high up his line of priorities. I define quality as a beautiful woman or, in beautiful picture, as a sensual thing. I cannot find quality in paper. Uniformity is the magic word for paper and I am sure it is so in all of your worlds. Most of you cannot make uniform paper and only two or three mills in U.S.A. can do so. I am defining uniform paper in this context by a given product from a given machine made week in and week out, month in and month out, that conforms to a specification.

Mr Hopkins' many colleagues ignore the psychological effects identified by Mr Stopper and, in his control of the production in the European area, he provides full-time inspectors in the firm's supplying mills. We have found this to be most satisfactory and we get a good quality paper of reasonable uniformity in this environment.

Mr Stopper We are continuing to compete, but I cannot fully agree with what you have said. One point that I would like to mention is the uniformity of the paper produced and I can only refer to the European mills (maybe some in the Far East), but for our product at least, because of the connection that we have with the papermills, they are able with a very few exceptions to make uniform paper. Here again, we disagree. I do not know your product, but with punched card stock, for instance, the quality level of European manufacture is comparable with the American level—in some cases, even better.

Mr McClaren I hope I have an international point of view, but the cost to the corporation is about ten times greater to have other people's paper used in our machines and this to me is truly a measure of uniformity.

ECONOMIC MODELS AND THE U.K. PAPER INDUSTRY

N. B. MORGAN, Scientific Control Systems Ltd., London

Synopsis Models of the behaviour of national economies have been developed and popularised for some time. The economic behaviour of whole industries has also been examined and statistical analysis has uncovered at least the main features of the working of a number of industries. There is no difficulty in principle in carrying this process of analysis on to the examination of individual products. The paper discusses these topics and illustrations are provided of national and industrial economic models and the behaviour of several product groups in the U.K. paper industry are examined. The results show that statistical analysis can reveal important economic influences at work, but that it is necessary to beware of special factors, technical, political and social, which can be of overriding importance for particular products at particular times.

Introduction

THE very concept of economic modelling needs more introduction than does most subjects. This arises from the fact that there is a good deal of disagreement on what an economic model is (or should be, if it existed) and of how an economic modeller approaches his work. It is, perhaps, a sign of the rudimentary state of economic science that this sort of philosophical debate should have such an important place in the activity of its practitioners: alternatively, it may be the intimate personal significance of economics that leads people to philosophise on it.

Recently, Coen, Gomme & Kendall(1) read a paper at the Royal Statistical Society, describing some striking empirically discovered relationships among a number of economic series, generally in lagged form. The object of the work was to obtain relationships that could be used for prediction, hence, for control. The evidence and the relationships were not disputed, but the approach was. As on a number of other occasions, the absence of an acceptable theoretical framework that can be utilised in interpreting the results was criticised. This kind of criticism raises the question of the proper nature of a model.

Under the chairmanship of M. I. MacLaurin

In practical work, the division between those who look for relationships first, then (as an option) go on to rationalise them and those who like to have the explanations first before setting up tests of the relationships implied by these explanations is often somewhat artificial. There will be no need to emphasise to those concerned with controlling systems that no model can or need be complete. A model, which is often set out in mathematical terms, sets out the relationships that determine the behaviour of the system to be studied. Only those relationships that are relevant to the purpose of the study (say, the control of the national economy) need to be set down and, if only a single relationship is needed, then the modeller is fortunate. *Explanation* is never complete and there is a sense in which the very well established and rigorously tested laws of physical science cannot be explained. Certainly, one cannot explain why the law of gravitation is what it is and not something else.

The economist, demanding a theoretical framework, is usually generous in his readiness to accept explanations: for example, naïve psychological suppositions about human nature suffice, so it is very doubtful whether the theoretical framework of economics is based on sound theories of individual and social behaviour. Even if the assumptions made were soundly based, it is certain that no prior reasoning indicates the functional form of economic relationships or the length of lag with which the relationships operate. Thus, the economist must rely on a statistical approach in order to bring his theoretical anticipations into a form of use for practical purposes (or tested, for that matter). Of course, once such a relationship has been set out, an unlimited number of explanations (in the sense of rationalisations) can be provided of why it exists in this form. No doubt, the reader will be able to rationalise the skewed pattern of income and wealth that has prevailed in many countries for as long as data are available.

Those interested in control theory, however, will be prepared to concentrate their attention on the relationships found to hold in the economic system and to relegate explanation to a lower place. It follows from these considerations that statistical methods are needed to obtain economic relationships in a form useful for control purposes. I shall not attempt to cover questions of statistical methodology or technique in this paper; I shall, in fact, refer almost entirely to regression methods. This reflects the concentration of statistical economic work in this area. The paper will proceed through a number of examples at different levels of economic generality, without engaging in a full discussion of the economic theories that the relationships so obtained suggest. The bibliography appended contains a list of books and articles that provide full descriptions of the statistics and economics to which the paper refers.

National economic control

THE most aggregate level for which I shall present an example is the national one. This is not because the international level is unimportant: recent events have shown us all how important these are and the paper industry has historically been open at the international level. There is not much of a record of international economic control and, however, for political rather than economic or even industrial reasons, attention has been focused on the problem of national economic control. Speculating, I would not expect the choice of relationships used for economic control to change much if multi-national bodies such as the European Economic Community were to become the units of economic management rather than individual countries.

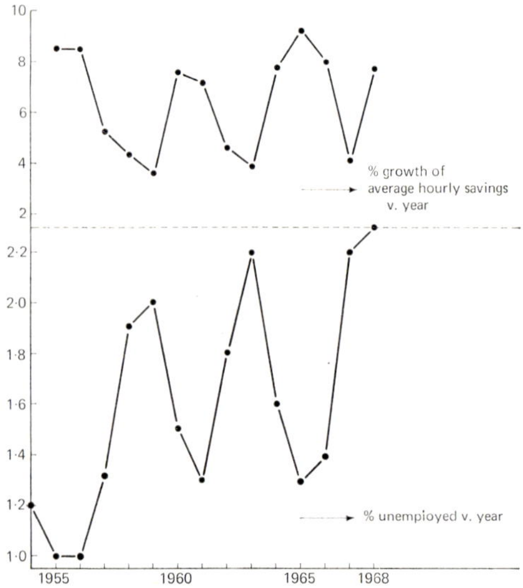

Fig. 1

Economic and statistical analysis of the national economy came to take up a large part of the attention of those interested in economic matters in the thirties. This was partly due to the academic and proselytising work of Lord Keynes at that time, also to the economic, social and political circumstances. Since the second world war and, most particularly, recently, attention has swung from the problem of combating unemployment to reducing the rate of inflation. A simple control mechanism has been suggested here: wage increases (among other things) cause price increases, the rate of wage inflation depends inversely on the level of unemployment; so, if the national economic managers create high levels of unemployment, they will slow the rate of inflation. New devaluation and tax increases raise prices even in the absence of wage increases, so in this paper I shall concentrate on the link between the level of unemployment and the rate of wage inflation. If such a relationship were established, the level of unemployment could be used to predict and control inflation. I believe that it is fair to say that such a relationship (of unspecified form) fits the theoretical framework of most economists. It is known as a Phillips' curve and the U.K. annual data on which an empirical study could be based are shown in Fig. 1 (the data are obtained from the *D.E.P. Gazette*). At first sight, a good relationship appears to exist between the movements of these series. A couple of years ago, the conventional wisdom held that the high level of unemployment prevailing in 1967 and expected to be maintained in later years would hold down increases of earnings and so help to reduce inflation and increase our international competitiveness. I do not know what the Cabinet (our economic managers) thought, but it is reasonable to believe that they had this control mechanism in mind.

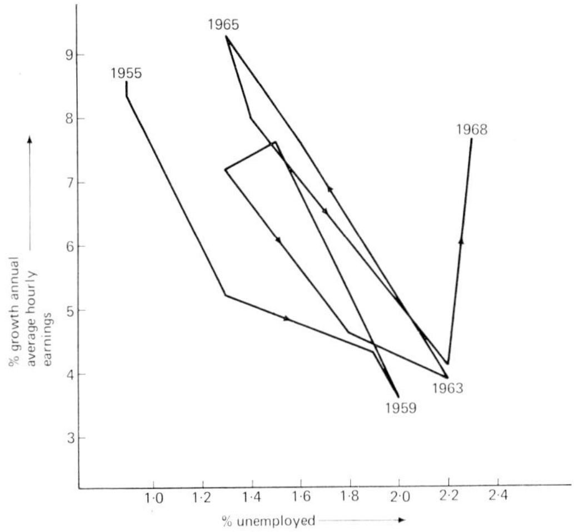

Fig. 2

An examination of Fig. 2, however, indicates that the control mechanism does not operate in the simple way suggested. The regression results relating earnings growth to the unemployment level are of no interest: the long-term factors driving up both the level of unemployment and the rate of wage inflation are sufficiently important to negate the apparent short-term effects of cycles of economic activity. For example, the short-term anti-inflationary effect of the economic downturn 1955–59 evident in Fig. 2, is not repeated so effectively in later cycles when higher levels of unemployment failed to keep earnings growth to the 1959 rate. A trend variable is needed to account for the persistent movement of the short-term Phillips' curve to the right. Now, the search for an economic variable that simply trends over time is all too easy: the greater proportion of all economic variables do this. Yet there are two that conform to notions acceptable to some economists anyway—the price level and the stock of money in existence.

Unfortunately, for this approach, the reader will recall that the object of controlling the growth of earnings was to control the growth of prices. A theory that reverses this order is no use to the price controller. Nor is he compelled to accept that it is the price level that plays this determining role, for, as has been suggested, a book of economic statistics will provide many other variables with the required characteristic of trending either up or down over time and their choice could be justified with an effort of rationalisation. At the time of writing, U.K. money supply data are not adequate to test this variable's performance as a controller of the rate of wage inflation, though recent theoretical and political interest in this aspect of economics means that preparation of such data is being given priority. Coen, Gomme & Kendall[1] have shown that U.K. money supply, as currently rather poorly measured, appears to control the behaviour of equity share prices on the London Stock Exchange, so one may hope that improved data will permit this line of research to be extended to more central fields of economic policy. Although inflation is an intensively studied area of economics and one in which a great deal of policy experience has been gained, no well-known and proven control mechanism yet exists for restraining the rate of wage inflation.

Industrial economic control

IT SHOULD not be assumed that all aspects of national economic control are as refractory as the control of wage inflation, though experience indicates that economic management is no easy task at the national level. It is possible that this is because the quantities being studied and (hopefully) controlled are very aggregate ones. Thus, one might reasonably expect the earnings of miners, hairdressers, civil servants and car salesmen to behave quite differently: it may be necessary to disaggregate to the industry level before progress can be made.

In the examples put forward here, the control implications that are drawn are of the most simple-minded type. For economic systems, this may not be a great disadvantage, as knowledge of the individual pieces of the system's make-up is not yet sufficiently refined to give one confidence in a control operation that relies on an intricate pattern of responses at several removes from the initial act. This section of the paper is split into two parts to discuss examples of two very simple kinds of control action suggested by the economic relationships obtained. I have called these *passive* or indirect and *active* or direct control and hope that these terms do not clash with conventional control theory usage. Active control situations are those in which the manager responds to a problem by influencing the variable of concern directly; passive situations are those in which the manager is unable or unwilling to influence the variable directly and is reduced to making the best provision he can for the inevitable. This will probably involve the active control of other parts of the system over which the manager has more power.

Passive industrial economic control

There is an almost unlimited number of aspects to the work of an industrial manager, varying in their relative importance from industry to industry. Yet the course of demand over the planning period ahead is of universal interest in setting the framework for wage bargaining, buying raw materials, determining the rate of production and the level of stocks held, price-setting and so forth. No example for one industry can be taken to apply as it stands to another, but the method and notions involved are often of general application. A couple of years ago, Mr Hughes and I wrote a paper for *The Statistician*,[2] in which we presented some results obtained from an examination of the monthly history of the registration of new cars in Great Britain over the period 1959–66; this is presented here as providing an example of when a simple economic model can be of great value.

The number of cars registered in a month varies for all sorts of reasons. We are still examining a quantity that is the result of aggregating the actions of hundreds of thousands of relationships, in the face of which any type of control would be quite impracticable, but it turns out that a very good fit can be obtained for the car registration data using only one equation, with three types of predicting variable—

1. Some auto-regressive terms, whose main function is to fit the very jagged seasonal pattern evident in the data. If the data had been de-seasonalised, most of these terms would have been unnecessary, though trend and constant terms may have remained. These may be taken as summarising the fact that no attempt has been made to examine the network of social, psychological and other factors that persistently tended to push up the rate of new car buying over this period.
2. Two lagged values of bank rate, which indicate that a 1 per cent rise in bank rate reduces the monthly rate of new car buying by stages by almost 10 000 cars, 10 per cent of the total registrations nine months later.
3. The effectively unlagged value of bank advances outstanding to personal and professional borrowers. An increase in the volume of such advances is reflected in increased new car buying: a result that corresponds with the general experience of many bankers.

The course of the actual car registrations made and those predicted by this model are shown in Fig. 3.

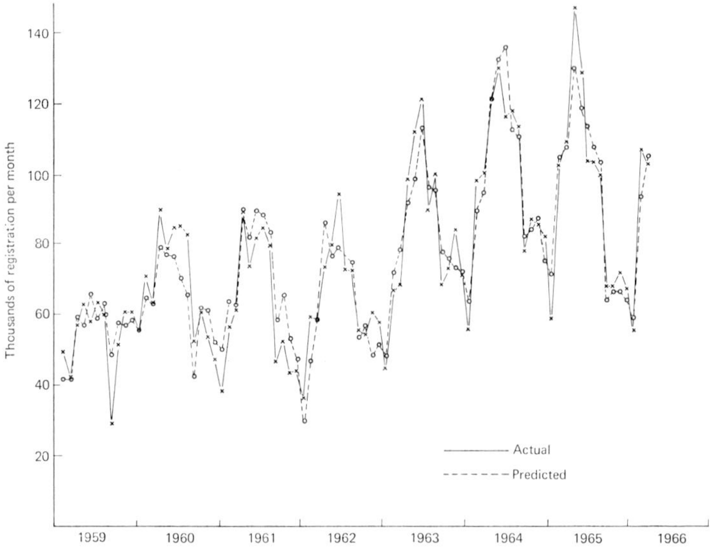

Fig 3.—New car registrations in Great Britain 1959-66 monthly (*source Monthly Digest of Statistics*)

From the point of view of national economic managers, an active control operation is possible. If the Chancellor of the Exchequer judges the level of domestic car buying to be too low or too high, this simple model suggests a way in which he can act to obtain the level he desires. The data to which the model has been fitted run up to February 1966. Let us say that the Government judged the level of new car buying to be too high in mid-1966 (as it did) and so raised bank rate and made a number of other financial and economic moves of a restrictive nature. It may be remembered that the imposition of a wage freeze was thought at that time to be something of a novelty, but Table 1 (largely reproduced from *The Statistician* article) shows that the model was able to predict the very substantial consequences of the 'July measures' in depressing the car trade the following winter without making any allowance for incomes policy or any other such innovations.

This table, incidentally indicates that national economic management is not a hopeless task, as the discussion in the section on national economic control may have implied. Nor is the individual industry's passive control problem

TABLE 1—FORECAST NEW CAR REGISTRATIONS OVER THE YEAR MARCH 1966 TO FEBRUARY 1967 (THOUSANDS): ERRORS COMPARED WITH ANNUAL CHANGES

Month	*Actual registrations*	*Predicted registrations*	*Percentage errors (actual–predicted divided by actual)*	*Percentage actual change from same month previous year*
March 1966	137·2	135·4	+1	−7
April	137·1	115·4	+16	+15
May	120·3	100·0	+17	+15
June	102·7	99·7	+3	−1
July	99·7	95·5	+4	−1
August	59·4	57·6	+3	−14
September	54·6	59·3	−9	−21
October	53·0	56·7	−7	−27
November	52·8	49·6	+6	−23
December	39·3	46·4	−18	−30
January 1967	103·6	89·3	+14	−4
February	92·9	92·9	=	−10

unmanageable. It may be that the car industry could do nothing to avoid the domestic slump of sales; certainly bank rate is beyond its control, but it could have been in a position to take defensive action. To determine the best form of such action would require a very thorough knowledge of the industry; each manufacturer and even each plant may require different strategies—to export more, to cut prices, to invest less, to be tougher in wage negotiations and so forth. The single economic relationship between the rate of interest and the rate of new car buying, however, sets the framework for all this, perhaps very complex, internal management analysis.

Details of the ways in which the car industry can react in the way which I have called *passive* to this change in their market situation are not of interest to an audience of papermakers, nor do I have the knowledge of the control systems operating in that industry that will differ anyway from plant to plant. Two general points hold—(*1*) nine months' notice is better than none and permits a wider range of responses to be considered and implemented; (*2*) it is not appropriate to make long-term decisions on the basis of the relationship obtained, for the Government can act to reverse or intensify their interest rate policy at any time. This has little immediate effect in the car industry, but the industry's one-year view should change at once. A very simple-minded view suggests that investment decisions should not be changed in such a situation; instead, more or less shift and overtime working and other short-term measures should be used. Of course, if an examination of the broader, national or international situation suggested that credit would remain expensive and in restricted supply for some years ahead, then investment decisions should be changed.

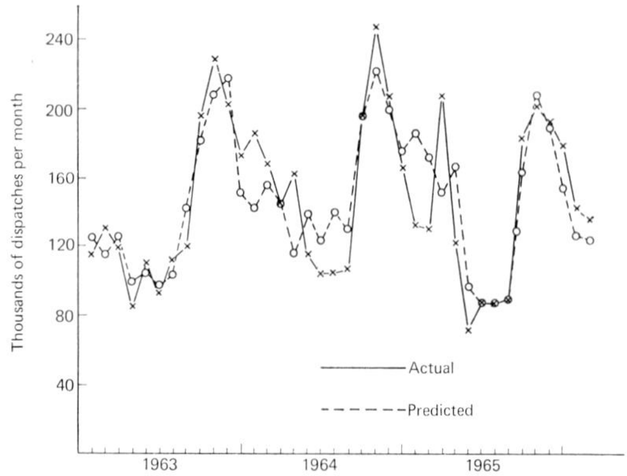

F g. 4—Comparison of T.V. dispatches

It is of some interest that similar results may be obtained from an examination of the history of manufacturers' dispatches of television sets. Once again, a 1 per cent increase in bank rate leads to a fall in dispatches of about 10 per cent of their total volume, some half a year later. In both these cases, cars and television, there were many technical innovations over the period studied and one might expect growing affluence and heavy advertising to have had an effect in changing consumers' tastes; nevertheless, in both these consumer durable industries, the national credit situation continued to set the framework for more detailed management policy-making.

Active industrial economic control

Although the behaviour of the car market may be predictable, there is less reason to believe that this is so of any particular manufacturer's range of cars, less still of any single model. At this level, economic factors often play a smaller part being swamped by technical and stylistic developments. Examples of an individual firm coinciding with the market arise in the cases of the U.K.'s television companies. Each of these companies is granted the temporary monopoly of the right to televise advertisements in a particular region and its income comes from selling this facility to advertisers. The amount of advertising time that can be sold to the potential advertisers is limited by law in a rather intricate way, but which may be taken as making it a constant quality. Nevertheless, the price at which this time is sold may be varied freely and a great deal of involved price changing and negotiation goes on. Although a large number of management problems exist for a television company, it is clear that determining the general price structure of the time that it has to sell is the key one. Conventionally, this structure is reset at six-monthly intervals (spring and autumn) and, since this variable is certainly under the company's control, the central management problem is simply formulated—set the general price level so as to maximise revenue.

Of course, the monopoly that a television company has is restricted. If it is too aggressive in its pricing, advertisers will tend to use the press or posters instead or to shift their expenditure to another region whose television company is offering better value, at least for a while. In economists' jargon, the problem is to estimate the position of the demand curve for the company's advertising time. Demand for this time was still trending upwards at the time of the study, as advertisers were still shifting from the older media—the press, magazines and so forth. Consequently, there was a steady tendency for all the television companies to raise prices during this period. In practice, a rather humbler objective than maximising revenue is acceptable: the company would be pleased to know whether it should raise its prices by more or less than the other companies, without necessarily setting the very optimum price.

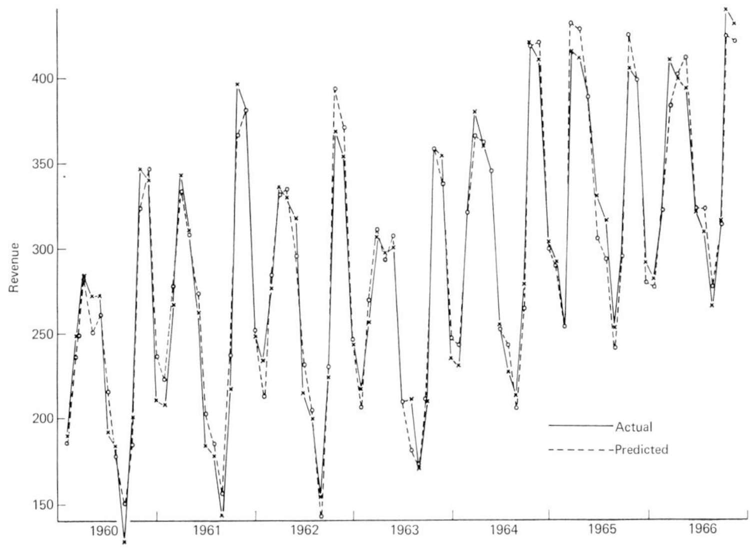

Fig. 5—Actual and predicted monthly revenue of a U.K. television company

After the previous discussions in this paper, the reader will not be surprised that a simple one-equation regression model was used as the basis of the approach to this problem. Seven years' monthly data on the company's revenue were available, together with corresponding information on the price of its advertising time, the number of television sets able to receive its programmes and advertisements and a number of economic variables measuring conditions external to the company. Analysis of these data obtained the fit for the revenue series shown in Fig. 5. The data are confidential and so the units are disguised and anonymity is preserved: the results are presented in proportional form.

The terms selected by the regression method used may be put into three categories—

1. Some autoregressive terms, whose interpretation is as for those in the car registrations equation discussed above—that is, they fit the sharp seasonal pattern and provide for the upward trend present in the series.
2. Some terms linking the prosperity of the company directly, but with a lag, to that of its region and the nation. These are of significance for the passive control problem discussed above, but not for active control.
3. Some terms linking revenue to variables that the company has at least some ability to control: its 'set count' and the price of its time. In the statistical analysis, the influence of these two factors is confounded, for the number of sets and the price of television time have both risen historically, though the number of sets had almost stabilised by the end of the period studied. There is scarcely any control relationship between the two: only a small proportion of advertising time is sold directly on the basis of set count. One approach would have been to conduct the analysis in terms of the revenue per set in the region rather than the revenue unadjusted in any way. Yet, as the analysis was carried out, the effect of a 10 per cent rise in the price of time was found, after allowing for the persistent growth in the set count, to build up over a year to a 4 per cent rise in company revenue. In the experience of the company, the demand for its time was inelastic and the control implication was obvious—to raise its prices rather more than the other companies at the next six-monthly round of price revisions.

In the two parts of this section, the two parts of the economic manager's problem have been discussed. First, the framework in which he must operate is predicted and, second, the effect of the items under his control is assessed. In the cases examined, there are substantial time lags before the full effects are felt of the initiatives taken. Of course, this may not always be the case, but the presence of such lags indicates that very substantial problems will exist in controlling such systems in a delicate way, though rough control may be more practical.

Application of these methods to the U.K. paper industry

THE field of paper manufacture is a new one to me, but, with the help of people in the industry, I have obtained a certain amount of background information and data. Only a small part of this is presented here, the analysis of which was carried out with the assistance of Mr Naqvi of Scientific Control Systems Ltd. A full survey of the whole industry has not been attempted; instead, three cases have been selected for presentation here. The U.K. experience has been studied of—

1. Consumption of food wrappings.
2. Consumption of kraft wrapping paper.
3. Consumption of good quality printing and writing papers.

Now, the paper industry has a number of features that make it somewhat different from those discussed above. It is an international market; it is a stable product; it is a key product in the economy and, more especially, the international trade of many of the important producers; it has a fast-growing output and heavy investment in technologically advanced and large plants; and is often in receipt of Government assistance and subsidy in one form or another; it does not have a large trade cycle. Although the full set of its characteristics is probably unique, it shares each of these features with some other industries: chemicals, man-made fibres and shipbuilding come to mind. In the remainder of this paper, it will be seen that some of the paper industry's features, at least, can be subjected to a statistical form of economic analysis.

Consumption of food wrappings

The first case to be considered from the paper industry is the consumption of food wrappings. At first, we tried to relate this to the growth of personal income after taking account of taxes and such items and inflation—the quantity known as real personal disposable income—but this proved quite inappropriate. Real personal disposable income increases year by year, whereas Fig. 6 shows that the consumption of food wrappings paper fell in the U.K. after 1963–64. This suggests the introduction of a technical or institutional factor at that period. Personal observation and discussion with people in this field suggested that the use of polythene bags and wrappings had become more widespread associated with the spread of self-service stores. The most convenient variable was found to be the number of supermarkets, which shows a strong growth from the early sixties.

The effect of fitting an equation using these two terms to the data on consumption of food wrappings for the years 1954–68 is shown in Fig. 6. Growth in disposable income has the effect of increasing food wrappings consumption, whereas a growth in the number of supermarkets has the opposite effect. Of course, the growth of supermarkets and of income are correlated, even if not causally related. Taking the two effects together, one expects a simultaneous 3 per cent growth of real personal disposable income and 10 per cent growth in the number of supermarkets per annum to have a balancing effect on food wrappings consumption at present levels of these variables. As this growth of income is slightly greater than usual and the growth in the number of supermarkets has often been of the order of 20 per cent in a year, one must expect the future trend of this class of paper consumption to continue to be downward.

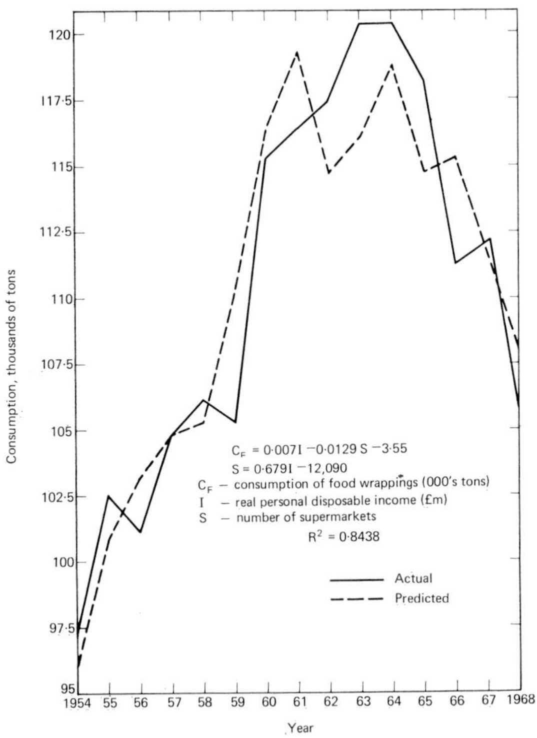

Fig. 6—Consumption of food wrapping paper—source B.P.M.A. Directory of Super Markets & Self Service Institute and National Income and Expenditure Blue Book

Presumably, supermarkets could play a larger part in the economy without increasing their numbers; for example, by increasing their average size or their turnover per unit of area. Furthermore, I see no automatic reason that they should be opposed to the use of paper for wrapping. The reader will be better able than I to judge what the future resultant of these conflicting factors discussed above may be and the export retailer whether there is a causal relationship between the growth of incomes and supermarkets. For interest, a very simple, two-equation linear model is given in Table 2. The second equation is derived from the period 1961–68 only. A fuller study could examine more complex functional forms and more variables. In particular, it appears that the actual behaviour is lagging behind that predicted in the last few years: this suggests that consumption of food wrapping paper now lags behind real personal disposable income by a year. We have not considered the very wide range of possible models, for the simple structure set out below is able to predict the 1968 consumption to within 7 per cent.

TABLE 2—CONSUMPTION OF FOOD WRAPPING PAPER MODEL

$$C = 0{\cdot}007I - 0{\cdot}0129S - 3{\cdot}55$$
$$S = 0{\cdot}679I - 12{\cdot}090$$

where C = Consumption of food wrapping paper ('000 tons),
I = Real personal disposable income (£m.),
S = Number of supermarkets.

For 1968, this gives $C = 113$, which is 6·8 per cent different from the actual value.

Consumption of kraft wrapping paper

The second case considered was another industry group, kraft wrapping paper, but this time one without any dramatic reversals of fortune in the U.K. market, at least. In earlier sections, the difficulty of analysing a smooth series has been discussed. As many economic series have this form, it is easy to achieve a very good fit with any of a wide range of explanatory variables and very hard to select which of this range is the best. A method of overcoming this difficulty is to take differences in the series to be explained (in this case, to take the annual differences in the consumption of kraft wrapping paper) and to perform the usual regression analyses with this form of the variable. The results of this transformation are displayed in Fig. 7, in which it can be seen that any trace of a trend in the series has been eliminated.

Consumers' expenditure with the effects of inflation removed was tried first for food wrapping paper. Since the kraft paper consumption variable has been differenced, annual consumption is differenced too. I think that the reader will agree that this gives a very good fit, in which all the turning points are identified (though the extent of the sharp down-turn in 1958 is substantially underestimated). Of course, kraft paper is not all sold to individual consumers directly; on an annual view, the evidence is that the firms and organisations buying kraft paper are geared in their activity to the level of individuals' spending.

An interesting feature of the result is that kraft paper consumption appears to have been more volatile with respect to general consumption in the fifties than it has been in the sixties. In each of the boom years of 1955, 1957 and

1959, the increase in consumption was greater than that predicted; in the intervening slump years of 1956 and 1958, it was even smaller or more negative than predicted. The fit is almost perfect for 1960–62, after which the actual behaviour of kraft paper consumption was less exuberant in the boom years of 1963, 1964 and 1968 than predicted. The reader may be able to provide an explanation of this phenomenon.

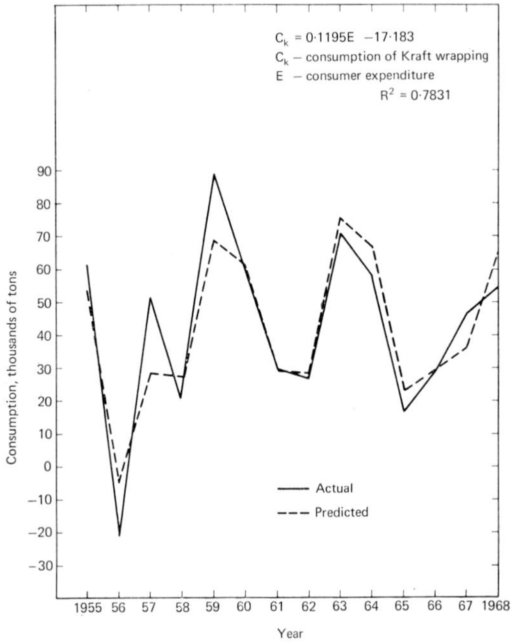

Fig. 7—Consumption of kraft wrapping paper—annual differences (Sources B.P.M.A. & National Income and Expenditure Blue Book)

Consumpt on of good quality printing and writing papers

The third and last case to be presented here was of a narrower group. It has been suggested above that aggregates often behave in a more regular way than their components, whose performance may be significantly effected by shifts of fashion and such factors. The case selected for discussion here represents a further stage of disaggregation and is another one that requires that the data be differenced to avoid having to analyse a trend-dominated series. The consumption of good quality printing and writing papers has risen in almost every year since 1957. On taking the annual differences in the consumption of these items, however, a trendless series is obtained, as can be seen in Fig. 8.

Once again, the annual differences in real personal disposable income were related to this series and it can be seen from Fig. 8 that a generally good fit is obtained with this single, simple explanatory variable. Although the turning points are still picked out, the quality of fit is somewhat reduced from the standard of the previous example. Furthermore, the analysis of the remaining qualities of printing and writing papers was unable to find a statistically significant fit with this explanatory variable. It may be that we have come to the end of the road with simple, economic variables and that more intricate marketing studies paying due attention to changes in taste, the impact of advertising campaigns and technological changes will be needed for studies of more detailed sections of the paper market.

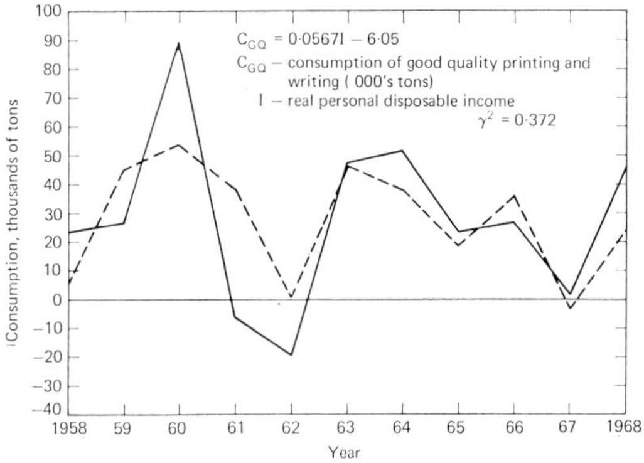

Fig. 8—Consumption of good quality printing and writing paper—annual differences (Sources B.P.M.A. & National Income and Expenditure Blue Book)

In each of the three cases from the paper industry that have been discussed, the control implications are of the variety I have called passive—that is, it is impossible for paper industry managers to influence real personal disposable income, consumers' expenditure or the growth of supermarkets to a significant degree. They must therefore simply adjust to the expected outcome for their industry. Long-term, this income variable is expected to grow by about 3 per cent per annum in the U.K., but this will vary from one year to another in the range of 1–7 per cent, partly as a result of Government action to curb inflationary tendencies in the economy. Consumers' expenditure is expected to grow at a similar, hopefully slightly slower rate. I have not the expertise to assess whether or when the current, rapid growth of supermarkets will reach its ceiling, but the trade association may be able to do so. Generally, the relationships obtained may be used to provide a broad framework for long-term planning and control.

The Government and a number of private agencies produce forecasts of many economic quantities for a year or so ahead at frequent and regular intervals. Although these have often been proved wrong in detail, they can be used to make up tentative medium-term plans. Then, as hard information from the national budget and other sources becomes available, these can be revised. For control purposes, it is unfortunate that no long lags exist to give warning of coming booms or slumps, though a study using quarterly or monthly data might be able to provide some help.

The techniques of economic model building have been and are extremely useful in other industries. In these cases, however, the economists working on the problem have had the opportunity of spending considerable time doing research and investigating the factors affecting the industry in question. The examples I have discussed from the paper industry were chosen from the point of view of readily obtainable data in the limited time available for preparation of the paper. For this reason, they are open to some obvious criticism and I expect this to become apparent during discussion.

The point I hope to have made is that, given a concentration of effort and availability of data, it should be possible to obtain some really useful results. This is despite the international nature of the paper business and its susceptibility to economic policies outside its control.

The way in which the paper industry should respond to this knowledge of their situation requires an understanding of the internal character of the industry. A later stage of a full rather than illustrative study of economic models applied to the paper industry would involve an examination of the detailed interrelationships among stock levels, production processes, investment plans and so forth. The paper by Mr Anderton is a contribution to this field. I hope that enough has been said in this one to indicate the extent to which very simple models can assist in understanding the economic framework within which the paper industry operates.

References

1. Coen, P. J., Cromme, E. D. and Kendall, M. G., 'Logged relationship in economics forecasting': *J. Roy. Stats. Soc.*, 1969, (A), **132** (2), 133–163
2. Hughes, P. A. B. and Morgan, N. B., 'The use of computers in forecasting': *The Statistician*, 1967, **17** (3), 279–299

Bibliography

Statistics

1. Johnston, J., *Econometric Methods* (McGraw-Hill, New York, 1963)
2. Draper, N. and Smith, H., *Applied Regression Analysis* (John Wiley & Son, New York, 1966)

Economics

3. Samuelson, P. A., *An Introductory Analysis* (McGraw-Hill, New York, 1967)
4. Lipsey, R. G., *Introduction to Positive Economics* (Weidenfeld and Nicholson, London, Second edition, 1966)
5. Phillips, A. W., 'The relationship between unemployment and the rate of change of money wages in the U.K. 1861–1957': *Economica* (New Series), 1958, **25** (100), 283-299

Discussion

The Chairman Your preprint has aroused considerable interest and, in order to contain the session to some extent, I will now call upon Mr D. L. Cooper of Pira, who has kindly agreed to present another point of view on your topic. After giving you a chance to answer Mr Cooper, I will ask Mr Anderton to speak, then we will take questions on the whole of this session.

Mr D. L. Cooper I think we should all thank Mr Morgan, coming as he does from outside the industry, for the work that he has put in to postulating models for some of the industry's products. He indicates in his paper that he may expect criticism from more informed sources. I have not the experience of economic modelling in the industry to qualify for such a description, but I should like to make one or two comments on some of the statistical aspects of his paper. I intend to discuss none of the technical aspects of the statistical models that are proposed.

I should first like to make a brief comment on the model of TV company revenue. I find it difficult to understand why the 'set count' should be a variable over which the company can be regarded as having control. I am not surprised that, when set count reaches saturation point, there is no correlation between it and revenue, but I would have thought that some term in the model might be worth keeping. At some stage, TV companies are going to be concerned with colour TV advertising and the charges for advertising are almost certainly then going to depend on set counts of colour TV reception. The experience of using set count in the black and white model might well be useful.

I should also say that I am not surprised to find a significant relationship between charges and revenue in situations where monopoly conditions for that medium exist.

Turning to the paper industry models and the consumption of food wrappings, although I can see some degree of relationship between real disposable income and use of food wrap, I find it more difficult to accept a model that postulates a causal relationship between food wrap and supermarkets. I am not disputing the fact that there is a statistically significant relationship, but is it meaningful? To me, a more reasonable explanation that the last few years is not such a good fit is not because food wrap has become out of step with real disposable income, but because the supermarket variable is not in fact a causal relationship. Indeed, even accepting that it may be, it is unlikely to be linear, since the rate of increase in supermarkets is likely to slow down as they account for more and more of consumer purchasing power. I suggest that any phenomena that accelerated in the year in which the down-turn of wrap production started would yield a correlation coefficient of appropriate magnitude.

I am also slightly concerned to know why Mr Morgan should choose real personal disposable income for explaining food wrap and consumer expenditure for explaining kraft wrapping. Why is it necessary to take out trends in one case and not the other?

I am pleased to see that he emphasises the impact of technological changes, marketing and other local variables. I can think of two examples from our industry in the recent past; tissue production and coated papers would have related more to these than to dependent variables from the national economy.

The point I want to make most strongly is the difference between statistical relationship and a causal relationship. Whereas regression techniques can establish the regression coefficients that give the best estimate of the value of a dependent variable from a set of data, unless there is a permanent causal relationship, we may lose our way if we use such techniques to predict the future. In such circumstances, I wish there were a way of indicating not only the degree of significance between relationships, but also the degree of irrelevance.

This paper contains some stimulating ideas and I should not wish to seem only destructive in my criticism. The problem of forecasting, particularly long-term forecasting of economic models, must be a particularly intractable problem and intuitively one feels that in the welter of data available such well-founded models must exist. The only advice I can offer is to establish very clearly the aims behind building the model at all—whether forecasts are to be for short-term, long-term, company-based, nationally or internationally based, then to settle for dependent variables for which some causal relationship can be reasonably advanced.

Mr N. B. Morgan I will try to answer your points in sequence. The first one deals with the set count variable in television. In the late fifties and early sixties, in the U.K., there was a persistent growth in the ownership of television sets, particularly to receive commercial stations. It was possible for television companies to stimulate such growth, firstly, by putting on appropriately attractive programmes and, secondly, by advertising. Even now, it is possible for companies (in this country at any rate) to expand their set count by putting in bids for other television transmitting masts. Various regional companies have done this and, consequently, they have expanded: thus, the set count has historically been under the influence of the television companies to some extent. This has become a control problem for them, because every now and again they get the opportunity of putting in a bid for a particular mast. There is competition among the various companies. This is a variable, they are making decisions. There are also a second channel ITV or a colour channel.

I am not really sure that I can say much about the causal relationship between food wrappings and supermarkets. I am sure that there may be confusing features of the data.

On the general question of causality, I must confess that the philosophy of the nature of cause always gives me difficulty. It seems to me that most people, by saying there is a cause and effect relationship, mean that there is a statistical relationship that holds in many different situations. Moreover, they mean that it holds over a reasonable period of time, in one part of the country and another, for one class of people to another, bad and good economic conditions alike and so on. I agree about the supermarket's case. Unfortunately, the data is not complete enough to test in full causal sense. It seems fair to say that some of the cases I have presented from outside the paper industry have a relationship that is more or less causal, tested over 15 years or so. With unemployment, money supply and earnings in slumps and booms, the relationship continues to hold. I think therefore that a reasonable causal relationship does hold in some cases. I am not saying that it could not be improved by better testing. The physical sciences have been working on causal relationships for 300 years or more, using techniques that are still improving and still being tested. We should not be surprised that economic models are still at a relatively rudimentary stage, though I have to confess that the paper industry models I have presented are extremely simple.

We use PDI in the food wrapping relationship, because it gives better results than the other explanatory variables tried, but I have no explanation why it should be better.

INDUSTRIAL DYNAMICS: A TECHNIQUE FOR GAINING UNDERSTANDING OF COMPLEX SYSTEMS

R. H. ANDERTON, IPC (Group Management) Ltd., Feltham, Middx.

Synopsis Control of a complex organisation requires that its response to decision inputs be predictable. Traditionally, this has been made possible by almost breaking down the organisation into very simply coupled subsystems. This allows decision-makers to use a relatively simple and usually informal model.

Size and environmental rate of change are now tending to make the decoupling hard to make effective. Industrial dynamics provides both a theoretical structure and a computer simulation technique designed to provide decision-makers with a richer model of the dynamic feedback processes generally involved in industrial operations. The structure defines a system in terms of a closed boundary containing feedback loops, which in turn are formed by the relation between level and rate variables. The DYNAMO compiler allows models of systems conceived on these lines to be programmed in a straightforward way so as to allow experimental investigation of the system variables. The paper discusses case studies and concludes with notes on model validity and the problem of data.

Introduction

THIS paper was prepared as an introduction to the subject of industrial dynamics. It is thus very largely an account of ideas developed by Prof. Jay Forrester and his associates at M.I.T. to whose original writings the interested reader is referred.[1–2, 5, 8] We can start, however by establishing the need for such a technique in a particular context.

New kinds of help are needed for managers charged with the organisation and control of complex industrial systems.

The tendency is for these systems to become larger, more extensive and often more diverse. At the same time, they are being required to respond more rapidly and more flexibly to change, both in technology and in the pattern of demand for their output. Although these tendencies increase the amount of

Under the chairmanship of M. I. MacLaurin

variety with which the manager must cope, there are rather definite limits to the amount of variety that an unaided human being can process. These limits are obvious enough in the straightforward case of sequential information processing such as reading, although the limits are rather high. They may be less obvious and more serious when this serial approach breaks down and relations between pieces of information are the essence of the matter. This occurs in risky situations when probabilities of separate events have to be compounded to estimate the probability of occurrence of a complex of events. It also occurs when the dynamics of situations determine the system behaviour, that is, when there are many strong couplings between parts of the system showing relationships developing over time.

In both these situations, the predictive power of the unaided human brain is rather weak. The matter is made worse because typically the weakness is not understood. It is the argument of this paper that—

1. These pressures on our mental powers are becoming increasingly felt both directly in our industrial systems and indirectly through failures in control in the economic and social systems that impinge on them.
2. The techniques of industrial dynamics developed by Forrester and his co-workers can provide powerful help in tackling the problems created by these pressures.

A managerial dilemma

IT WILL be interesting first to look at some of the ways in which the impact of these problems have been reduced by organisational conventions. We have said that there is a rather low limit to the number of interelationships we can handle together. (It seems that we have difficulty in predicting the behaviour of systems in which more than four or five variables dynamically interact.) As we have to have at least some competence in predicting system behaviour, if we are, in any sense, to control it and as industrial systems contain typically many more than five significant variables, we have little choice in organisational design. The subsystems must be largely decoupled so that the interactions between parts are simple. Putting it another way, the system must be almost broken down. We can give three examples of how this is done—

1. The use of buffering stocks to separate stages in a manufacturing process is a familiar one. This has a similar effect to that of a low pass filter in electrical engineering. Rapid changes are absorbed by the stock and only slow variations are transmitted. Thus, in the short term, the systems are decoupled.
2. The divisionalisation of large corporations by market or product or sometimes by function.

3. Less apparent, but of great importance, is the separation of activities according to the characteristic time that associated changes take to occur. Activities in an organisation can be thought of as forming a hierarchy in which higher levels have longer characteristic times.

   If these characteristic times are separated by factors of about 5–10, then the higher level is quasi-static from the viewpoint of the lower so that the higher level factors can be thought of as fixed policies or constraints.

These conventions have been (and are) remarkably effective in decoupling complicated systems and so making them manageable. They allow the manager to hold in his head a simple model that adequately represents the system he is managing.

Yet, we will now point out how, in each of the three examples given, current trends are making the solution less effective, in that the economic costs are increasing while the solution itself can at the same time be raising new dynamic problems.

TABLE 1—AN EXAMPLE OF THREE STOCKS IN CASCADE
(The figures are percentages)

	Retailers				*Wholesalers*				*Finished goods store*			
Week	*Out*	*In*	*Balance*	*On order from whole-salers*	*Out*	*In*	*Balance*	*On order from finished goods store*	*Out*	*In*	*Balance*	*On order from pro-duction*
1	100	100	100	100	100	100	100	100	100	100	100	100
2	110	100	100	130	100	100	100	100	100	100	100	100
3	110	130	110	110	130	100	70	190	100	100	100	100
4	110	110	110	110	110	190	150	70	190	100	10	370
5	110	110	110	110	110	70	110	110	70	370	310	Nil
6	110	110	110	110	110	110	110	110	110	Nil	200	20
7	110	110	110	110	110	110	110	110	110	20	110	110

Note; The order rule is to order sufficient to provide a balance equivalent to one week's sales as estimated from last week's sales

In the case of buffering stocks, the investment will increase if demand becomes more fluctuating, less predictable and more varied. If operations become more extensive, time lags and distortion in information and (possibly) naïve computer programming may tend towards the dynamic effect caricatured in Fig. 1.

In the case of divisionalisation, as products and markets change their identity, opportunities for synergy may be lost if rigid decoupling is maintained. Dynamically, too, inconsistent divisional targets can result in pendulum swings as, again, time lags and distortion in inadequate information links constrain the process of compromise.

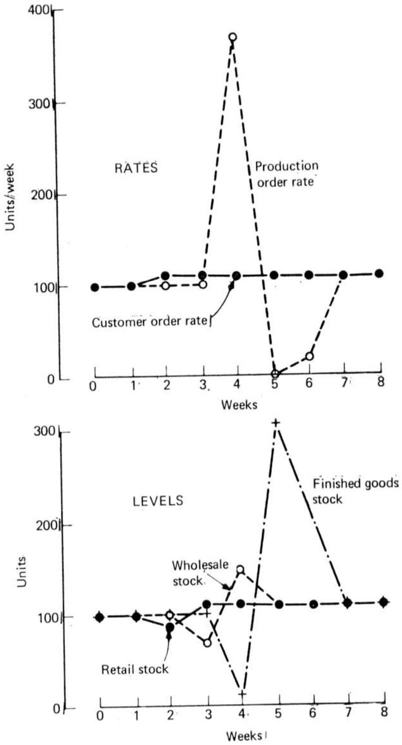

Fig. 1—Example of three stocks in cascade

In the third example, the time-separating hierarchy, the difficulties are more fundamental. The environmental rates of change are becoming such that the quasi-static assumptions of lower echelons are no longer valid.

For example, rapid external changes in market demand and in technology produce competitive pressures for internal changes at traditionally high levels of the hierarchy. These changes may be so frequent that the lower levels may not be able to increase the tempo of their activities in proportion and the time separation breaks down.

Without a stable policy framework within which to work, lower echelons are constantly attempting to adjust to or anticipate new situations. A steady state is almost never reached. A major decision input will now set off long chains of response and reaction, some links of which will feed back on themselves, positively or negatively. Intuitive predictions of results based on simple concepts of causality are now unlikely to be useful. Sometimes in fact the results will be counter-intuitive, nor is formalisation of the model through standard statistical and econometric methods likely to help. The presence of time lags, autocorrelation, non-linearity and large numbers of variables make these techniques extremely difficult to apply, to say the least of it.

What is needed is, firstly, a theoretical structure that provides a language and concepts, secondly, a simulation technique based on this theory that, together, will allow real complex systems to be modelled and explored. With these richer models, complexity can then be accepted and coped with, rather than artificially cut down.

A theory and technique that promises to meet these needs will be briefly described in the next section.

The industrial dynamics technique

The theoretical framework—Industrial dynamics provides a theory of system structure that can be outlined as follows. A system is a collection of parts

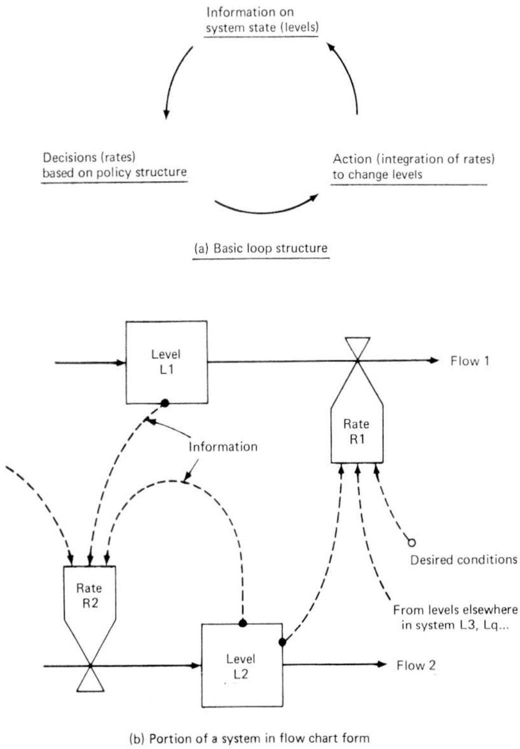

Fig. 2—System structure in industrial dynamics

connected to achieve a purpose. On this basis, a *system boundary* can be defined that includes the area of interest. System behaviour is generated inside this boundary. Within this boundary, the system is built up by the interaction of *feedback loops*. These are of two basic types, negative feedback loops (which are goal seeking) and positive feedback loops (which produce growth). These loops are constructed by connecting two types of variable. The first are called *levels* and are the accumulated quantities in a system (stocks, for example). The second are *rates* and fix the flow of quantities, which will be integrated to change a level. In a system, rates are determined by the 'policy' structure of the system, which in turn is defined by the goals and observed states, together with a procedure for converting the discrepancy between these two into a desired action. This system structure is illustrated in Fig. 2. This theory of system structure is held to be of very general validity, describing systems that extend beyond industrial ones[1,2] to include social and economic,[3] even biological situations.[4]

What then becomes possible is a very direct modelling of hypotheses on the *mechanism* of the system under study. In particular, it stresses the role of information in integrating an organisation and enables the importance of its quality and timeliness to be evaluated. A clear distinction is made between the design of policy that determines how information is converted into action and particular decisions made within a given policy structure.

Simulation procedures—If a system is described in accordance with this scheme, it becomes a straightforward matter to write sets of difference equations that define the system operation through time. Numerical solution of these equations then allows the trajectory of the system to be plotted, starting from some initially defined states. Time is quantised into discrete solution intervals made small enough for interpolation between solutions to be accurate enough for the purpose in hand.

During each solution interval, *rates* are calculated using sets of *auxiliary* equations supplied with the values of the levels *existing* in the previous solution interval. New level values can then be arrived at by adding to the old ones the appropriate rate multiplied by the solution interval.

The DYNAMO compiler[5] enables equations to be written very directly in a high level computer language. Graph plotting and tabulating output routines are included. DYNAMO 2 (a new version) allows FORTRAN statements to be included. Special DYNAMO functions include delay functions of various types, test functions such as steps and pulses, random noise generation and the ability to input functional relationships in the form of tabulated numbers. A time-sharing version of DYNAMO is also available.

Other systems modelling languages exist, for example, CSMP 360,[6] which have certain advantages.[7] The virtue of DYNAMO is the directness with which it relates to the theoretical structure just described.

The application of industrial dynamics

THE basic methodology of industrial dynamics is not hard to learn, but its successful application is less straightforward. All modelling is and must be an art in which judgment and experience play a large part. In industrial dynamic modelling, the necessary judgment is perhaps only rarely combined with the analytic ability also needed. Furthermore, it is seldom, if ever, easy to construct tests of validity for proposed models if they represent highly complex situations.

TABLE 2—OPPOSING VIEWS OF MATHEMATICAL MODEL-BUILDING

Conventional views	*The industrial dynamics position*
1. Data The main obstacle to the development of useful models of industrial operations is the lack of adequate data. The first step must be extensive collection of statistical data. Analysis of simple relationships between measurable variables should be the starting point of model building.	We usually start already equipped with enough descriptive information to begin the construction of a highly useful model. A model should come first. The model builder is more in danger from being insensitive to and unperceptive of important variables than from lack of information, once the variables have been exposed and defined.
2. Validity Objective tests must be available to establish the degree of correspondence between the model and the real world and the levels of significance of its hypotheses.	Model validity is a relative matter. The usefulness of a mathematical simulation model should be judged in comparison with the mental image or other abstract model that would be used instead.
3. Prediction The aim of model building and the criterion for its success and scientific status is its ability to predict the future states of the system.	System models should predict and reproduce the behaviour character of a system, not specific events or particular unique sections of actual system time history.
4. Solutions We should aim at analytic solution of models and this implies, in most cases, limitation to linear systems with constant parameters.	Most dynamic behaviour in social systems can be represented only by models that are non-linear and so complex that analytical mathematical solutions are impossible. For such systems only the simulation process using step-by-step numerical solution is available.
5. Generality The level/rate structure enshrined in the DYNAMO compiler is of limited applicability. It describes a small sub-set of industrial situations.	The structure will come to be recognised as having simple elegance, universality and a fundamental character common to a very broad range of systems running from physical devices through medicine and psychology to social ecological systems.

These factors, together with the large claims often made by industrial dynamics practitioners for the universality of their results have made the topic the subject of some controversy among management scientists. Table 2 sets out in extreme form some of the differences that emerge.

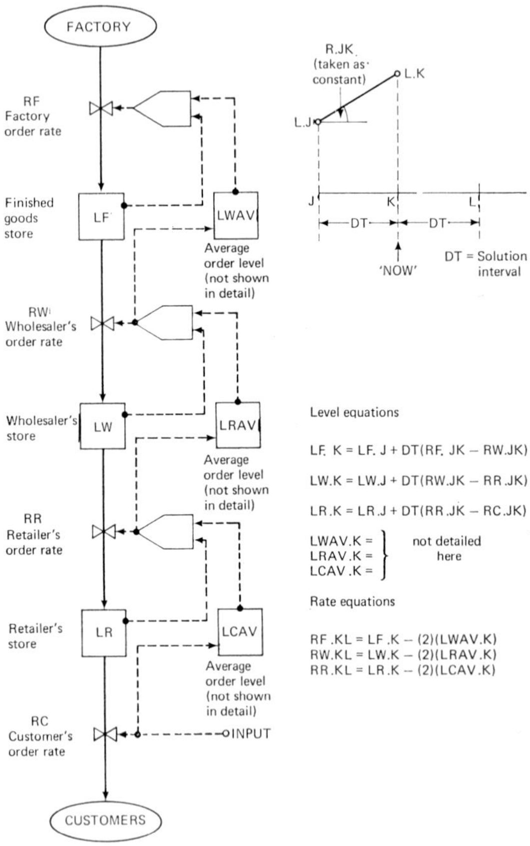

Fig. 3—Set of equations (simplified) for 3 stock example

Misunderstanding of the ways in which industrial dynamics is intended to be used accounts for much of the criticism and some of the difficulties in application.

An industrial dynamics study starts with a problem resulting from a system showing aberrant behaviour of some kind, a resistance to change perhaps or a persistent oscillation not obviously related to system inputs. It then attempts to show how the behaviour results from a defect in system structure. Finally, by experimentation on the simulation model, it suggests changes in system policy or structure that remove the self-generated disturbances.

The emphasis throughout is on organisational design or redesign made effective by an understanding of the system dynamics, particularly its sensitive points of leverage.

The work done at M.I.T. and elsewhere has covered many fields extending from the early work on production-distribution systems to broader studies of corporate growth and market dynamics most recently to large-scale research on social systems.

The lessons from this work confirm that—

1. Systems behaviour is frequently remarkably insensitive to parameter change sometimes even by factors of up to five.
2. There are a few sensitive points where small changes can produce radical changes in system behaviour.
3. Prediction of behaviour is beyond unaided intuition and requires a disciplined structural investigation using computer simulation techniques.

It seems likely that only by improving our understanding of system structure in the way these conclusions suggest will we resolve the classic problem of the control engineer, now shared by practising industrial managers. The problem is—How can we create organised systems that are at once stable and responsible to rapidly changing inputs?

There will be no simple solutions, but we may learn how to produce complicated ones.

An example from the paper industry

THESE ideas can be made more concrete by sketching a possible ID approach to a practical problem, which could emerge in the paper industry. It has to be made clear that what follows is in no sense the result of an actual study; however, I have taken the circumstances of an actual situation so as to make the material plausible. What I have chosen to do is to take the production-distribution system of a papermaking company and to examine the possible implications of a decision to change it in a fairly radical way by the introduction of a central warehousing function.

Let me start by listing five points I want to illustrate by means of this example—

1. The aspects of interest in a dynamic study differ from those considered in the more usual static one.
2. There are differences as well as similarities in the characteristics of industrial management systems and engineering systems.

3. Understanding the characteristics of systems of this kind often requires bold use of 'soft' data. (The further point that this is much less hazardous and justifiable than might be thought cannot be demonstrated without an actual study.)
4. The definition of the system's boundary is crucial in modelling.
5. Light can be thrown on the transient effects of changing company situations that tend to preoccupy senior management are often perceived in terms of crisis, usually receive no systematic attention and are, in fact, closely related to problems well studied by control engineers.

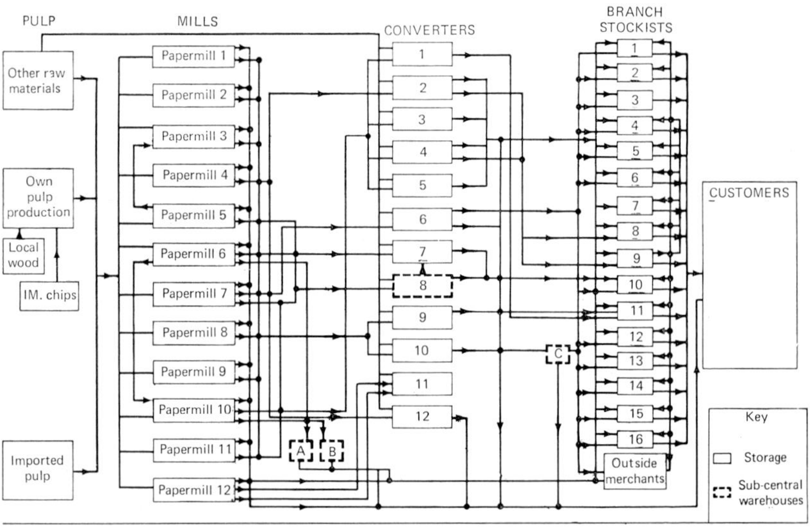

Fig. 4—An existing production-distribution system

Fig. 4 shows in schematic form the configuration of an existing production-distribution system. Typically, a chain in this system works as follows. Papers of a particular grade, but of different weights and finishes are produced at a mill in a repetitive sequence of runs. A stock of the various grades of paper is held in reel form to feed the converting process and a stock of finished paper products allows the response to orders from the branches to be independent of the production cycle. Large customer orders can be slotted directly into the production cycle. Neither of the two stocks is designed to smooth the production process as stock reorders are calculated to replenish the stock to a predetermined level on each cycle, based on an average of past sales. This system allows considerable flexibility in that it can react very quickly to changes in demand pattern, order mix or customer priority. Its disadvantages are that it is unlikely to provide the theoretical savings possible with a few centrally sited warehouses that would allow optimum use of bulk transport and economic order quantities. The use of central warehousing might also be expected to result in reduced delivery times. We shall return to this in a moment.

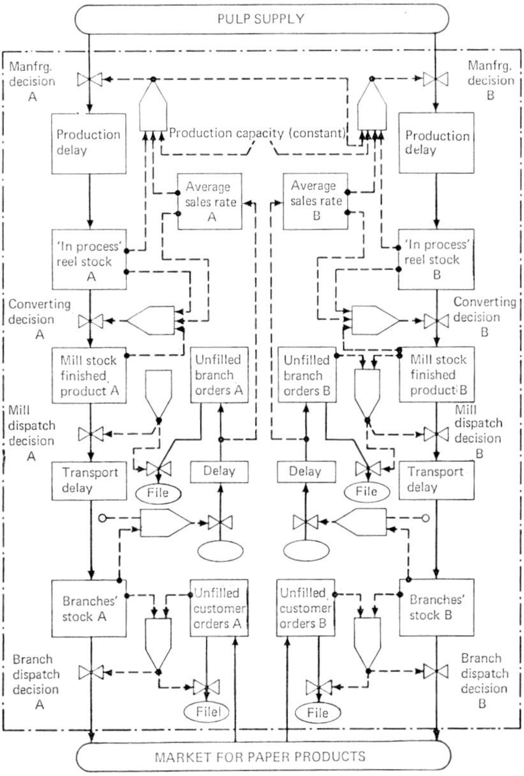

Fig. 5—Simplified flow diagram for production-distribution system (two paper product groups only)

Firstly, let us look at the way in which industrial dynamics could represent such a system. Fig. 5 is a flow diagram based on the model developed by Forrester[(1)] drawn for just two paper products. It shows how the rate of flow of product from pulp to customer is determined by decisions based on information about the state of affairs in other parts of the system. It allows the introduction of delays at various stages in the process, including lags in information flow. The equations that convert information into rate decisions define the policy structure of the system.

Notice two things—

1. We have drawn systems boundaries at particular points. Pulp buying and storing decisions, plant capacity decisions and the market are all outside this boundary.
2. We have limited the model to two product lines.

Decisions of this type are not arbitrary, but they can be made sensibly only if there is a clearly defined purpose (usually the solution of a specific problem) associated with the model. The point that there is no such thing as *the* model of the system is an obvious one, but nevertheless often forgotten.

For example, we might investigate the following. If the capacity of the papermachines is almost fully utilised, preference is likely to be given in the manufacturing decision to those grades of paper that allow higher tonnage rates of production. The effect of this would be to increase average delivery delays on the less favoured products. What system behaviour might result from this? The model builder would clearly have to include some representation of the market in order to show how order rate would react to the longer delivery time. He might want to include decisions on the increase of papermaking capacity. He would certainly have to consider how many different lines he should include in his model and an effective representation of the priority production decision. This could lead perhaps to studies of a long-term realignment of customers following fluctuations of sales linked with the state of the economy.

The model that would result from this would be totally different from the one limited as in Fig. 5.

Let us return to this simpler model. Fig. 6 shows some typical order data. A high level of noise is imposed on long-term trends. A good deal of this noise may be the result of the discrete nature of branch ordering, perhaps times to coincide with known production schedules. Three points can be made here—

1. An important difference between managed industrial systems and many physical systems is that there is in the latter often a fairly clear separation of signal frequencies, noise frequencies and the natural frequencies of the system, whereas in management systems this is only rarely so.
2. In industrial dynamics, attention is focused on the continuous nature of the systems it studies. This is not to say either that the discrete characteristics of systems are unimportant or that they cannot be repeated in an industrial dynamics formulation, only that most of the interesting dynamic behaviour can be modelled with continuous approximations to the system elements.

3. In industrial dynamics studies, systems behaviour generated in noise-free decision processes is emphasised. This is in contrast with event-based simulation methods, which try to generate the stochastic behaviour of the system. Yet noise signals are often used to test the sensitivity of the system to random inputs.

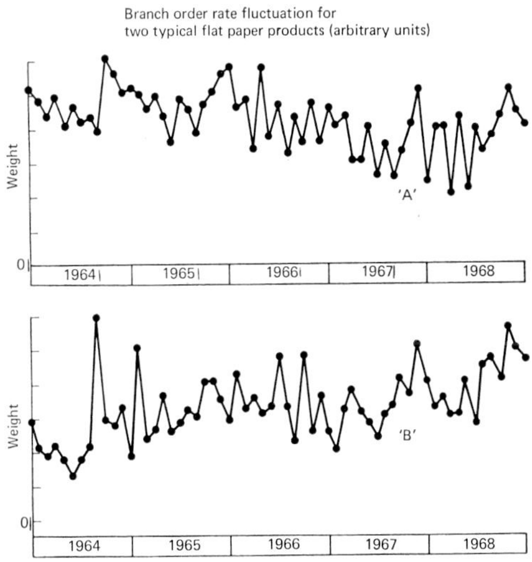

Fig. 6—Some typical order rate patterns

It might appear with demand data of this kind that the dynamic response characteristics of the system were of minor importance. Certainly, in industries subject to rapid swings in fashion or frequent production innovation, inadequate dynamic response is more immediately noticeable. Yet no industry—however static it may appear—can neglect its dynamic behaviour. There are few industries, for example, unaffected by the external state of the economy. Devaluation, lowering of trade barriers, changing cost of capital, changing levels of consumer expenditure can in their various ways set off chains of reactions that, if inappropriate, can be unnecessarily expensive. The need to detect change in the presence of noise entails an averaging process that itself introduces a lag.

The other source of dynamic effects nearly always present is competition. One of the ways in which companies compete is by offering better customer service and, in particular, shorter lead-times on delivery. In our example, we can take it that pressures to improve delivery provide the impetus for the central warehouse proposal.

The 'best' configuration of production and storage locations can be worked out straightforwardly, which is not to say easily, by operational research techniques fairly well established. Such a study might show an impressive saving both in cost and in delivery time, but it would be very likely to be a static study rather than a dynamic one. It would thus say nothing about the effect of change in the new system, whether this change be an external effect in the economic or market environment or the internal impact of the new system itself.

To end this paper, a short sketch is given of how industrial dynamics might study these aspects of the proposal.

We noted in the account of the existing production-distribution system that the production decisions were made in direct contact with incoming orders, in some cases from customers or at least from the branch level. If central warehousing were introduced, this contact would be lost and production control would infer correct production rates and mixes from the data available to it. This would require the setting up of a new information system. Alternative systems could be studied using a development of the model of Fig. 5, modified to include both the central warehouse and the new information handling procedures. The tendency of these systems to amplify noise inputs in the way we have referred to would be examined using a model of this type. One might look also at the effect of using information on other variables such as delivery delay on system operation. If it were favourable, new measurements might be worthwhile making.

We could turn next to the study of possible reactions to the warehouse introduction itself. One of the characteristics of large systems we have remarked on is their ability to resist change. This may not be the result of some natural conservatism among the people running them, but be caused by the hidden closed loop structure of the system. For example—

1. There could be a change in customer ordering patterns corresponding to a lower stockholding by them when they perceived the shorter lead-time on delivery.
2. Failure to signal a clear change in goals corresponding to the potential of the new system may result in self-regulating actions in related areas that nullify much of the value of the change.

Detailed examination of effects of this kind could be conducted by a model outlined in Fig. 7, based on one developed in detail in Forrester's paper.[8]

We can see from this feedback loop representation how central a part delivery delay plays in the operation of the total system. Firstly, it decreased the effectiveness of sales effort, hence decreases orders. Secondly, if it differs from the company norms' for delivery, delay pressures will be created for additional plant capacity. If we add the positive feedback loop that tends to increase the sales force as sales rise (or decrease the force if the sales fall), we

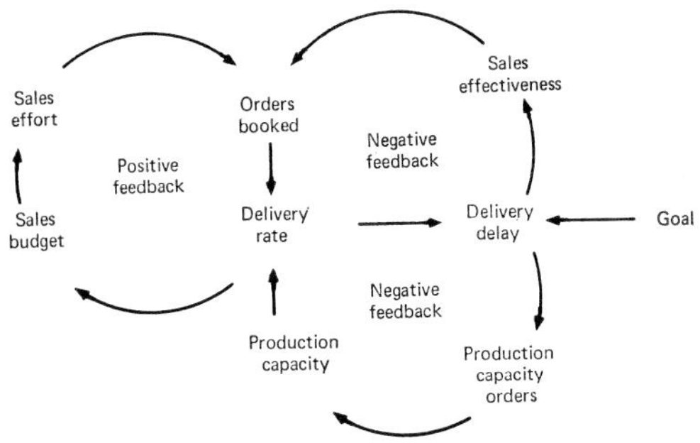

Fig. 7—Three coupled feedback loops

have three interacting loops that can exhibit a variety of behaviour difficult to predict without simulation. The effect of differential delivery delays on products of varying profitability provides a further dimension of complexity.

All we have been able to do is to sketch an outline of certain speculations about system behaviour. Because the actual study has not been done, they have been based on models developed in other situations by Forrester and his associates. In the nature of things, the complexity of system behaviour cannot be described verbally in a few sentences and this has not been tried. Indeed, to succeed would have contradicted the central theme of this paper.

The purpose has been simply to stimulate control engineers and others to widen the boundaries of the feedback systems they study to include factors beyond physical ones. The work will be difficult, but the pay-off may be high. More should make the attempt.

References

1. Forrester, J. W., *Industrial Dynamics* (The M.I.T. Press, Cambridge, Mass., 1961)
2. Roberts, E. B., *The Dynamics of Research and Development* (Harper and Row, New York, 1964)
3. Forrester, J. W., *Urban Dynamics* (The M.I.T. Press, Cambridge, Mass., 1969)
4. Street, P. R. and Dent, J. B., 'Industrial dynamics and an approach to its use in farm management research': *The Farm Economist*, 1968, **11** (8), 345–353
5. Pugh, A. L., III, *DYNAMO Users Manual* (Second edition, The M.I.T. Press, Cambridge, Mass., 1963)
6. Brennen, R. D. and Silberberg, M. Y., 'Continuous systems modelling programs': *IBM Systems J.*, 1967, **6** (4), 242–266
7. Charlton, P. J. and Street, P. R., 'Some general problems involved in the modelling of economic systems on a digital computer' (A.R.C. Symposium on the Use of Models in Agricultural and Biological Research, February 1969)
8. Forrester, J. W., 'Market growth as influenced by capital investment': *Ind. Management Rev.*, 1968, **9** (2), 83–105

Discussion

Dr J. Grant I have had occasion to make a number of market studies, mostly in connection with my work in developing countries, where the problems are rather different from those outlined by Mr Morgan. Usually, there is no existing papermaking industry and one must try to ascertain in advance what sort of paper the one-machine mill that is going up should make from the materials available. In the course of years, I have developed a technique of my own, making use of such elementary statistical expertise as I possess and, in general, it has not worked out too badly. Often, I do this with some trepidation, partly because the information I receive may be unreliable and partly because the installation of just one converting factory can completely upset my deductions. If I decide, for example, on a general purpose mill for making cheap writings and printings, then, if a cement factory starts up, the demand for sack paper could be far in excess of the total other paper and board requirements of the country; a kraft mill would then obviously be the best type of mill to choose. Is this ad hoc approach really the best one in the circumstances or would it be better to try to apply Mr Morgan's more sophisticated methods?

On the kind of long-range forecasts referred to, if they become generally known, they can often trigger off a set of circumstances that make the forecasts quite wrong. A classical example occurred in Europe about eight years ago when a well-known international organisation forecast a great increase in woodpulp consumption over production three or four years later. As a result, the woodpulp mills put in much plant to increase production; when the three or four years had elapsed, there was a surplus instead of a deficit. The market has not really settled down after that episode even now. If long-range forecasts of that kind are given too much publicity, therefore, they can become a menace.

Mr N. B. Morgan I agree with your first comment, for I am all for ad hoc methods. I know nothing about developing countries' economies, but, if you take the very simple situation you referred to, then the kind of methods I have been talking about are probably not very valuable. It is much better merely to ask the consumer how many rough paper sacks he wants.

I am rather against long-range forecasting in economics, except in the very broadest sense. One believes a growth rate for this country is going to be 3 per cent per annum, for example, instead of 10 per cent. Long-range forecasts for small sectors of the economy must be very doubtful. The outcome for a particular industry or plant can be affected by other people, quite ignorant of it, probably quite uninterested in it, operating on some quite different sector of the conomy with entirely different objectives. Thus, the Chancellor of the Exchequer may have nothing against motor cars; nevertheless, with other considerations in mind, he can make tax and other changes that throw all forecasts out of line.

Dr R. A. Holm The previous statement on pulp capacity has a direct bearing on the point that was made about the necessity of studying the detailed dynamics of economic systems and, proceeding in this pattern, I think the second paper pointed the way very clearly. How do we logically lay down what we believe to be the particular pattern of action or reaction of a system? Does Mr Anderton have any particular methods that he can recommend to determine the non-linear coefficients? It is obviously impracticable and perhaps impossible to determine what these functions are by simply looking up all the past results of similar actions. How do you, in fact, get the main factors and their relationships?

Mr R. H. Anderton There is no simple answer to that question. You arrive at a model by doing rather painstaking informal investigations like interviewing, by data collection in the various parts of the company that seem to be important and by slowly trying to build up an understanding of what is happening. There is a certain kind of man who will do nothing until he has all the data and will only then proceed to analyse it. Industrial dynamics takes the position that usually there is enough known in a descriptive, verbal way about the operations of the company to make a first attempt at making a model. (There is often confusion here between precision and accuracy. One can study something precisely—that is, in numerical terms—without knowing about it very accurately.) A preliminary model is then constructed and one performs experiments with the model to find out which parts of the model (gain factors, non-linear relationships, etc.) really matter—that is, to which the model is sensitive. Having worked that out, you can then go back to collecting more data in that sensitive area. One of the most remarkable results of Forrester's work is that the behaviour of very large systems seems to be quite insensitive to the values of very many parameters, even through factors of five. What it is usually very sensitive to is the structure of the system, in particular, the information that is available and being used at the various decision points. Some people tend to say that they know nothing about that particular variable, therefore they leave it out. In fact, by doing so, they are putting it in, giving it a rate of zero—and you can usually do much better than that! You may know it is seven (maybe eleven), but certainly not zero. Usually, it is more important to put all major influences in, even roughly, than to put emphasis in the early stages on an elaborate empirical investigation of the data.

Mr H. B. Carter On my way to this conference, I ran into a trades union official and this very problem came up. He accused the profession that I represent and of which I am proud to be a member of not being able to do anything about the wage/price spiral. I am sure we have enough brains in the profession now to be able to predict the result of any of the various actions and control activities that are embarked on generally by governments. He claimed that we need a deal of statistics and all the facts. I said that all we need are sampling techniques and we can then arrive at all the factors, until we come to a very pertinent point. What happens if, as a result of this control activity, the trade unions are told that the way to stop this is to first of all put a freeze on wages. He answered, 'That would be political suicide; we could never accept that.' So the point of this comment is how do you program such a reaction into an appropriate control when 50 per cent of the labour force who are to be directly affected will reject it anyway?

Mr Morgan First of all, there seems to me to be a reasonable chance that a number of ways of controlling the situation exist—indeed, any situation—and reasonable therefore that, of these ways, some of them will have results worse than the complaint for at least some of the population. Many people regard a wage freeze as being worse than having a stop/go cycle.

Mr Anderton There is a very interesting class of control system in which what is controlled is very complicated and is, in fact, much more complicated than the complexity of the controller. The variety of response available to what is controlled is thus greater than the number of control actions available to the controller. This happens very much in national control systems, when the Chancellor of the Excheuqer introduces a new tax system, for example. Then there may be thousands of people who are trying to find ways of evading tax. There are more of them than there are of him. This raises a fundamental point that can perhaps be tackled only by the mathematical theory of games. It is a whole new class of control system.

Mr A. J. Ward May I return to Mr Carter's question. I think, there was some confusion in his comments. These techniques do not generate control strategies for you, but enable one to demonstrate the dynamic behaviour of the system studied.

On the second part of his question, engineers tend to fight very shy of problems that are not physical and are not easily quantifiable such as marketing, selling, promotion and investment. These are often dominated by dynamic phenomena and control systems engineers have just the right background to formulate and analyse them. Even if the numerical accuracy of the answeer is poor, a system study will often show marked differences in the quality of the system's general behaviour for different varieties of control action. Here is a real opportunity for engineers to make a contribution to the non-technical aspect of the business.

Mr P. E. Wrist I would like to say to Mr Anderton that I think the techniques he is discussing are probably very effective in periods of stability or in periods of continuing expansion. They tend to break down during transition periods. I would like to mention three examples.

One is the departure from normal behaviour during periods of national uncertainty about the future economic climate or a loss of public confidence. During these periods, for example, inventories may either be run down to a very low and dangerous level in anticipating a recession or run very high in anticipating a shortage or a strike. Most of the models of which I have been aware are not adequately able to predict and incorporate these things in time. They can usually analyse what happened after the event, but they often cause misleading projections of demand for the short term and a distortion of demand over the long term. If we followed the predictions, we might conclude that a new expansion of capacity was required immediately only to find a few months later that the assumptions on which it was based have disappeared. Looking back, we can analyse why we were misled, but the model did not prevent us from being misled.

Another factor that can distort demand projections is a changed ordering pattern that happens during times of shortages. We have found that during these periods there is an excessive amount of double and triple ordering. As soon as shortages begin the ease, these duplicate orders vanish overnight and arrears that previously appeared to be as firm vanish in a very short time.

Another effect that makes economic models of limited value is their inability to predict the unusual. Events that happen only once would never be incorporated into the model, because of no previous experience and the probability that they will not happen again. By way of example, if we look at the book industry in U.S.A., 1967 was a year of very rapid growth and forecasts for continued growth were made. Many plans were made on that assumption. The following year, the expansion did not occur and, on investigation, it was found that the government towards the end of the previous fiscal year had a large sum of money left for education and the only thing that people could buy in a short time was school books. As a result, in the summer of 1967, an excess of school books was bought. In 1968, the stock was drawn on and fewer new books were bought. This kind of action will probably not happen again for a long time; when it did, it had a dramatic effect on a system such as you described.

Mr Anderton I agree with most of what you say. No technique can prevent you making mistakes in prediction, but if you have a technique for understanding the mechanism of events, even post hoc, then you will at least not make the same mistake twice. With an increasing understanding of the way the systems are working, I think we will be able to anticipate events more frequently.

THE FUTURE OF CONTROL ENGINEERING

J. F. COALES, O.B.E., M.A., F.I.E.E., F.Inst.P., F.I.E.E.E., F.R.S.
Professor of Engineering, University of Cambridge

Introduction

UP TO a few years ago, all industrial processes were either batch processes or were operated on a steady state basis, although some (like papermills, automatic looms and steel-rolling mills) operated in steady state only for a limited period until the specification or grade of the output had to be changed. Thus, industrial processes can be divided into three classes—

1. Batch processes.
2. Continuous processes.
3. Quasi-batch processes.

Batch processes

In the early days, all processes were batch processes and this meant that they were all time-varying and therefore difficult to control accurately. They were in fact controlled by time schedules based on previous knowledge of the process, but this is in general possible only when both the desired product and the raw materials are closely specified. Any physical process can be represented by the simple box on the left of Fig. 1 in which the raw materials are represented by $\mathbf{v}$, which includes composition, quantity and such other relevant variables as temperature, viscosity and physical dimensions, depending on the nature of the process. These raw materials are operated on by the plant that has certain characteristics such as physical dimensions, temperature and pressure, which appear as coefficients in the set of differential equations, by which they may be represented mathematically.

In addition to the characteristics of the plant, the characteristics such as rates of reaction and catalyst activity of the process must also be known and they also appear as coefficients in the set of differential equations. All these coefficients are denoted by $\mathbf{a}$. The raw materials having been operated on by the plant and process, their composition, temperature, etc. (or in the case of

Under the chairmanship of D. Attwood

mechanical processes, physical dimensions) change with time and these quantities can be represented by **x**, which is called the state variable of the system. Certain of the components of this state variable correspond to those components of the product which appear in its specification designated **z** and the object is to control the process by the application of heat or chemical additives in the case of chemical and metallurgical processes or by moving a tool in metal-working, designated by **u**, to change the state variable so that the relevant components of **x** approximate to the desired output **z** in the shortest possible time. It should be noted that the **u** values, called the controlled variables, always imply the application of energy or material to the process.

All physical processes can be represented by a set of simultaneous differential equations of the form—

$$\dot{\mathbf{x}} = F(\mathbf{x}, \mathbf{a}, \mathbf{u}, \mathbf{v}, \boldsymbol{\upsilon}, t) \qquad . \quad . \quad . \quad . \qquad (1)$$

in which $\boldsymbol{\upsilon}$ represents unwanted disturbances that cannot be measured.

If the **a**, **u** and $\boldsymbol{\upsilon}$ terms are known, then the process is completely defined by equation (*1*) and the **x** can be calculated for any *t*. In the case of a batch process, **v** will be the initial value of **x** at time $t = 0$ (that is, $\mathbf{x}_0$) and the **v** will be functions of time. In general, the material throughout the plant will not be homogeneous and diffusion processes will be involved, resulting in a set of partial differential equations, which makes the derivation of a mathematical model in the form of equation (*1*) extremely difficult, if not impossible. Fortunately, however, continuously variable systems of this type can frequently be broken up into sections that can be represented by lumped parameters, but this inevitably increases the number of state variables **x** and plant coefficients **a**. At the present time, it is impossible to solve large sets of partial differential equations involving time even with the largest computers available and so from now on we shall consider only sets of ordinary differential equations and assume that any distributed parameter system in which we are interested can be approximated by a lumped parameter system.

It will now be apparent that, in order to model a batch process, it is necessary to know how the state variable **x** changes with time for different values of **u** over the whole range of **x** met in practice. If the **u**, **v** and **x** values can be measured as functions of time, then it is possible to hillclimb on the coefficients **a** and so determine them. The coefficients **a** will then truly represent the plant characteristics. Unfortunately, two difficulties usually arise. Firstly, there are unwanted disturbances, some of which may arise from the ambient conditions that cannot be measured and that are represented by $\boldsymbol{\upsilon}$ in both Fig. 1 and equation (*1*). Thus, the values of **a** that can be obtained by regression analysis or hillclimbing can be no more than estimated values designated by $\hat{\mathbf{a}}$. Secondly, it is not usually possible to measure all the compoents of **x** required

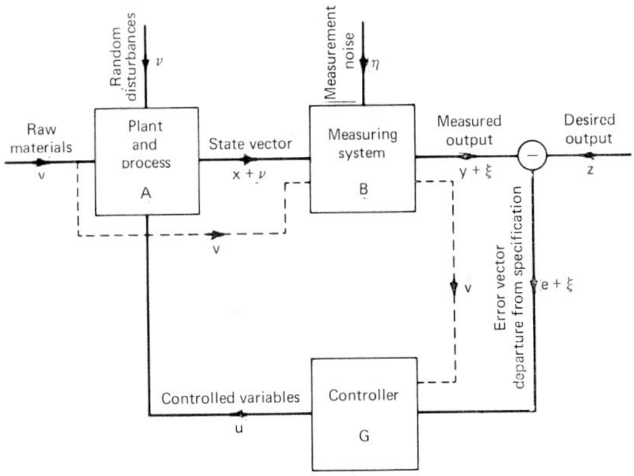

Fig. 1

to represent the state of the plant and process by a set of first-order differential equations of the form of equation (*1*) and, in any case, they cannot be obtained directly, but only by means of measuring instruments. These have to be included in the system and are represented by the box on the right of Fig. 1. This means that we must write—

$$\mathbf{y} = B(\mathbf{x}+\boldsymbol{\upsilon})+\boldsymbol{\eta} \qquad (2)$$

where $\mathbf{y}$ is the measured values of those components of $\mathbf{x}$ that can be measured and B is called the measurement matrix. All measurements are to a greater or lesser extent noisy and so we have to add a noise vector $\boldsymbol{\eta}$ in equation (*2*). By combining equations (*1*) and (*2*), we get—

$$\dot{\mathbf{y}} = \varphi(\mathbf{y},\mathbf{a},\mathbf{u},\mathbf{v},\boldsymbol{\upsilon},\boldsymbol{\eta},t) \qquad (3)$$

from which theoretically the estimated values $\hat{\mathbf{a}}$ of the plant parameters can be determined. This rather formidable equation can fortunately very often be simplified, because the noise vectors $\boldsymbol{\upsilon}$ and $\boldsymbol{\eta}$ are uncorrelated with the other variables. In this case, equation (*3*) can be rewritten in the form of equation (*4*)—

$$\dot{\mathbf{y}} = \varphi(\mathbf{y},\mathbf{a},\mathbf{u},\boldsymbol{\upsilon},t)+\xi \qquad (4)$$

where ξ is the combined noise vector.

When a number of the components of $\mathbf{x}$ cannot be measured, however, equation (*4*) is not a true representation of the system and there are insufficient equations to include all the values of $\mathbf{a}$. In this case, we have to start with a structural model of the system in form of a set of higher order differential equations of the form of—

$$\mathbf{y}^n = f(\mathbf{y},\mathbf{y}^1,\ldots,\mathbf{y}^{n-1},\boldsymbol{\alpha},\mathbf{u},\mathbf{v},t)+\xi \qquad (5)$$

in which $\mathbf{y}^1 \ldots \mathbf{y}^n$ are the first and higher derivatives of the $\mathbf{y}$ vector and the $\boldsymbol{\alpha}$ values are the plant parameters—that is, the coefficients (some representing time constants) and indices, if the system is non-linear, appearing in the equations. The $\mathbf{y}$ to $\mathbf{y}^n$ being measurable, it is now possible to determine $\hat{\boldsymbol{\alpha}}$, the estimated values of these coefficients and/or indices and so an approximate structural mathematical model is obtained. In general, this will be a set of non-linear equations, but most industrial processes are fortunately operated over a narrow range of variables and the equations can usually be piecewise linearised with a small number of linear zones. In this case, the structural model can be simplified to the normal* linear form of equation (*6*), where A' is the plant matrix and C' is the control matrix—

$$\dot{\mathbf{y}} = A'(\mathbf{y},t)+C'(\mathbf{u},t), \quad \mathbf{y}_0 = \mathbf{v} \qquad (6)$$

Theoretically, it is now possible by the calculus of variations to calculate the u in terms of the error matrix, equations (*7*) and (*8*), which will bring $y(t)$ to the desired value z with minimum expenditure of energy or any other criterion J on which it may be required to optimise the system. The criterion J is frequently called a *cost function*, since it should include, besides the measure $g_1(\mathbf{e})$ of the divergences of the output product from the desired specification, the cost of applying the controls $g_2(\mathbf{u})$ and the fixed costs such as labour and capital charges $g_3(\mathbf{w})$. The form of the cost function J will be as given in equation (*9*), in which the kernel function is usually made up of the algebraic functions $g_1(\mathbf{e})$, $g_2(\mathbf{u})$ and $g_3(\mathbf{w})$. By minimising J, the $\mathbf{u}$ are obtained as functions of time as given in equation (*9*)—

$$\mathbf{u} = G(\mathbf{e},\mathbf{v},t) \qquad (7)$$

$$\mathbf{e} = \mathbf{z}-E\mathbf{y} \qquad (8)$$

$$J = \underset{T\to\infty}{\text{Limit}} \int_0^T [g_1(\mathbf{e})+g_2(\mathbf{u})+g_3(\mathbf{w})]dt \qquad (9)$$

In the case of a batch process, equation (*6*) must be integrated to give equation (*10*), which completely defines it.

$$\mathbf{y} = H(\mathbf{a}, \mathbf{v}, \mathbf{u}, t), \qquad \mathbf{y}_o = \mathbf{v}, \ \mathbf{y}_T = \mathbf{z}/E \qquad (10)$$

where $t = 0$ and $t = T$ are taken to be the times at the start and completion of the process, respectively.

Continuous processer

With the growth of the chemical and oil industries, the need to process large quantities of fluids led to the development of continuous processes that proved much easier to control than batch processes. The great advantage of

* The term *normal form* is used for a set of first-order differential equations

continuous processes is that, because the product is being removed and raw materials are being added continuously, the rates of reaction remain constant and so the process can be operated in steady state and is therefore time invariant, provided the plant itself does not change and there are no changes in ambient conditions to affect the process. Time can be dropped from equation (*1*) and it can be represented by a set of differential equations of the form of equation (*11*)—

$$\dot{\mathbf{x}} = F(\mathbf{x}, \mathbf{a}, \mathbf{u}, \mathbf{v}, \upsilon) \quad . \quad . \quad . \quad . \quad (11)$$

Furthermore, provided the plant can be approximated by a number of sections with lumped parameters, no partial derivatives will be involved and equation (*11*) will be a set of simultaneous ordinary differential equations. This makes the process very much easier to control and it will be seen that, provided the raw materials do not change, the $\mathbf{v}$ are constant and so when the plant has been run up to the state when the relevant components of $\mathbf{x}$ are equal to the desired values $\mathbf{z}$, the control variables $\mathbf{u}$ should be kept constant, provided that $\mathbf{a}$ is also constant. Thus, provided there is no change in the plant characteristics, due to wear and tear or changes in catalyst activity or in the composition and flow of raw materials, there is no control problem while the plant is running in steady state.

Unfortunately, it is often impossible to maintain the composition and flow of raw materials exactly constant and some adjustment of the controlled variables is necessary, but the required variations are usually small. This means that we have to retain the $\mathbf{u}$ in equation (*11*). Changes in the plant characteristics due to wear and tear (for example, fouling up of pipes and valves) does occur and changes in the ambient conditions may also cause unexpected fluctuations in the operation of the plant. The unwanted fluctuations and measurement noise may be represented as before by the addition of a random variable ξ as indicated in equation (*12*).

$$\dot{\mathbf{x}} = F(\mathbf{x}, \mathbf{a}, \mathbf{u}, \mathbf{v},) + \xi \quad . \quad . \quad . \quad . \quad (12)$$

Thus, industrial plants have traditionally been designed for the manufacture of closely specified products from closely specified raw materials. On the basis of laboratory and pilot plant experiments, the plant has been designed and manufactured, then it has been set up and maintained to do just this. Under these conditions, if the product fails to meet the specification, no attempt is made to correct this automatically nor can the product be modified, except by modifying the plant or setting it up afresh. If there were no wear and tear of the plant and if the raw materials were always closely within the specification, there would be no need for any feedback control at all, provided the plant was unaffected by external conditions. There are always uncertainties in the operation of the plant, however; raw materials do change from time to time and these result, if they go uncorrected, in changes of the product specification and therefore of spoilt work.

It is to overcome these difficulties that feedback control is used, first of all to maintain such plant parameters as temperature, pressure and fluid flow constant and, secondly, to enable them to be changed to new values should there be changes in raw materials or in the external conditions under which the plant is operating. If the physics and chemistry of the manufacturing processes were fully understood and if the essential physical and chemical properties of the raw materials could all be measured continuously, there would be no need for feedback control, provided the new operating conditions required could be calculated quickly enough to correct for the changes in raw materials. Before the days of automation, this is exactly what the craftsmen used to do. By long experience, the turner knew that, if he got an unusually hard bit of steel, he had to turn it at a lower rate and with a smaller cut and so he adjusted his turning accordingly. There was some feedback, of course, because he had no instruments as a rule to measure the hardness of the steel and it was only by observation of the way it cut that he knew that it was harder than usual.

Similarly (apart from inherently unstable processes), in the control of industrial processes, it is due to ignorance of the characteristics of the materials and the physics of the processes that negative feedback is used to give adequate control. In most processes, it is relatively easy to discover by empirical observations over a long period the effect on the finished product of small changes of temperature and pressure on the one hand or of changes of cutting speed and depth of cut on the other. Thus, by observation of the quality of the finished product, adjustments can be made to correct for small changes in the raw materials or, indeed, of tool or plant wear and tear, even when relatively little is known about the physics and chemistry of the processes. This may be considered the first step in quality control and has resulted over the last 10–20 years in increasing demand for what are known as quality control instruments.

The first step is always to set up local control loops to keep the operating conditions constant, then to add quality control instruments, observations of which are used to alter the operating conditions over a small range to correct for changes of plant parameters and raw materials. Fortunately, these changes have usually been small and relatively slow and it is for this reason that the three-term controller has proved so successful. A change in raw materials implies, of course, a change of the $\mathbf{v}$ in equation (*11*) and changes in the plant parameters a change in the $\mathbf{a}$ of equation (*11*) so that the $\mathbf{x}$ will no longer correspond with the desired output $\mathbf{z}$. Thus, the controlled variables $\mathbf{v}$, the set points in conventional process control, have to be changed in order to achieve a new and satisfactory steady state.

In conventional control, no account is taken of the route by which the plant is brought to the new operating condition nor of the amount of spoilt work that results during the change from one steady state to another. The controllers are set up in such a way, however, that the new steady state is reached in the shortest time possible without appreciable overshoot. With the advent of on-line digital computers, the operation of the three-term controllers can be replaced by calculations carried out in the digital computer as has been done in the early applications of direct digital control. The use of a computer enables the best operating conditions (that is, the set points) for different raw material compositions and plant characteristics to be computed at regular intervals so that, when materials or plant characteristics change, the set points can be corrected automatically. In general, this is the best that can be done with direct digital control until a dynamic mathematical model of the plant and process in the form of equation (*11*) has been developed.

Direct digital control

Many continuous unit processes are now provided with data-loggers by means of which the operating conditions can be recorded and stored and the three-term controllers have been replaced by algorithms in a digital computer program. For many of these processes, very simple algebraic models of the form of equation can be developed—

$$\mathbf{x} = f(\mathbf{a}, \mathbf{u}) \qquad . \quad . \quad . \quad . \quad . \quad (13)$$

in which, because it is a continuous steady-state system, the $\mathbf{v}$ are included in the state variable $\mathbf{x}$.

We can rewrite this—

$$\mathbf{y} = f(\mathbf{a}, \mathbf{u}, \mathbf{v}) \qquad . \quad . \quad . \quad . \quad . \quad (14)$$

where $\mathbf{y}$ includes the output quantities that are specified, if the functions f are known and, during a specified period of time, the $\mathbf{u}$, $\mathbf{v}$ and $\mathbf{y}$ are measured and stored, the plant characteristics $\mathbf{a}$ can be calculated by hillclimbing or regression analysis. Then, assuming the $\mathbf{a}$ and $\mathbf{v}$ remain constant during the ensuing period, the $\mathbf{u}$ for optimum operation in accordance with some predetermined criterion can be calculated. The $\mathbf{u}$ are then taken to be the required set points for the ensuing period and the plant is brought to the new steady state by means of the controller algorithms. During this ensuing period, the $\mathbf{u}$, $\mathbf{v}$ and $\mathbf{y}$ are again recorded and stored so that the $\mathbf{a}$ can be recalculated at the end of this period to take account of any changes in the plant characteristics and new values for $\mathbf{u}$, the required set points for the next period, can be calculated and applied.

In this way, the set points are updated at the end of any period in which there is a change of the raw materials; alternatively, small changes of the desired output specification can be accommodated. The system therefore operates as a feedback control system, although only an algebraic model is used. Furthermore, the system is truly adaptive in that it takes account of changes in plant parameters and so it is comparable with a time-varying dynamic system, provided the sampling periods are short compared with rate of change of plant parameters, raw materials, etc. Further non-linearities in the plant model present no special difficulties.

This simple system has limitations, of course, the most important being that the route by which the plant goes from one state to the next is predetermined by the three-term controller algorithms and will not be optimum. Provided the required changes are small and/or infrequent, however, this is of little importance; in any case, the computer will almost certainly give better steady state operation than a human operator. Another limitation arises from the difficulty of carrying out hillclimbing proceedings with a large number of variables even with very powerful computers. At the present time, the maximum number of components of $\mathbf{u}$, $\mathbf{v}$ and $\mathbf{y}$ that could be accommodated if the $\mathbf{a}$ are to be calculated in reasonable time is 20–30, depending on the form of the equations; with computers rapidly becoming smaller, more powerful and cheaper, this limitation should disappear in a few years.

If the $\mathbf{a}$ cannot be calculated sufficiently quickly with the available computer, they can be assumed constant over longer periods, in which case calculation of the $\mathbf{u}$ (which usually has relatively fewer components) for the new values of the $\mathbf{v}$ and/or $\mathbf{y}$ is quite simple, since no hillclimbing or regression analysis is required. This simple method of computer control has considerable advantages when unwanted random fluctuations and/or measurement noise are severe, because the mean values of the variables can be taken over each recording period. Provided the noise is of higher frequency than that of the changes in raw materials or plant parameters, it is only necessary to choose the longest recording period compatible with adequate control, an exchange relationship that is always present in noisy systems.

When the noise spectrum is such that simple averaging does not give adequate rejection, more sophisticated filtering techniques must be used. Statistical methods are particularly well developed for determining algebraic models from noisy data, so noisy data is unlikely to be a serious limitation in the application of this simple DDC to continuous processes. If a dynamic model is required (as it will be, if optimum control of the plant is wanted while going from one steady state to another), parameter estimation in the presence of noise is far more difficult. The optimisation procedure for linear filtering and prediction developed by Wiener has been extended to multi-variable systems by Kalman. If the system is linear, these methods can be applied fairly easily, using a digital computer for up to about 10 variables; for more complex systems, very little can be done at the present time.

Quasi-batch processes

Other than in oil refineries and large chemical works, there are very few continuous processes that operate in steady state producing the same product from the same raw materials year in, year out. The majority of industrial products are still manufactured by batch processes. For some industries such as the motor industry, an attempt has been made, when production quantities are large enough, to achieve steady state conditions by the introduction of transfer machines and flow production. In these cases, there are a large number of parallel operations producing intermediate products, which are then subjected to a second set of operations producing another set of intermediate products and so on until the final product is complete as indicated in Fig. 2.

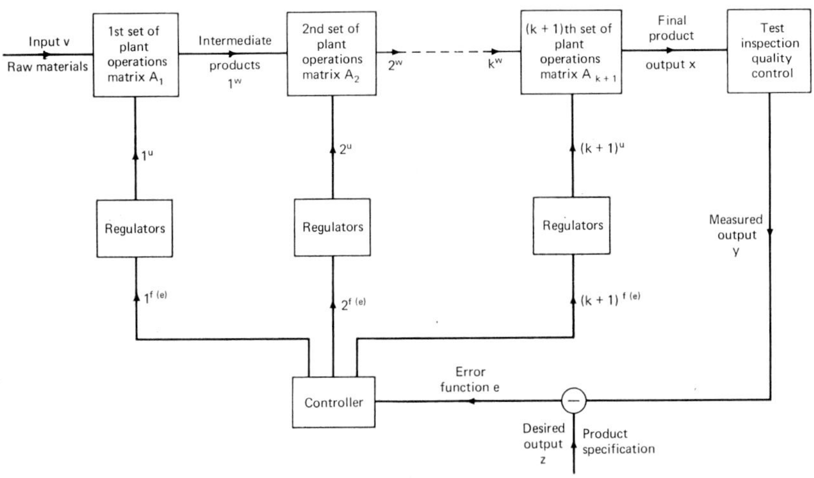

Fig. 2

This is exactly analogous to a chemical plant in which a large number of reactions go on in one unit operation (say, a distillation column), the products from which go to other unit operations for further processing.

There are other industries such as papermaking, glassmaking, steel and textiles in which the plant may operate in steady state for considerable periods while the specification is kept constant, but, because the demand for a particular grade of paper or type of cloth is limited, the specification of the output product from the same machine may have to be changed several times a day. In other cases, although the process is inherently continuous, the plant may for practical reasons have to be frequently stopped and restarted. This applies in a steel rolling mill, where the size of the slab rolled is limited by the size of the coiler. In this case, the ingot has to be rolled and cut into slabs of a certain size to match the coiler. Since the beginning and end of each slab is a gross discontinuity, it is effectively a batch process, although, for the middle 95 per cent of each slab, the mill is operating in steady state.

Similarly, a papermill, when it has been set up correctly to manufacture a particular grade, is in steady state operation and can be accurately represented by equation (*12*). When there is a change of grade, the mill must be brought to a new state in which the **x** must have different values and so different controlled variables **u** are required. In general, there is little difficulty in controlling the mill in the steady state condition, provided the raw materials can be held constant. As is well known, when a change to a new grade is required, it is difficult to achieve the new steady state in a reasonable time and without a great deal of spoiled product. It is evident that the greatest immediate savings under these conditions can be made not by improving the control of the steady state conditions, which are already relatively satisfactory, but by optimising the 'route' by which the plant is taken from one steady state to the new required steady state so as to minimise the amount of spoilt products.

This, therefore, becomes a start-up and shutdown (as distinct from a continuous control) problem and, over the period of grade change in a papermill or at the nose and tail of each slab in a steel mill, it must be considered a batch process. It is for this reason that I have called these processes quasi-batch processes. Under these conditions, we must revert to a mathematical model of the form of equation (*3*) and we shall find that, although the plant characteristics may not change with time, the **v** values are changing and so time must be retained as a dependent variable in the equations. It is also usually found that the **a** values are in the form of partial derivatives with respect to the initial conditions **v** and this is a further reason for retaining time as a dependent variable. Thus, in order to determine an optimum control strategy, the integrated form of equation (*10*) will have to be used.

Industrial processes

There are very few finished products that are produced from the raw materials by one piece of plant or unit operation. More often than not, a number of intermediate products or components are produced by a number of operations in series and parallel, then assembled together into a finished product as indicated in Fig. 2. Such is the situation in a steel works, the main processes of which are shown in Fig. 3. Even in a papermill, there are a large number of processes in series before the final product is ready for dispatch. Each of these processes can be represented by a mathematical model of the type indicated in equation (*10*), but it is found that the number of variables in each identifiable process is 30 or more. Thus, if all the models are combined to make a complete dynamic model of the whole process, the number of variables may well be several hundred. This means that there is no possibility at the present time of developing an optimum control strategy for a complete factory.

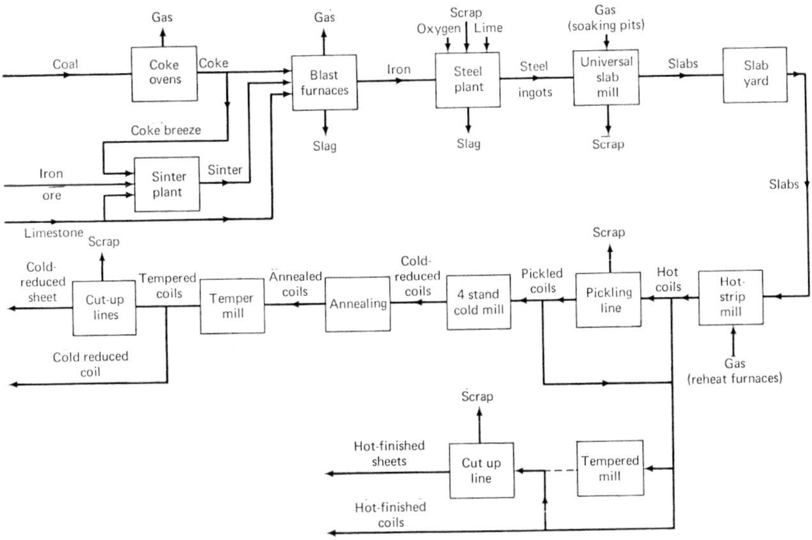

Fig. 3—Main processes of steel works

The present position

Unit operations

There is no doubt that, given time and adequate instruments to make the necessary measurements, also large enough computing facilities, a mathematical model can now be obtained for any industrial manufacturing process or unit operation. Furthermore, given a mathematical model of a process, theoretically, an optimum strategy for the control of the process can be developed by the calculus of variations (such as Pontryagin's maximum principle) for deterministic inputs. That this has been done in only very few cases is accounted for by the difficulties encountered. It has been found that, even when industrial processes are divided up into the smallest possible unit operations, these unit operations have 30 or 40 variables, the **u**, **v** and **x** of equation (*1*)—and, as mentioned above, the **x** vector includes intermediate variables that more often than not cannot be measured.

At the present time, most industrial plants are not sufficiently well instrumented for the data to be obtained and, in many cases, sufficiently accurate instruments have not yet been developed. When plants are sufficiently well instrumented, observations must be made over a considerable period and, at least, thousands of pieces of data will be required, so that some form of data recording equipment is desirable. When the data are available, a structural model of the type described above can be tested and, if it proves to be correct, the **a** of equation (*1*) can be calculated, provided a fairly large computer is available. Unfortunately, the structural model is seldom correct in every detail and modifications have to be made so that the development of an adequate model is usually both tedious and costly. I estimate that the cost of getting a model for the simplest process with about 30 variables is at least £10 000. Although the number of different processes being used in industry must run into tens of thousands, I doubt if one hundred of them have yet been modelled satisfactorily.

The development of a dynamic model is only the first step towards optimum or even improved control, because—even for deterministic systems—by use of the calculus of variations, it is possible to solve the two end-point boundary problem for a linear system only with about 10 or at most 20 variables and, if it is non-linear, it is far more difficult. The number of variables must therefore be reduced to those that are significant for control, which it must be hoped will not exceed about 10.

It is for this reason that linearisation of the model becomes important, since equation (*1*) can then be put in the matrix form for equation (*5*) and methods have been developed for reducing the plant matrix A. As mentioned above, continuous processes, because they work over a very large range, can usually be linearised, but batch processes prove to be much more difficult and approximations and changes of variables must usually be made in order to reduce the number of non-linearities. After development of the structural model, this is the next stage in the process. When the model has been linearised and reduced to a reasonable number of variables, it must again be tested, but this can be done of course with the data already collected, provided this covers the full expected range of operation of the plant. The next step is to calculate the optimum control strategy for some appropriate deterministic change in the operation of the plant such as small step changes in the controlled variables or in the composition of the raw materials according to some predetermined criterion. This strategy must then be tested either on the actual plant or on a computer using the original structural model to replace the plant. If a large hybrid computer is available, this can be done fairly quickly; if only a digital computer is available, it will have to be large and the computing time may be considerable.

Such a control strategy should ensure that within the expected range of operation the plant will be controlled in the best possible manner to take account of small variations in the raw materials and some variation in the desired specification of the product can be accommodated, but, theoretically, it will not correct for changes in the plant parameters or ambient conditions. In order to correct for these changes, some form of adaptive control will be required. Methods for doing this have been devised and used successfully in aerospace applications for which well-defined mathematical models of the systems are available, although they have not as yet been applied to industrial processes except as described in the section on DDC. Any adaptive control system almost certainly requires a model reference of some form, which, for practical reasons, should not be too large, so the development and reduction of a dynamic model is almost certainly essential if adaptive control is required.

No account has yet been taken of unwanted fluctuations in the operation of the plant or of measurement noise. Fortunately, in many processes, these random fluctuations are small compared with the accuracy to which the plant must be controlled. If they are not, deterministic control is all that can be achieved by the use of the calculus of variations, which will be certainly sub-optimum and may be far from satisfactory. If sufficient data are available, the effectiveness of the control strategy can of course be tested and sensitivity analysis can be used to predict the effect of unwanted disturbances, but more efficient control can be achieved only by the use of dynamic programming or special filtering techniques, which greatly add to the complication and the amount of computing equipment required if these methods are to be used for on-line control.

The control of complete factories

There are very few factories that produce a single product day in, day out and these are seldom, if ever, made up of few enough unit processes for the development of a dynamic model of the whole factory to be possible. As described above, individual processes, if given sufficient time, can be modelled and near-optimum control strategies can probably be developed. Even if all the processes have been modelled and can be combined, however, the number of variables will almost certainly be hundreds (if not thousands) and there will be no possibility of reducing this to the sort of number, probably fewer than 20, for which a reasonable dynamic control strategy could be developed. If the factory were producing the same product year in, year out (as does an ammonia plant), then it might be possible to develop a satisfactory control strategy to take account of changing raw materials, changing ambient conditions and wear and tear of the plant, provided a fairly large digital computer was available on-line. This is the exception, however, rather than the rule and most factories have to cater for a production programme that includes products of several different grades, colours and physical dimensions. Under these conditions, although the range of product specification may be relatively small, at least some of the processes will be in the quasi-batch category and the dynamic control of the whole factory will be out of the question. Under these conditions, the only possibility for control of the factory as a whole is either to take customers' orders and collate them or to make a future sales forecast on the basis of prediction from previous sales.

The prediction of future sales is just as much a control problem as the modelling of unit operations and there are well-established methods of doing this, but it should be pointed out that this prediction may depend on what the economists call exogenous variables arising from changing economic, climatic or demographic conditions. The prediction of these exogenous variables in turn depend on a knowledge of the behaviour of the economic, climatic and demographic systems in which the factory operates and so, as more sophisticated control of individual factories is required, the control or systems engineer is continuously forced to consider bigger and bigger systems. Assuming that a reasonable forecast of the desired output of the factory can be made, it is possible by operations research techniques (such as linear programming) to determine the best production schedule for the unit processes and this can be updated from day to day or at whatever intervals are required.

The production schedules for the individual processes having been determined, it should now be possible to use for their control dynamic strategies developed in the way described in earlier sections. If all the processes behaved exactly according to plan and the raw materials were always exactly according to specification, there would be nothing more to be done. In practice, of course, the outputs from the unit processes will not be exactly according to schedule in quantity or quality and the discrepancies must be fed back so that the future production schedule can be corrected. Changes in raw materials and faulty operation of early processes result in input changes to the later processes, which need to be optimised as far as possible to correct these discrepancies. Thus, a hierarchical system of control is required with information flowing up the line and orders flowing down. Fig. 4 shows such a system in operation at a steelworks.

Future developments

THERE are seven distinguishable areas in which active research and development is required—

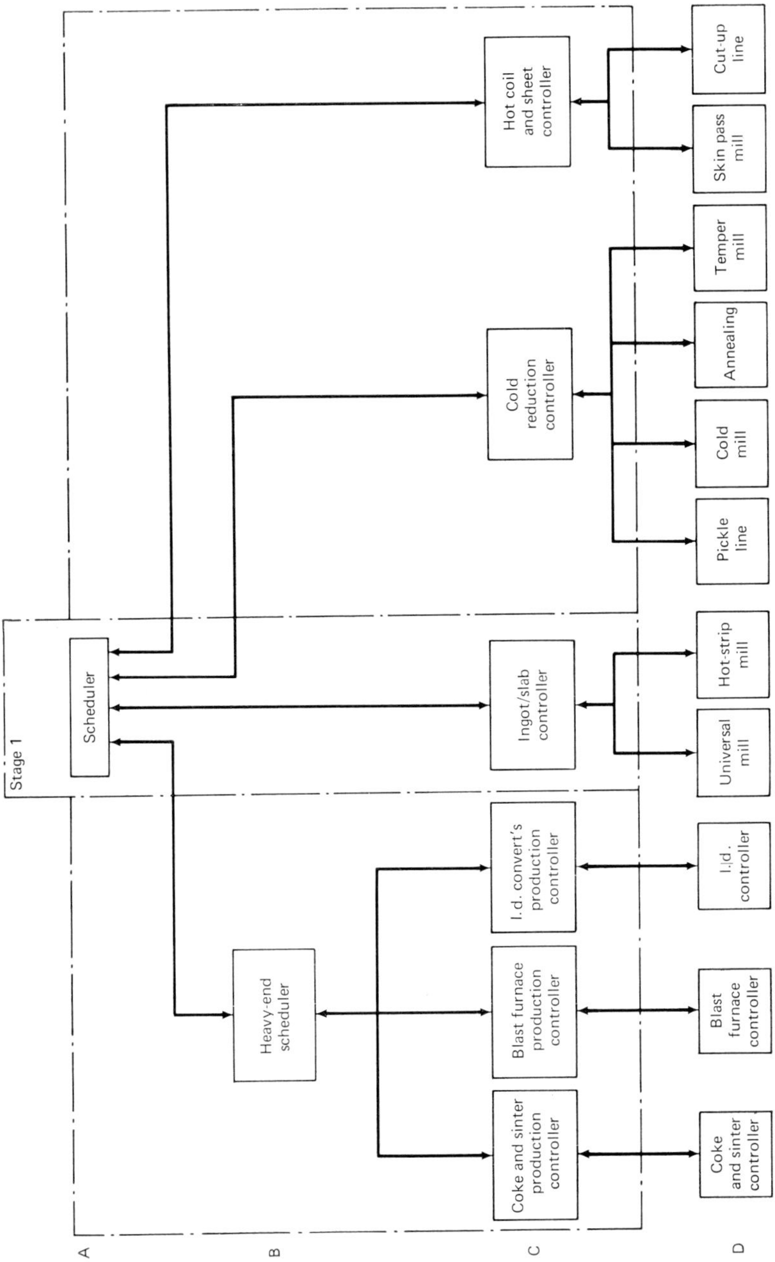

Fig. 4—Interconnections between computer systems—

Level A Plant scheduling B Plant optimisation
C Production control D Process data logging plus process control

1. Improved adaptive control of continuous processes.
2. Improved dynamic control of batch and quasi-batch processes.
3. Computer control of start-up and shutdown of complex systems.
4. The development of hierarchical computer control systems for complete factories.
5. The modelling of very large systems—for example, national economies, meteorological and demographic systems, transport and communications systems and large business and industrial groups.
6. The development of stochastic control systems.
7. Approximation of non-linearities and development of sub-optimum control strategies.

Continuous processes

So long as the range of products required from a single process is restricted and raw materials can be reasonably well controlled, it is probable that direct digital control of steady state is all that will be required. The urgent need is to improve our hillclimbing techniques for parameter estimation in systems that have a very large number of variables or cannot be linearised, so that adaptive control as described in the DDC section becomes possible.

Batch and quasi-batch processes

There can be little doubt that during the next few years the biggest economic gains will come not from improvement in steady state control of continuous processes (which is usually within 1 or 2 per cent of optimum anyway), but by increasing the throughput of batch processes and reducing the amount of spoilt product, when changing grade and at other discontinuities, in quasi-batch processes. As described in the sections on batch and quasi-batch processes, any major improvement almost certainly demands a time-varying dynamic mathematical model of the process and very little has been done as yet in this area.

Five steps are necessary in the development of optimum or near-optimum control of a batch or quasi-batch process—

1. Collection and recording of sufficient data.
2. Development and proving of an accurate structural model.
3. Simplification of this model by finding suitable approximations to allow it to be piecewise linearised and the number of variables reduced.
4. Development of a near-optimum control strategy.
5. Development and test, using the accurate model (*2* above), of the control algorithm.

At the present time, not one of these steps can be carried out for any but relatively simple systems and new and improved techniques will have to be developed.

Start-up and shutdown of complex systems

There are many very big systems such as electrical power generating plants, sugar extraction plants and chemical plants, in which for economic reasons the time taken to get into full production after a shutdown should be as short as possible, but which, for safety or other reasons, has to be carried out in accordance with a carefully worked out schedule—for example, the launch, landing on the moon and return of Apollo 11.

In an electrical generating system, the boilers and generators must be run up in such a way that components are not subjected to thermal shock and no dangerous temperature differentials arise. This may require nearly 10 000 separate commands during the start-up period and can be done in the shortest time only by means of a computer monitoring the system and issuing the commands in accordance with a predetermined program. This is a scheduling problem closely allied to optimum control, which must be considered as coming within the sphere of control or systems engineering.

In the case of a sugar factory, it takes about a week at the start of a campaign to get it into full production and a reduction of a day in this time would save many thousands of pounds. This problem should be treated as a quasi-batch process, but there are far too many variables (more than 200) for this to be possible, so the best compromise is probably to treat it as a problem of scheduling a number of unit processes as described in the next section.

Hierarchical systems

As explained in the section on factory control, a complete factory is far too complex for total dynamic control to be possible. Hierarchical systems have to be used with two or more levels of control. At present, little is known about the behaviour of multi-level systems and stability may be a problem. Relatively simple systems are in operation and research is going on with a view to a better understanding of hierarchical systems, finding principles for designing them and for comparing their performance, but very much more needs to be done.

Very large systems

As mentioned in the section on the control of complete factories, the optimum control of a whole factory inevitably depends on knowledge of a number of exogenous variables, which almost certainly have to be predicted from a knowledge of the environment in which the factory operates. A study of environmental systems such as the national economy therefore becomes necessary. Besides, prediction and control of public utilities such as electric power generation and distribution, gas distribution, transport, communications are essential to the efficient running of a community and, when these have been nationalised or are carried out on a national scale, they become very large systems indeed. The behaviour, design and control of these large systems are now being actively studied under the title of *systems engineering*, but it is a control problem, albeit a complicated and difficult one. All the techniques developed for the control of smaller systems are likely to be required and control will almost certainly be hierarchical.

Stochastic systems

As more accurate control of an industrial process is required, either to improve its efficiency or to make it conform better in a hierarchical system, the effect of measurement noise and unwanted fluctuations becomes more marked. If the system is non-linear, there comes a time when it must be treated as a stochastic system. The analysis of multi-variable linear stochastic systems is difficult enough: when they are non-linear, the mathematical techniques required for their analysis have not yet been developed.

At the present time, the best that can be done to control them efficiently is by the use of dynamic programming. Effectively, this consists of estimating the present state of the system and, by integrating forward along a number of controlled paths to the required end point and by comparison, finding the path that is optimum in some predetermined sense. Since many of the variables are noisy or subjected to random variations, only the most probable (expected) optimum path can be determined, but this is the best that can be done under the circumstances. Such a procedure requires a vast amount of computation and storage of information, so very large computers are necessary; even then, it can be done in reasonable time only for a few variables. Whereas for deterministic systems the time taken to compute the control variables goes up in proportion to the number of state variables, for dynamic programming of stochastic systems, the time required for computation goes up as the square of the number of state variables.

Rapid advances in computer technology are making very much more powerful computers both available and economically justifiable for on-line control of industrial processes; even so, dynamic programming is unlikely to be feasible for most unit processes until new methods have been developed.

Approximat on of non-linearities and development of sub-optimum control strategies

In view of the difficulties to be overcome before truly optimum control can be applied in an on-line adaptive form to any but the simplest processes, probably the most valuable results will come during the next few years from the intelligent use of approximate linear models and sub-optimum control strategies, of which predictive control is an example. When a process has been modelled, it is often possible to find new variables by transformation and combination of the measured variables; these result in approximately linear relationships, which in turn allow the model to be put into the normal form of equation (*6*). The successful linearisation of the structural model often calls for considerable ingenuity and may require different variables to be measured; if the necessary instruments are available or can be developed, it is likely to be worth the effort. Since the first step must be to develop a structural model of the process and the development of new instruments may be required, the most urgent requirement is for the development of structural models of many more unit processes.

Conclusions

DIRECT digital control can now give reasonable adaptive steady state control for continuous processes, but it is in better control of batch processes and of grade changes and unavoidable discontinuities in quasi-batch processes that the greatest gains are likely to be made during the next few years. The most urgent need is for the development of mathematical models for many more manufacturing processes so that dynamic control can be applied to batch and quasi-batch processes and adaptive DDC to continuous processes. In parallel with this, hierarchical control needs to be developed for the better control of complexes of unit processes and complete factories and the study of very large systems should continue. These is little hope of optimum dynamic control of whole factories or of any but the simplest stochastic systems in the foreseeable future, but research into the analysis of stochastic systems is of great importance.

List of symbols

$\mathbf{a}$	Plant and process characteristics: the coefficients, indices, time constants, etc. appearing in the true mathematical model
$\hat{\mathbf{a}}$	Estimated values of $\mathbf{a}$
$\boldsymbol{\alpha}$	The coefficients, time constants, indices, etc. in the true structural model of the system
$\hat{\boldsymbol{\alpha}}$	The estimated values of $\boldsymbol{\alpha}$
$\hat{A}$	The estimated plant matrix when the model is in the normal form of a set of ordinary linear first-order differential equations
$\hat{B}$	The measurement matrix
C	The control matrix
E	The error matrix
F	The function representing the physical structure of the process as a set of first-order differential equations

f	The function representing the physical structure of the process in terms of measurable quantities
g_1, g_2, g_3	The kernel functions of the performance criterion
G	The control function
J	The performance criterion that is to be minimised
$\boldsymbol{\upsilon}$	Unmeasurable process fluctuations
ξ	Combined noise $\xi - \mathbf{v} + \boldsymbol{\eta}$
$\boldsymbol{\eta}$	Measurement noise
t	Elapsed time
T	The time of completion of a batch process
$\mathbf{u}$	The vector representing the controlled variables—that is, movement of tools, flow of steam to heat exchangers, fuel, chemical additives, etc. that are used to control the process
$\mathbf{v}$	The vector representing the composition, quantity, etc. of the raw materials that are not continuously controllable nor used to control the process
$\mathbf{w}$	Fixed costs unaffected by the mode of operation of the plant, such as labour costs and capital charges, which have to be included in the cost function
$\mathbf{x}$	The state variable of the plant and process—that is, the vector representing the state of the plant and the materials in it at time t
$\dot{\mathbf{x}}$	The first derivative of $\mathbf{x}$ with respect to time
$\mathbf{y}$	The measured values of the state variable $\mathbf{x}$
$\dot{\mathbf{y}}$ or y^1	The first derivative of $\mathbf{y}$ with respect to time
$\mathbf{y}^n$	The nth derivative of $\mathbf{y}$ with respect to time, $d^n y/dt^n$
$\mathbf{z}$	The vector representing the desired output of the plant—that is, the specification of the product plus such other variables as the quantity required

THE FUTURE OF PROCESS CONTROL IN THE PAPER INDUSTRY

P. E. WRIST, B.A., M.A., M.Sc., The Mead Corporation, Dayton, Ohio, U.S.A.

History and philosophy of control systems

THE past decade has witnessed a major advance in the use of process control systems by the paper industry. Before that time, very few variables were controlled automatically, many instruments were either indicating or else, if controlling, were operated in the manual mode much of the time and the degree of instrument reliability and accuracy in the average mill did not promote operator confidence or use. Pneumatic controllers performing simple control loop strategies were the norm. Making paper was an art and frequently different papermakers would operate a machine in quite different ways to make the same given grade of paper. Fairly broad tolerances in paper specifications were accepted in the trade, since few machines could achieve or maintain very stable operations.

The economic pressures that developed during the fifties created two incentives to achieve improved process control—the trend to lower basis weights and the increased quality demands by the customer on the one hand and the need for greater operating speeds and machine efficiencies on the other. The rapid advances in electronic instrumentation, control system reliability and computer technology stimulated to a large extent by the space programme appeared to offer the elements of a solution. In several companies, this need for improvement in uniformity was broadly recognised and the early promises of utopia offered by the computer manufacturers captured the imagination of process engineer, paper salesman and company executive alike. In such a climate, optimism overstated the positive and ignorance overlooked the negative.

In 1961, the first installation of a digital process computer on a papermachine was heralded with much publicity, then quietly abandoned two years later. Another abortive installation was made in 1962 and seemed to this observer to have been primarily conceived as a paper sales promotional tool.

Under the chairmanship of D. Attwood

Several systems, coming on line in 1963, emerged during a period of scepticism and only after a radical departure from the initial control strategies suggested by the suppliers were they finally able to demonstrate the potential economic value of the process computer. Most of these installations are still in operation today. By 1969, over 70 digital control computers have been placed in operation or are on order by the paper industry—over 50 in North America and the balance in Europe. The majority are used to control papermachine operation although there are now several installations in the pulping and bleaching plant areas.

The most conclusive evidence that computer-based control technology will be the dominant theme in the future of process control in the paper industry is the growing number of companies operating multiple installations. Even though many accountants still debate the partition of economic benefits between systems engineering, improved instrumentation and computer control, the tangible improvements from their combined use have been recognised by operator and customer alike and have made computer-based or computer-compatible process control systems an accepted feature of most recent high-speed machine and pulpmill installations.

In spite of this proliferation of applications in an industry that, in the past, has been slow to embrace major developments with a high technological content, the present day state of development still exhibits very many signs of immaturity. The high cost of engineering analysis and software generation has shown to date little sign of reduction on successive installations. Instrumentation, hardware, software and control strategy are still rapidly evolving and the earlier ignorance of process technology that created the art of papermaking mythology has now given way to a belief by some that each new computer installation requires a brand new control strategy and an original software package. The long and expensive gestation period that precedes an installation, together with the shortage of skilled process and computer analysts to carry it out, have been the two major restraints on the growth rate of this new technology.

As a means of looking into the future, I have chosen to review the recent development of process control in the paper industry and to project some of the emerging trends and control concepts.

The early philosophy of the computer process control exponents was 'measure everything possible, feed the results into the computer, statistically analyse the signals, then calculate the appropriate corrective control changes'. It was assumed that these indiscriminate statistical models derived by regression and correlation analyses would substitute for our lack of understanding of the true process variable relationships. Perhaps the first significant finding of the early workers in computer process control was that, on a typical machine for much of the time, there was such a great amount of variation of a random or unpredictable nature that this type of statistical model was useless for control improvement. In computer lingo, excessive white signal noise gave garbage in, garbage out. One of the first criteria the Mead team established in the data-collecting phase of their work was to calculate an instantaneous material flow balance for the system and only if the balance closed within 5 per cent was the system sufficiently stable for the readings to be of value in developing a statistical control model. At this stage of the program, we became so submerged in data that it proved beneficial to re-evaluate the objectives of a process control system.

Objectives of a process control system

THE objectives of any process control system are threefold—(*1*) a uniform product, (*2*) a product with certain given quality specifications and (*3*) manufacturing it in the most economical manner possible—in short, stabilise-standardise-optimise, in that order.

A paper or pulp mill process contains some elements with very short transportation lags, some with very long ones and, in addition, some controllable process variables have very fast response times and others very slow ones. To achieve product uniformity under these conditions requires a system designed to operate in a *stable* manner—that is, we desire changes in a variable to occur only as the result of a deliberate control decision and that there be no—or, in practice, very little—uncontrolled or random variation. We cannot design a control system that can change process variables quickly enough to offset rapid, random variations in the process feed flows or composition. Stability for the most part must be designed into a system. It can usually be achieved more easily and at lower cost in a new installation than by modification of an existing one. The most appropriate first use of a computer on an existing machine is to locate and determine the causes of instability so that they can be corrected. This work is the role of the process engineer and must precede that of the control engineer.

Product specifications are frequently based upon tests simulating final use requirements. In these and many other instances, it is impossible continuously to measure the product specifications during production. Measurement and control of these properties depends therefore on the identification, measurement and control of other variables that have a high degree of correlation to the specified properties. The absorption coefficient for beta-rays, for example, is substituted for basis weight. There are two classes of variable—*process variables* such as roll and wire speeds, flow rates, material ratios and temperatures and *product variables* such as opacity, colour, surface texture, strength and porosity. Control theory assumes that, for a given machine, there is a relationship between a given combination of process variables and the resulting set of product variables—establishing the former set determines the other. Any set of process variables creating the desired product specifications is called an operating *standard* for the grade. The role of the control system is to adjust the process controls about the desired process standard to compensate for varying material inputs or other changing boundary conditions.

If a given grade can be produced by a number of different *standard* conditions, the final function of the control system is to select the most economical *standard* set taking into account the economic parameters associated with the specific order. Usually, a given grade can be made over a range of speeds, at each level of which there will be one or more operating standard sets. The economic parameters will include such factors as component raw materials costs, machine efficiencies as a function of speed, the cost of alternate ways of achieving a desired property and the costs associated with getting on and off the grade.

The unique role of the digital computer, in contrast to other control instrumentation and hardware, is its ability to contribute towards the attainment of all three control objectives. It can aid in identifying instability, memorising and achieving process standards, checking measured product characteristics against stored specifications and, finally, in selecting the optimum process economics.

The progress to date in the paper industry has been mostly towards achieving the first two objectives. In many cases, the early improvements achieved were a result of improved process stability rather than of computer control *per se*. In the control field, standardisation programs have been quite successful. Control algorithms (mathematical models) designed to imitate the behaviour of an experienced operator have also been successful and acceptable to the operators. Advanced control strategies based on sophisticated control theory have had limited application to date and it is only in the last few years that they are beginning to find a useful place as we gain a better understanding of process dynamics and the interrelationships of process variables. Process optimisation is beginning to appear in the bleaching plant area, where the incremental steps of brightness obtained from successive stages can be varied according to the prevailing cost of the different chemicals used. In no case, however, have we achieved the sophistication of the optimising function found today in certain food and petrochemical plants.

Process design—the pre-process control

THE important role that process design makes to subsequent control success has been emphasised above. A major change is taking place as a result of this realisation in the design of new mills. The prior emphasis of the chemical engineer on treating a system as a succession of essentially unrelated unit processes has given way to the systems engineer and an emphasis on process dynamics, interactions between variables and total systems behaviour. In this respect, the control system is no longer 'hung on' the process after the basic design is completed; instead, the system is designed with specific control strategies in mind.

How has this influenced paper or pulp mill design? In many ways—original layouts are made with instrumentation in mind, piping layouts and flow meter locations to ensure accuracy, gaps in the papermachine to accommodate traversing hardware, modification of flumes, silos, suction box separators, etc. to facilitate flow measurement. This trend will increase as we begin to control for additional product specifications. The advances and flexibility of control achieved by magnetic flow measurement has made this a preferred method for metering and proportioning feed flows. The improved control of flow and mixing that the magnetic flow meter has permitted has allowed the reduction or elimination of intermediate process chests, whose purpose was not that of storage, but of variable smoothing. This trend has the advantages of reducing transportation lags, elimination of many random consistency variations caused by inadequate agitation and economic savings both in capital, operations and pollution treatment, since the intermediate chests were either dumped to sewer at a grade change or increased the amount of off-quality paper produced at a grade change.

Recent findings in dryer research are drastically changing the approach to moisture control. The previous approaches of overdrying and remoistening or of segmented cross-machine drying at the end of a machine for uniform moisture profiles are being replaced by the use of controlled crown wet presses and more carefully designed dryer ventilation schemes. The prevention of product profile non-uniformity throughout the process is a more satisfactory and economical solution than the use of controlled cross-correction devices after non-uniformity occurs.

Other major advances have been made in the control of hydraulic instabilities from flow valves, pressure screens, head boxes and other mechnical and hydrodynamic devices in the system, all of which in the past have produced variations beyond the corrective ability of a control system.

System analysis has changed the design approach to such auxiliary components as whitewater saveralls and broke systems. Both have been shown to cause previously unsuspected disturbances in the uniformity of the product, when operated in a conventional manner and disturbances in filler, drainage and porosity attributable to these sources persist for a long time after the original upset—a break, for example—has been corrected. Process design changes have minimised these variations and, in one case at least, resulted in a substantial increase in effective saveall capacity.

I believe the impact of a system approach to pulp and papermill design has only just begun and that the increasing political and social pressure for both air and stream pollution abatement will end up by requiring the paper or pulp mill design engineer to include these operations in his design strategy. Stated in an unconventional manner, a prime role of the systems engineer is to design out variation and cost and to reduce the need for sophisticated process control systems before the process is built.

Instrumentation—sensors and control elements

THE present trend in instrumentation is from the mechanical and the pneumatic towards the electronic. The sales of electronic instrumentation in 1968 exceeded those of pneumatic for the first time and, by 1973, it is forecast that the switch to electronic will be close to 90 per cent. The major advantages favouring electronics have been compact size, flexibility, reliability and decreasing prices. The use of solid state and now of integrated circuitry has been a significant factor in this switch. The reduction in cost of integrated circuits—approximately 50 per cent/year—has been so significant that new approaches to instrument maintenance have become possible. The present difficulties of converting pneumatic instrument men to electronics will be overcome in the future by reducing the cost of electronic assemblies to the point that it is more economical to replace entire assemblies with spares than it is to try to correct the failure. This approach, with manual back-up, appears to have prevailed over the alternate approach of 100 per cent circuit redundancy. This trend will lead also to an increased use of standardised electronic plug-in modules.

Monolithic circuits have gone a long way toward overcoming another papermill problem—atmospheric corrosion of electronic circuitry. This factor will permit the gradual relocation of much of the control hardware closer to the process, particularly as the hierarchical approach to computer control comes into use. The traditional process instrument had either a pneumatic or an analog output. Even many digital sensors such as pulse tachometers displayed their readings in analog form. One of the major cost items and continuing source of trouble on the earlier computer systems was associated with the conversion and multiplexing of analog input signals. Fig. 1 shows a typical computer control loop today. In this instance, a digital process (namely, the absorption of beta-rays) is converted to an analog signal, then electronically massaged to take care of source decay, standardisation, etc.

Because of the high cost of analog to digital conversion and a requirement of some computer manufacturers to keep this function closely tied to the remote computer, the signal is multiplexed along with the other inputs. On the

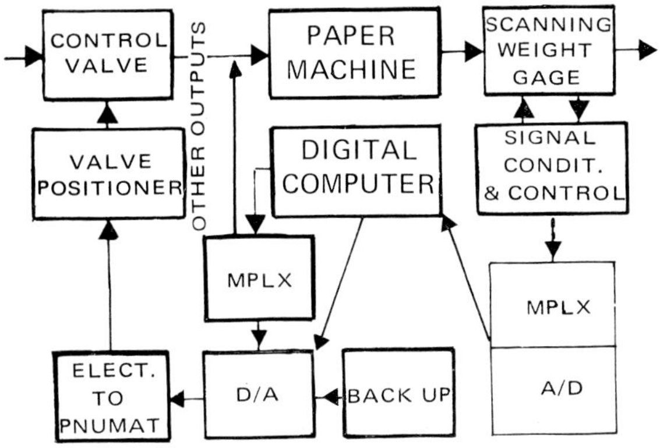

Fig. 1—A typical computer loop for analog instrumentation showing conversion and multiplexing steps required

output side, the reverse steps are taken and extended in this example to an extra pneumatic conversion in order to operate the valve positioner. One company is already offering a beta-gauge in which digital logic and circuitry are used through the input stages and many of the instrument functions are taken over by the computer. Fig. 2 shows such a simplified control loop. Similar approaches are also being developed for moisture and optical measurements.

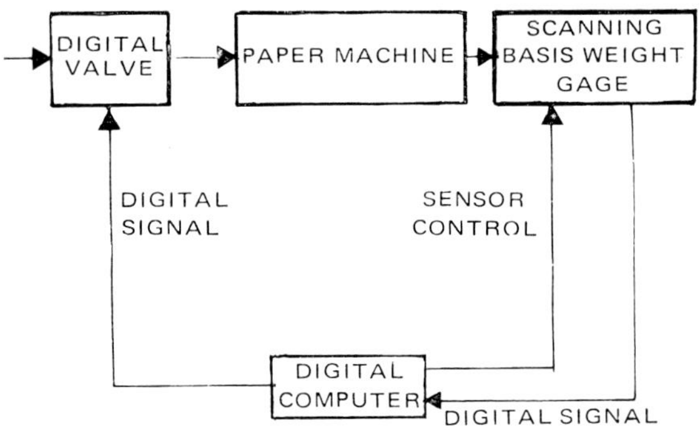

Fig. 2—A digital control system illustrating the simplification possible with digital instrumentation

This seems a logical direction and will probably expand, particularly in association with direct digital control.

Some process variables are unique to the paper industry and some have proved difficult in the past to measure reliably. Examples include freeness, consistency, formation and, in addition, product variables such as on-line porosity, printability and strength parameters. The ability of the computer to calculate often makes it possible to deduce some variables not readily measured directly from other more available measurements, also to make corrections to the results for errors caused by known process interactions. The impetus of the computer has generated a market that is attracting a steady stream of new or improved sensors for the paper mill. Taken together, these factors have greatly reduced control limitations imposed by the lack of suitable instrumentation.

Pneumatic power is still preferred for primary motive power on most valves. The inherent problems with electric control motors in a mill environment has limited their application. Digital and electronic valve positioners are finding use in DDC systems, however, so that the control loop can now be digital right up to the final stage. There would seem to be no particular economic incentive to eliminate the pneumatic power function.

Computers

LOOKING now to computer trends let us consider the stages through which we have progressed.

Stage 1—A conventional analog control loop is shown in Fig. 3. It could be either electronic or pneumatic. Its control modes are limited to gain, reset and rate, which are manually set, usually by the instrument engineer. The control point is manually set by the operator.

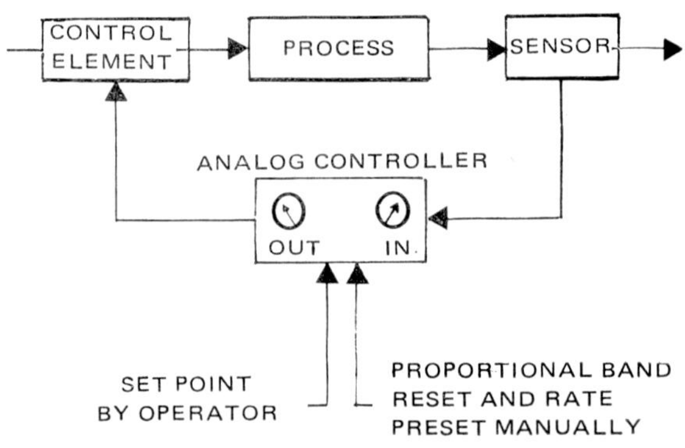

Fig. 3—A conventional analog controller with manual set point and three-mode control

Stage 2—The first application of the computer substituted the computer for the operator (Fig. 4). The analog control mode adjustments were still set manually and performed their normal task, but now the computer adjusted the set point according to a preprogrammed strategy. This approach is called *supervisory control.* Economics dictated a single computer to handle all controllers and the computer design required a remote air-conditioned location. Fig. 1 illustrated the problems of coupling each control loop to the computer and the economic compromises made to solve them. Early improvements in computer design occurred as the early modified electronic data processing units were replaced with units designed expressly for process control. Better performance and lower costs were the result.

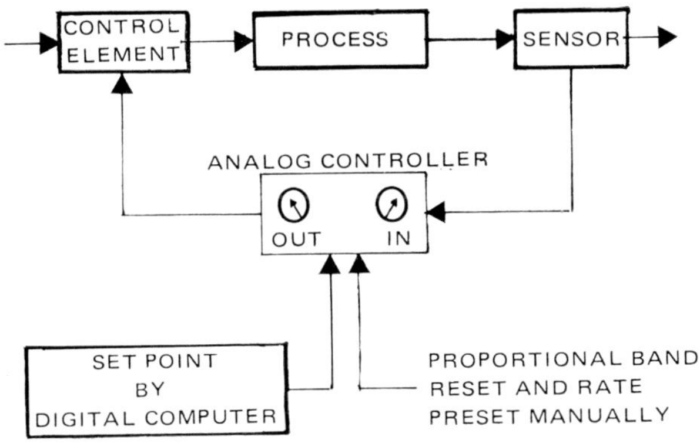

Fig. 4—Supervisory computer control

The first successful strategies were aimed at operator imitation—for example, recording the operator's manual settings when the machine was producing paper to specification and recalling these automatically whenever the same grade was run again. Immediate advantages were realised from this 'standard' approach, since the computer had an infallible memory. The next step was to analyse the operator's strategies for grade and speed changes, then to program the most successful ones into the computer. During computer failure, the system returned to conventional operation. There are still many computers operating in this manner and it is probably the most straightforward system for an existing machine, since it may be added to the existing control system incrementally. Besides, it allows two or more machines to be handled by a single computer.

Stage 3—Type A, DDC (Fig. 5). The next major step has been to direct digital control. In this mode, the function of the analog controller is replaced entirely by software within the computer. The counterparts to the gain, reset and rate controls are now defined within the computer program and the control point is entered into the computer by the operator or generated internally by a programmed strategy. In its simplest form, type A, the control constants are preset by the programmer. Interconnections between control loops are made internally by software and can be easily modified by a

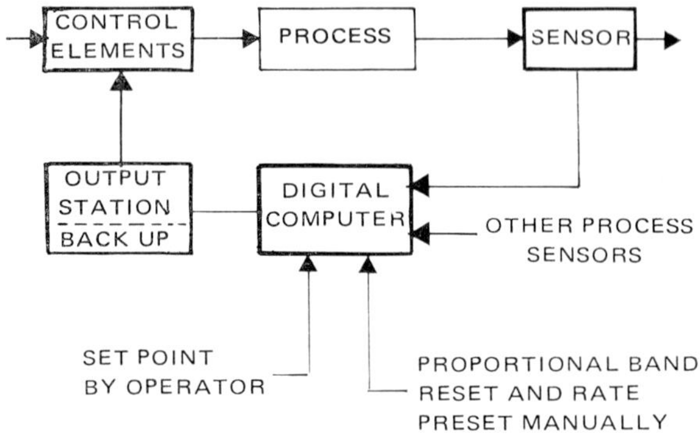

Fig. 5—Basic direct digital control: a digital equivalent to the analog centre of Fig. 3

program change. Instrument maintenance is reduced, since the hardware is greatly simplified.

There are two basic disadvantages to this approach—

1. Since only a single computer substitutes for all the analog controllers, the control action in every loop is now non-continuous. Each loop is served sequentially by the computer on a programmed cycle of one or two seconds. DDC is practical, because the high speed of the computer allows all the control functions to be serviced in a fraction of a control cycle, usually 30 per cent or less and so leaves a substantial portion of each cycle for other computational purposes.
2. Since all loops are dependent on one computer, a failure of it affects all control loops. This possibility is remote in practice, since computer reliability has proved greater than that of the papermachine itself. Yet DDC control valves and other DDC output devices are designed to retain their last position in case of a failure and an auxiliary analog back-up system is provided on some systems for a few key variables in addition to manual operation on all variables.

A type A system is the control equivalent of a mixed stage 1 and stage 2 supervisory system. The ratio is commonly 80 per cent stage 1 and 20 per cent stage 2. Although it was initially projected that a type A, DDC system over a minimum size would be cheaper than a conventional analog controller system, this does not appear to be true in practice because of the provision of back-up systems. It is unlikely that a type A system could be justified economically, unless it was later intended to modify it to a type B system.

Stage 4—In a type B, DDC system (Fig. 6), the full potential of the DDC approach is achieved. The 'frozen' control system of type A with fixed control parameters is replaced by a 'fluid' control system, in which not only the parameters, but the control linkages as well can be selected dynamically to suit the operating conditions. Thus, a different control configuration could be used during breaks for grade changes, even for different grades. Most of the new computer systems over the next five years will probably be of this type.

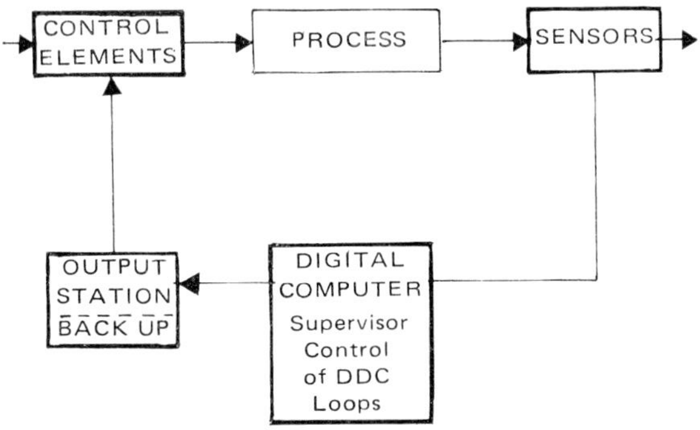

Fig. 6—Direct digital control with supervisory control of the DDC loops—that is, the DDC equivalent of Fig. 4

The most obvious feature of a DDC system is the absence of a control panel covered with controllers, limit warning lights, etc. They are replaced by a single display unit that on a command to the computer displays all the appropriate information of the selected loop. Loop parameters are changed at this same console. Since controllers can be viewed only one at a time, a computer-operated program must in the case of emergency record the identity of each loop in the sequence it goes out of limits to assist the operator in re-establishing control. This departure from familiar procedures necessitates a careful operator-training program. Current developments in cathode ray tube display systems will undoubtedly become commercial in the next few years. This will allow a wider choice of display features and, with the use of colour, a graphical differentiation of critical information. Such systems have been widely used in the space program and cost is the controlling factor in their use for industrial installations.

In stage 2, the role of the computer was strictly supervisory. It could organise its operating priorities according to their importance. Its active memory was largely available for computational procedures and the transfer time of data between active memory and disc storage was not a critical factor. In a DDC mode, a considerable portion of the active memory is set aside for the DDC control function, since there would be insufficient time on a large system to move the DDC package back and forth to disc or drum storage during each control cycle. This has two disadvantages—a given computer cannot handle as large a system as if it were used in a stage 2 system and, secondly, it requires the computer to give priority to the control function at the start of each cycle. This latter fact means that computations not completed in the open time of one cycle must be interrupted and possibly moved in and out of storage for completion during the next.

Stage 5—Hierarchy of computers (Fig. 7). The continued reduction in computer hardware costs promises that the next stage of computer system will consist of a central supervisory controller, much like the role of the computer in stage 2 served by one or more dedicated mini-computers located close to the process, each one of which will provide the DDC control function for a given subsection of the system. All computational operations associated with the supervisory function—models, standards, strategies, etc.—will be performed in the central computer and its output will adjust the control parameters and control points in the DDC units. Such a arrangement will have added value in those subsystems containing digitally based instruments, since the slave computer will also be able to handle the computional processing of the raw signal and standardisation routines. A further possibility is that it will eventually be possible by careful design of subsystems to standardise these subsystem hardware and software packages and so permit a distribution

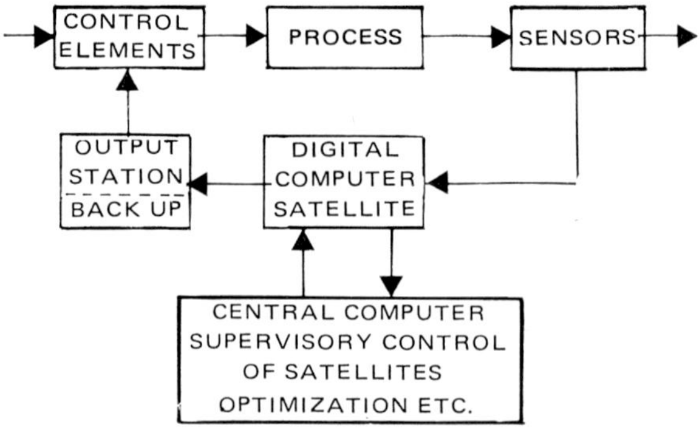

Fig. 7—Hierarchical computer control—a central computer exercising supervisory control over one or more satellite computers handling the DDC function

of development costs over many installations. The specific control models and strategies that define a machine's individuality and a company's proprietary technology would then all be contained in the central computer.

The biggest unknown in anticipating the rate and extent to which these new systems will be adopted is the relative economic return that arises from each new level of control complexity. The first gains in computer control are the largest and come from a few simple control strategies. Further gains are smaller and more costly to achieve. The cost picture is changing as hardware costs decrease rapidly and software costs stay relatively constant. This trend will increase the incentives for seeking ways to standardise subsystems of a control package so that they are suitable for a large number of customers and so justify the lower cost of a common hardware control package.

On many older machines, where major speed increases are not available to justify the costs of a computer system, it is probable that preprogrammed mini-computers will be used as part of a packaged control subsystem. The customer will have no special preparation or software maintenance to contend with and the system will be bought as a complete standard instrument control system. A basis weight, moisture content control package of this type has already been offered and other suitable subsystems would include head boxes, continuous digesters, savealls and refiners. These packages could be designed to be compatible with the hierarchy system in order to achieve wider application.

Mathematical modelling and control strategies

IN THE previous discussion, frequent reference has been made to models and control strategies. Control actions and their effects do not take place instantaneously. There are some delays in the control system itself and much larger ones in the process. Under these conditions, Attwood has shown that manual and analog control actions tend to be too frequent and drastic, resulting in a permanently unstable condition.

At least two kinds of model concept are used in process control studies. One approach attempts to describe the process mathematically in such a way that the interdependence of all the significant variables is explicitly stated. It is assumed that such a model (if available) would provide the basis for developing control actions unambiguously. The large number and the complicated interrelationship of papermaking variables makes such a *variable model* an impractical dream at this time. The second approach develops empirical relationships to describe the way in which small *changes* in one or more variables effect *changes* in the other significant process variables—that is, it is a *variable derivative model.* If the independent variables are subject to process control action, then such a control model describes the expectation of the process' response to the control actions. In manual operation, the model is a mental one and imprecise. In an analog system, it is usually a pneumatic or electronic network. In the digital computer system, it is defined in mathematical terms. The value of the model is the closeness with which it predicts the actual response of the process, *not* the degree of resemblance it bears to the physical system. It is used to plan a program of control actions that will move the process in the shortest possible time to a new control level or restore it from a random disturbance and minimise longer term oscillations.

Most of the variables on a papermachine are interrelated to several others. This means that a change in one variable will cause a change in several others. Unless these interactions are taken into account by the control system, a control level change in one variable will cause disturbances in several other variables and may cause the entire process to oscillate for a long period. Analog control systems are able to deal only with simple interactions. Two analog techniques that have been used are cascade and ratio or product control. Examples of their use have been in dryer controls and stock metering, respectively.

The computer has permitted more sophisticated use of models. Dahlin has described a technique to decouple the interactions between two variables and applied it in a basis weight and reel moisture control system. In another version, Rounsley combined decoupling with a predictive strategy. A mathematical model in the computer keeps a real time record of the system's response to all previous control actions and can therefore recognise new disturbances by comparing the model response to the actual response and, if a difference exists, take appropriate action immediately. Moisture control with these techniques gave a fivefold reduction in variation compared with a more conventional analog control system.

To date, little success has been reported with feedforward control. For example, feedforward basis weight control based on thick stock consistency, stock flow rate and stuff gate control has proved too insensitive, owing to poor consistency measurement and an inability to model all the variables influencing the fibre retention on the Fourdrinier machine and the presses.

Significant improvement in basis weight control has been obtained on one of our machines using an approach that is possible only with the aid of the computing ability of the digital computer and is illustrative of the technique of adaptive modelling. The example is based on three observations—the extremely reliable and accurate readings possible with modern basis weight gauges, the relatively slow changes in retention factors on the wet end and the relatively faster fluctuations in consistency of the thick stock supply to the machine. The equation relating the major variables (but ignoring time delays) is—

$$\begin{bmatrix} \text{thick stock consistency} \times \\ \text{thick stock flow rate} \times \\ \text{wet end retention factor} \end{bmatrix} = \begin{bmatrix} \text{oven-dry basis weight} \times \\ \text{machine speed} \times \\ \text{deckle factor} \end{bmatrix}$$

In this equation, basis weight is measured accurately and is a critical product property. Machine speed, deckle factor and stock flow rate are all relatively constant and usually fixed during a production run. The short-term basis weight fluctuations that occur are primarily a result of random consistency variations, which, because of the time lag, are measured too late to be corrected by a feedback control loop. The retention factors usually drift slowly as a result of slow system build-up, etc.

The control strategy developed was to decouple total basis weight and moisture content at the reel and use the oven-dry basis weight reading in a feedback control loop to set the thick stock flow control valve or stuff gate. This is conventional so far. The consistency reading signal is measured by the computer and corrected for known interactions of flow rate and of temperature by means of statistically derived models—say its corrected value is C_1. An apparent retention factor is calculated by substitution of measured values in the above equation. The instantaneous retention factor calculated will fluctuate as a result of the errors caused by the time displacement of consistency and basis weight measurements. A smoothed average retention factor calculated from successive readings, however, will give a very accurate estimate because of the slow time dependence of the factor with respect to the sampling frequency. This average value can then be substituted in the equation and a corrected average consistency reading C_2 calculated.

The consistency controller is now arranged so that the control set point is adjusted to maintain C_2 at a desired level and the instantaneous correction signal to the dilution valve is determined by the error signal (C_1–C_2). In this way, the basis weight signal in effect provides a continuous recalibration of the system and compensates for drifts in the consistency gauge itself and the machine retention factor. Such a strategy is a form of adaptive control.

This type of updating of model constants by on-line calculation can be done by use of time average readings as in the above case or by off-line control test readings taken intermittently and entered into the computer by the control laboratory, if the variable concerned has a slow rate of drift.

Strategies such as this offer several possibilities not previously available—

1. It is possible to control accurately while using a slowly drifting primary sensor.
2. It allows a desired variable to be calculated indirectly from the measurement of another correlated variable, for which the proportionality factor may vary slowly.

The advantages so obtained will lead to improved control and cheaper measuring devices, as well as to the measurement of variables for which accurate sensors had not previously been available.

The economics of process control

A DIGITAL computer system of the type widely used today is an expensive item. A computer may cost approximately \$250 000 (£100 000) and a like amount will be divided between software development and instrumentation. The cost is somewhat the same whatever the application or capacity of the plant. I have already indicated some of the reasons that these figures have not declined as fast as would be expected for a new technology. There are signs that the situation is beginning to change, but I am sceptical that total costs will decline very rapidly, since savings will be largely balanced by increased sophistication for a number of years.

Perhaps the earliest improvements will be for marginal installations, by which I mean those of smaller size or older vintage that could not justify a standard system. Here, we have seen that dedicated mini-computer subsystems may be the most economic purchase with control capabilities restricted to a few key variables and the use of standard control programs already built in. On new larger installations, however, the added capabilities of a more complex system will continue to be justified.

Specialised software houses are claiming that, through the sharing of software programs over many installations, they will be able to offer cost savings not available to the inhouse staffs of individual paper companies. To date, the technology has not slowed down sufficiently for many duplicate sales of the same model system, but this could become increasingly a factor as time passes. They also claim the additional advantage of being unbiased in equipment selection, which will lead to more competitive pricing, although to date the lack of software standardisation between computers, etc. has made it costly to change from one supplier to another for successive jobs. The feast or famine situation within a paper company once the initial installations are made will probably cause a decline in complete inhouse capabilities, particularly as these services become available outside. Smaller internal specialist staffs to update installed systems will be a more economical solution.

Although the pulpmill and bleaching plant have not received as much attention to date as has the papermachine, the indications are that these applications will yield substantial returns from chemical cost savings and they have the added advantage that these systems are not in general as sophisticated or costly and their product mix does not vary as much as for an average papermachine.

Conclusions

IN SUMMARY then, I expect to see the following major trends continue—

1. Increased use of computer-based control systems.
2. Improved and new instrumentation, with wide use of integrated electronic circuitry, digital techniques and a trend to move many of the functions of the instrument into an associated digital computer.
3. A continued expansion of DDC, although for a number of years hybrid systems mixing DDC and analog controller may represent the optimum economical balance.
4. Increased use of dedicated mini-computers either as stand-alone installations or as slave units tied to an hierarchical computer control system.
5. Extensive use of sophisticated models and control strategies, particularly the use of the decoupling, predictive and adaptive varieties. I do not expect any extensive use of feedforward strategies in the near future, because of the highly interactive nature of most of our operations and the successes achieved by other means.
6. A gradual extension of computer use towards the third objective of process control—the economic optimisation of the manufacturing process.
7. The generation of routine production statistics, etc. by the computer system for use by the management information system and a gradual flow of information back from this source for setting up the process variables.
8. A slow trend to standard subsystems of hardware and of software and another trend away from software into hardware for many of the routine and repetitive operations.

References

1. Dahlin, E. B., 'Integrated control of Fourdrinier machines', Second Symposium on Water Removal at the Presses and Dryers (Technical Section CPPA, Mont Gabriel, 1968)
2. Rounsley, R. R., 'Predictive control applied to a dryer section': *Pulp & Paper Mag. Can.*, 1969, **70** (2), 59–62
3. Dahlin, E. B., 'Computational methods of a dedicated computer system for measurement and control on papermachines' (TAPPI 24th Engineering Conference, September 1969)
4. Moore, J. F. and Gardner, N. F., 'Process control in the 1970s': *Chem. Eng.*, 1969, **76** (12), 94–106
5. Birkett, R. B., Henderson, R. L., Hobart, D. G. and Tsai, T. H., 'Process control for the 1970 mill': *Tappi*, 1968, **51** (10), 83A–89A
6. Attwood, D., 'The interaction between human and automatic control', *Papermaking Systems and their Control* (this volume), 69–87
7. Keyes, M. A. and Temple, R. H., 'Automation trends in papermaking systems': *Papermaking Systems and their Control* (this volume), 417–436

THE FUTURE OF CONTROL PROCEDURES FOR MANAGEMENT

Prof. S. EILON, D.Sc., Ph.D., D.I.C., F.I.Mech.E., F.I.Prod.E., Management Engineering Section, Mechanical Engineering Department, Imperial College of Science and Technology, London

---

Introduction

THERE is no difference, at least in principle, between control of administrative systems and control of other systems. The purpose of a control procenure is to specify the way in which the behaviour of a system can be affected. The controller of the system scrutinises information about the performance of the system and every so often he has to make a decision, namely, to choose between several courses of action open to him. This decision is transmitted to the system with the expectation that it will react in a certain way. This cycle of events—monitoring and evaluation of the system's behaviour, followed by a decision for corrective action—is the essence of the control process, irrespective of whether it is control of inanimate systems or managerial control of industrial enterprises.

There are, of course, many facets of the industrial environment that can make managerial control very complex. For one thing, information about the performance of the system is rarely complete, but, as managers often have to act within very severe time constraints, they may be unable to afford the time required for further data collection. Then, the system's response may be uncertain, as is very often the case in economic and social systems both on the macro and the micro scale; so the decision maker has to act in the knowledge that there may be serious discrepancies between his expectation and the system's actual subsequent behaviour. A third important feature that management of an industrial system has to contend with is that the time lag between decision and response can be quite substantial. This is certainly the case at the higher echelons of the management hierarchy, when questions about the allocation and conversion of resources are contemplated. Add to that the fact that there are many levels at which control can be considered, that a system to describe an industrial enterprise can be regarded as a plethora of

Under the chairmanship of D. Attwood

systems within systems and that any one manager may have several roles to play in more than one single system and we begin to realise the enormity of the task of analysing management control in a general way.

One approach is to start with very simple systems and to inquire into control relationships within them; in particular, to examine the interactions among controllers when several are assigned to look after the same system. Such an examination may well result in certain general concepts and definitions about linkages between controllers and the way that they can affect the state of the system. These concepts may then be helpful in the study of complex procedures in large systems.

The fewest states that a controller or a system can assume is two, since by definition a single-state system cannot be controlled and, similarly, a single-state controller can have no effect on the system. There are numerous examples of systems operating in a two-state mode—lighting circuits (the light is either ON or OFF), go or not-go gauges in inspection of components, appointments by selection boards (a candidate is either accepted or rejected), jury decisions in court (guilty or not guilty) and so on. The digital computer, which conforms to the rules of binary logic, is another such example. For the sake of convenience, let us label the two states as 1 and 0 or ON and OFF, respectively. There is no special merit in these names, except that 1 and 0 are used in binary arithmetic and ON and OFF describe the two states of a current flowing through electrical circuits. As the two-state operation is fundamental to the understanding of the control function, we shall now examine the two-state control process in some detail.

The two-state single controller

IF THE controller may be either in the state ON or in the state OFF and, similarly, if the system may assume either of these states, then all the possible control procedures may be listed in what is generally called a *truth table* (Table 1). Each procedure expresses the correspondence between the states of the controller and the states of the system. In procedure 2, for example, when the controller switches to ON, the system moves to state ON; when the controller switches to OFF, the system reverts to OFF.

TABLE 1—FOUR PROCEDURES FOR A TWO-STATE SINGLE CONTROLLER

		System			
Controller	*Procedure*	*1*	*2*	*3*	*4*
ON		ON	ON	OFF	OFF
OFF		ON	OFF	ON	OFF

Of the four possible procedures, clearly the first and the last have no need for a controller, since the state of the controller in neither case affects the state of the system. Procedure 2 calls for *affirmation*, namely, it requires the system to assume the state indicated by the controller. Procedure 3 is that of *negation* (or *inversion*) and here the state of the system is the opposite of that of the controller. If in procedure 3 the labels of the states of the system are reversed and the labels of the state of the controller are left as they are, then procedure 2 is obtained. This is reminiscent of the NOT operator in Boolean algebra and the NOT gate in circuit theory: when an output is connected to an input through a NOT gate, the output is ON when the input is OFF and vice versa.

We see, therefore, that for a two-state system with a single controller A, only two control procedures are feasible—

Procedure	*Controller*	*System*
Affirmation	A	As A (or A for short)
Negation	A	Not A (or $\bar{A}$ for short

In administrative systems, the NOT gate has the connotation of management by exception. The controller may be described as a passive component in the system; as long as he remains in a state of passivity (denoted as OFF), the activity under his charge continues unabated. When he perceives an exception to normal working, he takes action (he switches to the ON position) and causes the system to stop; this is precisely how a NOT gate is defined.

Linkages among several controllers

SUPPOSE that several controllers are assigned to the same system, so that there is complete correspondence between the states of each controller and the states of the system. It is important to note that it is not the state of the individual controller under such conditions that necessarily determines the state of the system. If that were the case, a one-to-one correspondence would have to exist, then either the roles of the controllers are incompatible (when one controller requires it to be in a different state) or the controllers must all act in unison to ensure that no conflict occurs. When the controllers are set to work in unison, then a one-to-one correspondence exists not just between the system and each controller, but among all the controllers as well.

In the absence of a one-to-one correspondence, the state of the system is determined by the combination of the individual states of the controllers. The way in which the controllers are connected defines the control procedure for the system, namely, the correspondence between the composite states of the group of controllers as a whole and the states of the system.

Time-independent gates

THE concepts of AND gates and OR gates,[1, 2] which are used in designing electrical circuits, are helpful in considering the fundamental ways in which controllers can be connected to form control procedures. To these concepts, we add the ANDOR gate, which can be constructed as a combination of the basic AND, OR and NOT gates, as we shall see later. The ANDOR gate is not commonly identified as a separate gate in circuit theory, but it is a convenient concept to employ, particularly in the study of administrative control procedures. The three gates (henceforth, the terms *gate* and *link* are used interchangeably) are defined as follows—

AND gate — If there are several inputs connected to an output through an AND gate, the output is ON when all the inputs are ON and the output is OFF when one or more inputs are OFF.

OR gate — If there are several inputs connected to an output through an OR gate, the output is ON when one or more inputs are ON and the output is OFF when all the inputs are OFF.

ANDOR gate — If there are several inputs connected to an output through an ANDOR gate, each input can cause the output to be switched ON or OFF.

To these (see also Fig. 1), we should add the NOT gate, which was discussed earlier. It does not define any linkage between controllers, but acts as a negation operator—

NOT gate — If an output is connected to an input through a NOT gate, the output is ON when the input is OFF and vice versa.

The operation of the AND, OR and ANDOR gates is summarised in Table 2, which shows the state of a system with two controllers A and B, where each controller has two states designated as 1 and 0 (or ON and OFF or TRUE and FALSE), respectively, in the form of a truth table.

Changing the state of the system

THE results in Table 2 reveal some interesting facets of the way in which each controller can effect a change in the state of the system. Suppose that the system is in state 1 and that the two controllers are linked by AND. If an event occurs that causes either controller to want to change the state of the system from 1 to 0, he is able to effect such a change.

TABLE 2—THE TWO-STATE TWO CONTROLLER SYSTEM

Controllers A	B	AND	*System* OR	ANDOR
1	1	1	1	1
1	0	0	1	0
0	1	0	1	0
0	0	0	0	1

If both controllers want the system to change, then again the control procedure AND allows both to act and to achieve that aim.

If, on the other hand, the system is in state 0, then the action of one controller (even of both) may not be effective in changing the state of the system from 0 to 1 when the controllers are linked by AND. It all depends on the initial

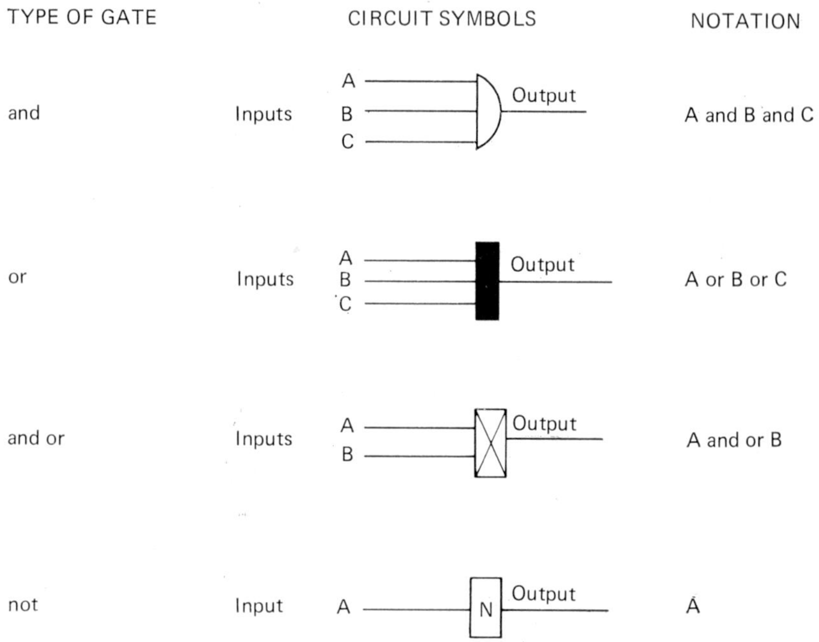

Fig. 1—Four types of gate

state of the controllers. For example, if one is at 1 and the other at 0, then only the latter can effect a change in the state of the system; if the former acts (by changing his state) or if both act simultaneously (each reversing his own state), no change in the state of the system will result.

In the case of the OR link, the situation with regard to change is a mirror image of AND. If the system is at 0, either controller or both can effect a change; if the system is at 1, then the result depends on the initial state of the controllers (see Table 3).

The ANDOR link allows either controller to trigger off a change; if both controllers act simultaneously (each reversing his own state), then no change in the state of the system can take place and this result holds irrespective of the initial state of the system or the initial respective states of the controllers.

TABLE 3—ONE OR TWO CONTROLLERS ACTING TO CHANGE THE STATE OF THE SYSTEM

System's initial state	*Controllers' action*	*Effect on the system* AND	OR	ANDOR
1	One acts	Change	?	Change
	Both act	Change	?	No change
0	One acts	?	Change	Change
	Both act	?	Change	No change

Notes

1. A controller acts by switching his initial state (from 1 to 0 or from 0 to 1).
2. The symbol ? in the table means that the effect on the state of system depends on the initial states of the controllers.

The NOT gate

THE combination of these three gates with a NOT gate causes the system to behave in the opposite way described in Table 3 and the control procedures to result are—

NAND = NOT AND

NOR = NOT OR

NANDOR = NOT ANDOR

The contrast between the AND and the NAND procedures is shown in Table 4 and similar relationships between between the controllers and the system for NOR and for NANDOR are also given in the table.

TABLE 4—NAND, NOR, NANDOR

Controllers A	B	AND	NAND	*System* OR	NOR	ANDOR	NANDOR
1	1	1	0	1	0	1	0
1	0	0	1	1	0	0	1
0	1	0	1	1	0	0	1
0	0	0	1	0	1	1	0

It is not difficult to see now how ANDOR can be described in terms of AND, OR and NOT. If an AND procedure and a NOR procedure are linked by an OR operation, the ANDOR procedure is obtained, hence—

ANDOR = (AND) OR (NOR)

similarly, NANDOR = (NAND) AND (OR)

Examples

SEVERAL examples of electrical circuits are shown in Fig. 2. The three switches in (*a*) control the current flowing through the line; they are connected in series and form an AND gate. The switches in (*b*) are in parallel and form an OR gate. In (*c*), the output is ON if either A and B are ON or if C is ON, whereas (*d*) shows an example in which A and either B or C should be ON for the output to be ON. In (*e*), when the solenoid C is energised, it moves the switch to an OFF position; when the solenoid is not energised, the switch moves to an ON position: the output for the whole circuit is ON when A and B are ON or when D is ON or when C is not ON. In (*f*), an ANDOR gate is shown; if the switches A and B are as shown (call this position ON) or if both are switched to the other position, the output is ON; if one is switched OFF, the output is OFF.

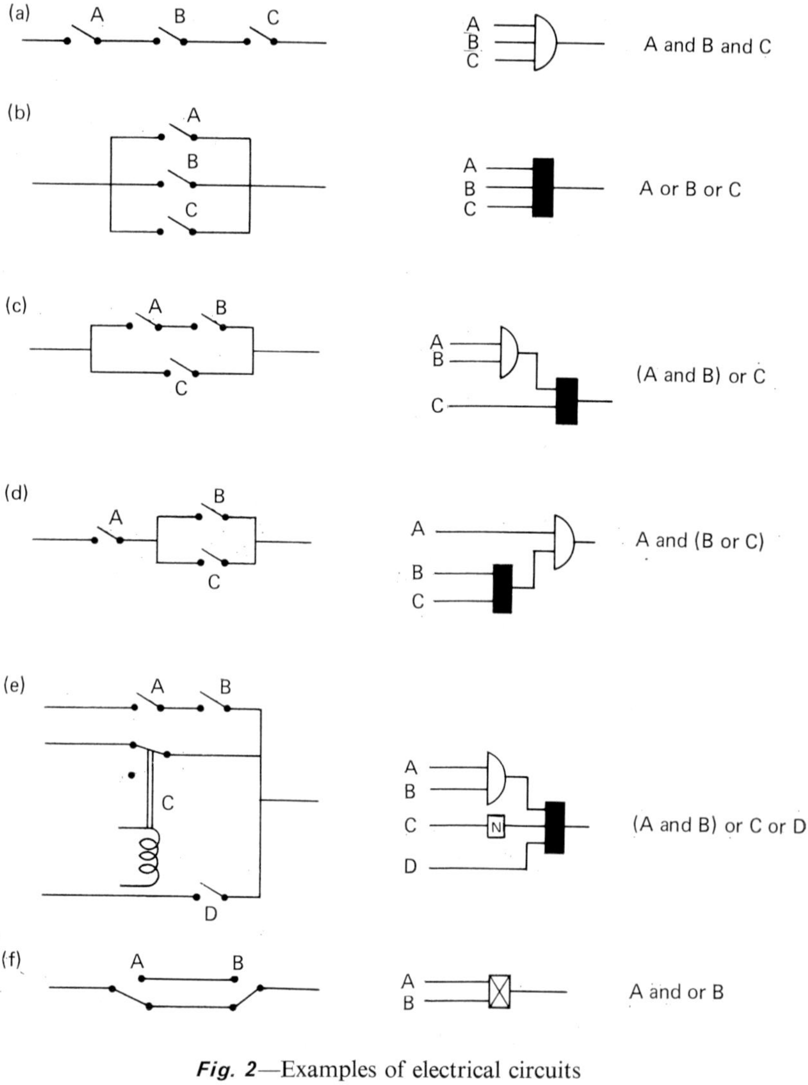

Fig. 2—Examples of electrical circuits

Control by members of a set

THE notation of Fig. 1 does not provide a convenient way of describing activities authorised by a majority vote or by agreement between a given number of members of a group of controllers. As these forms of control are common in administrative systems, a suitable notation for this purpose would be an advantage. For example, if two signatures of any three executives A, B, C are needed to validate a cheque, the procedure may be described as—

$$(A \text{ and } B) \text{ or } (B \text{ and } C) \text{ or } (C \text{ and } A)$$

but more conveniently as Z(2), which means that authorisation may be given by two members of set Z, where Z = {A,B,C}. Thus, the notation Z(x) means that only x members of a set need to agree on a given action and Z(m) requires a majority of the members of the set to agree.

Control for administrative systems

THE linkages forming AND, OR and ANDOR gates conform precisely to the three control procedures discussed in another paper[(3)]—

Control in series All the controllers connected in series need to agree on a course of action for it to take place. If one controller objects, he overrules the others. This procedure is designated as the AND gate.

Control in parallel When controllers are connected in parallel, they all have to agreed before an activity can be stopped. This procedure is a mirror image of control in series, but here each controller can overrule the others if he objects to the activity being terminated—this is, in fact, the OR gate.

Conjoint control Each controller can decree that action be taken or that an activity be stopped. In other words, when a controller acts, he activates all the other controllers to whom he is connected in parallel. If he switches OFF, it is as if all the others switch OFF at the same time; if he switches ON, they all switch ON. Here, the last controller has the last word and annuls all previous actions of the controllers. This procedure is the ANDOR gate.

Other possible linkages

TABLE 2 lists only three control procedures resulting from various linkages between two controllers, but there arre many other possible procedures as can be seen from the exhaustive list in Table 5. Each procedure is concerned with four outcomes for the system, the outcome in each case being either state 1 or state 0.

There is one procedure for which all the outcomes are state 1
$(^3_4)$, namely, four procedures, with three outcomes as state 1 and one outcome as state 0.
$(^2_4)$, namely, six procedures, having two outcomes as state 1
$(^1_4)$, namely, four, having one outcome as state 1 and
one procedure for which state 1 does not emerge.

The total number of possible alternative procedures is $2^4 = 16$.

A close examination of these procedures reveals that some include various operations in Boolean algebra, such as *negation* (the NOT gate, as in alternatives 10, 11 and several others), *conjunction* (the AND gate in procedure 12), *disjunction* (the OR gate in procedure 2), equivalence (this is the ANDOR gate in procedure 8) and *exclusive disjunction* (which is the negation of ANDOR,

TABLE 5—SIXTEEN PROCEDURES FOR A TWO-CONTROLLER SYSTEM

Controllers A	B	*System* 1	2	3	4	5	6	7	8	9	10	11	12	13	14	15	16
1	1	1	1	1	1	0	1	1	1	0	0	0	1	0	0	0	0
1	0	1	1	1	0	1	1	0	0	1	1	0	0	1	0	0	0
0	1	1	1	0	1	1	0	1	0	1	0	1	0	0	1	0	0
0	0	1	0	1	1	1	0	0	1	0	1	1	0	0	0	1	0
			↑ OR			↑ NAND	↑ A	↑ B	↑ ANDOR	↑ NANDOR	↑ $\overline{B}$	↑ $\overline{A}$	↑ AND			↑ NOR	

namely, NANDOR, as shown in procedure 9). Furthermore, all the other procedures may be described in terms of the basic AND, OR and NOT gates. Further observations are summarised below.

Procedure

1. The system has only one state and has therefore no need for controllers.

2. This is the OR gate (namely, disjunction) so that the procedure in this column may be described as A or B.

3. If the labels for B were to be reversed, the result in this column would be identical with that in the previous column, namely, the procedure may be described as A or $\overline{B}$.

4. $\overline{A}$ or B.

5. $\overline{A}$ or $\overline{B}$. This is also the NAND procedure, namely—

$$\overline{A} \text{ or } \overline{B} = A \text{ nand } B$$

6. Here, the system adheres to the state dictated by A; controller B is redundant.

7. The system is controlled by B; here, A is redundant.
8. This the ANDOR gate; also, note that—

$$\text{A andor B} = (\text{A and B}) \text{ or } (\bar{\text{A}} \text{ and } \bar{\text{B}}))$$
$$= (\text{A and B}) \text{ or } (\text{A nor B}).$$

9. This is exclusively disjunction and may be described as—

$$\text{NOT(ANDOR)} = \text{NANDOR},$$

also A nandor B = (A nand B) and (A or B).

10. This is the reverse of column 7—that is, $\bar{\text{B}}$.
11. This is the reverse of column 6—that is, $\bar{\text{A}}$.
12. Here, we have conjunction, namely, the AND gate.
13. A and $\bar{\text{B}}$; this column is also the reverse of column 4—that is, the procedure may also be described as NOT ($\bar{\text{A}}$ or B).
14. $\bar{\text{A}}$ and B; alternatively, this procedure is NOT (A or $\bar{\text{B}}$).
15. $\overline{\bar{\text{A}} \text{ and } \bar{\text{B}}}$ = A nor B.
16. As in procedure 1, the controllers are redundant here.

Boolean operations

TABLES 2, 4 and 5 suggest that Boolean algebra may be helpful for formal descriptions of control procedures and relationships between controllers. Consider the following Boolean operators[1, 2] involving A and B—

Operator	*Symbol*	*Meaning*
Conjunction	A:B or AB	If A has the value 1 and B has the value 1, then AB has the value 1, otherwise AB is 0
Disjunction (also called 'alternation')	AVB or A+B	If either A or B has the value 1, then A+B has the value 1, otherwise A+B is 0
Implication	A⊃B	If either A is 0 or B is 1, then A⊃B is 1, otherwise A⊃B is 0
Equivalence	A≡B or A/B	If A and B have the same value, then A/B is 1, otherwise A/B is 0

The operators *conjunction* and *disjunction* are also referred to as *Boolean* (or *binary*) *multiplication* and *addition*, respectively. The rules of these operators are—

Boolean multiplication (conjunction)	*Boolean addition (disjunction)*
1.1 = 1	1+1 = 1
1.0 = 0	1+0 = 1
0.1 = 0	0+1 = 1
0.0 = 0	0+0 = 0

We can now construct a truth table to find the value of several expressions consisting of A and B (Table 6). The results are obtained by simply performing the operations of Boolean addition or multiplication of the respective states of A and B (or their complements $\overline{A}$ and $\overline{B}$) in each row, as required by the operator in each expression.

TABLE 6—VALUES OF SOME BOOLEAN EXPRESSIONS

A	B	$\overline{A}$	$\overline{B}$	A+B	$\overline{A+B}$	$\overline{A}+\overline{B}$	AB	$\overline{AB}$	$\overline{A}\,\overline{B}$
1	1	0	0	1	0	0	1	0	0
1	0	0	1	1	0	1	0	1	0
0	1	1	0	1	0	1	0	1	0
1	0	1	1	0	1	1	0	1	1
Procedure in Table 4		11	10	2 ↑ OR	15	5	12 ↑ AND	5	15

AND, OR, ANDOR

COMPARING the results in Tables 5 and 6, we find that—

$$A\ B\ = \text{A and B}$$
$$A+B = \text{A or B}$$

The AND gate is equivalent to conjunction (that is, the Boolean multiplication of the individual states of the linked controllers) while the OR gate is equivalent to disjunction (namely, it involves the Boolean addition of the individual states). Thus, the Boolean expression AB should be read as 'A and B' and A+B should be read as 'A or B' (rather than 'A plus B').

If we examine the definition of *equivalence*, we find that it coincides with the ANDOR gate and, if this operator is denoted by a stroke /, then A/B is read as 'A andor B' or as 'A stroke B' (but not 'A divided by B'). From procedure 8 in Table 5, it is evident that—

$$A/B = \text{A andor B}$$
$$= (\text{A and B}) \text{ or } (\overline{A} \text{ and } \overline{B})$$
$$= AB+\overline{A}\ \overline{B}$$

and this is easily verified by performing the Boolean addition of the columns for AB and $\overline{A}\ \overline{B}$ in Table 6.

Operations involving several controllers adhere to the usual algebraic conventions, so that—

$$A+B = B+A$$
$$A\ B = BA$$
$$A+(B+C) = A+B+C$$
$$A(BC) = A\ B\ C$$
$$\text{also } A\ (B+C) = AB+AC$$
$$\text{similarly } (A+B)(C+D) = AC+AD+BC+BD$$

Alternative procedure for the two-controller two-state system

ALL THE 16 alternative procedures for two controllers handling a two-state system are listed in Table 5. This list is derived by enumerating all the possible outcomes for the states of the system and relating them to the corresponding states of the two controllers. It is not difficult to see that this is in fact an exhaustive list and covers all the possible combinations of linkages between the controllers. One way to enumerate all these various combinations is as follows—

	Alternatives	*Procedure in Table 5*
(*a*) Outcome for the system is independent of the states of the controllers	1	1
	0	16
(*b*) Outcome for the system depends on the state of one of the controllers	A	5
	$\overline{A}$	11
	B	7
	$\overline{B}$	10
(*c*) Outcome for the system depends on disjunction of the controllers	$A+B$	2
	$A+\overline{B}$	3
	$\overline{A}+B$	4
	$\overline{A}+\overline{B}$	5
(*d*) Outcome for the system depends on conjunction of the controllers	AB	12
	$A\overline{B}$	13
	$\overline{A}B$	14
	$\overline{A}\overline{B}$	15
(*e*) Outcome for the system depends on disjunction of the linkages in (*d*)	$\overline{A}\overline{B}+AB$	8
	$\overline{A}B+A\overline{B}$	9

All the other combinations in (*e*) are already covered in (*b*) and (*d*) as can be seen from the results in Table 7.

TABLE 7—RESULTS FOR DISJUNCTION OF THE LINKAGES IN (*d*)

	AB	$A\overline{B}$	$\overline{A}B$	$\overline{A}\overline{B}$
AB	AB			
$A\overline{B}$	A	$A\overline{B}$		
$\overline{A}B$	B	$\overline{A}B+A\overline{B}$	$\overline{A}B$	
$\overline{A}\overline{B}$	$AB+\overline{A}\overline{B}$	$\overline{B}$	$\overline{A}$	$\overline{A}\overline{B}$

Each cell in this table is derived by performing the Boolean addition for the corresponding row and column linkages. For example, the result for the second row and the first column is $A\overline{B}+AB = A(\overline{B}+B) = A$. The ten results (the cells above the diagonal are a mirror image of the cells below) include the four listed in (*b*) and the four listed in (*d*), leaving only two outcomes for (*e*). It can be shown that the outcomes for the system that depend on disjunction of the linkages in (*c*) have already been listed; similarly, that the outcomes that depend on conjunction of the alternatives in (*c*) or in (*d*) have all been accounted for.

Sequential links

A NOTABLE feature of Boolean algebra is that all the elements in an expression are assumed to act simultaneously, so that there is no merit in the order in which they appear in a given gate, thus—

A and B = B and A that is, AB = BA
A or B = B or A that is, A+B = B+A
A andor B = B andor A that is, A/B = B/A

In circuit theory, the sequence of operation of controllers is generally of no importance, since it is the final state of the controllers that determines the state of the system and this is illustrated in the examples in Fig. 2. Besides, there is no question of hierarchy in a procedure such as A and B; both controllers are equal in status and in their ability to affect the state of the system. This is not to suggest that they necessarily use this ability equally well or that they act with equal frequency. If A is quicker than B to discern changes in the environment and reacts accordingly, then the procedure A and B results in B behaving comparatively sluggishly and A doing most of the work of switching ON and OFF. Nevertheless, both have *equal opportunity* in this control procedure and this is what is referred to in an earlier paper[4] as first-order control.

When we examine administrative systems, we find that time-independent gates are not adequate to cover the variety of procedures that can be designed and it is therefore necessary to add the time dimension or the sequence in which controllers are required to perform their duties. It is suggested that the requirement for such a sequence can be denoted by an asterisk in the control expressions (namely, and*, or*, andor*). This notation has the following meaning—

A and* B — A acts before B; if A is ON, the matter is referred to B and, if he is ON, the output is ON, otherwise it is OFF; is A is OFF, however, there is no need to refer the matter to B, because his action will not affect the outcome.

A or* B — A acts before B; if A is ON, the output is ON and there is no need to refer to B; if A is OFF, however, then the matter is referred to B, who by switching to ON will overrule A.

A andor* B — A acts first, but, whatever he decides, the matter is referred to B, who may acquiesce or reverse A's decision; the state of A signifies the state that he wishes the output to be at (that is, he is ON when he wishes the system to be ON and OFF for the system to be OFF); the state of B signifies his reaction to A's decision: B is ON when he approves and OFF when he disapproves. The resulting state of the system is then derived from Table 2, in which the state ON is denoted by 1 and OFF by 0.

Sequential links can be described by Boolean expressions similar to those used for time-independent links, except that an asterisk is added to denote the sequence in which the controllers have to perform—

$$A{*}B = \text{A and* B}$$
$$A \overset{*}{+} B = \text{A or* B}$$
$$A/{*}B = \text{A andor* B}$$

The examples in Table 8 perhaps illustrate the way in which such a language can be used to describe administrative procedures. In these examples, $Z_1 = \{n_1 \text{ known members}\}$ and $Z_2 = \{n_2 \text{ known members}\}$; the two sets need not be mutually exclusive.

It should be noted that, for a given control procedure, the correspondence between the composite state of the controllers and the state of the system is not dependent on whether the gates in the procedure are time-independent or sequential. For any given set of states of the controller, the result for the system would be the same, as can be verified by comparing the summary in Table 9 for sequential gates with that in Table 3 for independent gates. In other words, the algebraic rules for computing the outcome of any procedure that involves sequential gates are precisely the same as when the notations for sequential performance of controllers are deleted. The sequential procedure allows a controller to be ignored, however, when it is evident that he can have no effect on the outcome.

TABLE 8—EXAMPLES OF CONTROL PROCEDURES

Example	*Activity*	*Authorisation*	*Notation*	*Boolean notation*
1	Payment of accounts	Director A or B or anyone of the group Z_1	A or B or $Z_1(1)$	$A+B+Z_1(1)$
2	Appointment of executives	3 members of committee Z_2, then approval by director B	$Z_2(3)$ and* B	$Z_2(3){*}\ B$
3	Requisition of materials	Any member of Z_1 or two of Z_2 or B	$Z_1(1)$ or $Z_2(2)$ or B	$Z_1(1)+Z_2(2)+B$
4	Dealing with a customer's complaint	Any two members of Z_1: if the complaint is dismissed, it is then referred to B, who may decide otherwise	$Z_1(2)$ or* B	$Z_1(2){*}+B$
5	Major capital expenditure	Majority agreement by Z_1, but either A or B may overrule this decision	$Z_1(m)$ andor* (A or B)	$Z_1(m)/{*}(A+B)$

Consider, for example, the following three procedures—

A and B . . . . (*1*)
A and* B . . . . . . (*2*)
B and* A . . . . . . (*3*)

where the probability of A being ON is $p_A = 0{\cdot}8$ and the probability of B being ON is $p_B = 0{\cdot}6$. Let us assume that the states of the two controllers are independent of each other. The probability of the system being ON is 0·48 for each of the three procedures. In (*1*) and (*3*), B is required to be ready to act all the time, whereas he is required in (*2*) to act only 80 per cent of the time (since for the rest A is OFF and then the system is OFF, whether B wishes it to be at OFF or not). As for A, in the first two procedures, he acts all the time; in the third, only 60 per cent of the time. In reality, of course, the states of the controllers may be highly correlated, so that the probability of the system being ON may be much higher than 0·48, but the essential argument remains valid—namely, that the sequential linkages AND* and OR* dispense with the services of a controller when these services are not expected to be effective.

The definitions of the sequential links AND*, OR*, ANDOR* and the summary in Table 9 contain an interesting allusion to the way in which the state of controller B is to be interpreted. *In* AND* *and in* OR*, *the state of B signifies the state that he wishes the system to be in.* From Table 9, it is clear that whenever B points to 1 the system is at 1 and whenever B points to 0 the system follows suit, although—as we have seen—there are circumstances in which B is not called upon to act at all. *In* ANDOR*, *however, the state of B signified his agreement* (denoted by 1) *or disagreement* (denoted by 0) *with A's decision.* These observations are summarised in Fig. 3.

TABLE 9—RESULTS FOR SEQUENTIAL LINKS

Link	*Controller* A	B	*System*
A and* B	1	1	1
	1	0	0
	0	–	0
A or* B	1	–	1
	0	1	1
	0	0	0
A andor* B	1	1	1
	1	0	0
	0	1	0
	0	0	1

It should be recalled that the ANDOR (and likewise the ANDOR*) link allows *either* controller to change the state of the system. If, irrespective of the initial state of the system, B's will must prevail in a sequential link, then ANDOR* proceeds as follows—

1. Suppose that initially the states are—

A	B	System
1	1	1

An event occurs that requires A's decision, which can be signified by 1 or 0, depending on which state he wishes the system to be in after the event. When A has made his decision, B can either agree (denoted by

Controller	AND*	OR*	ANDOR*
A	State of controller corresponds		
B	to state he wishes the system to be in		1—if he agrees with A 0—if he disagrees with A

Fig. 3—The meaning of the controllers' states

state 1) or disagree (state 0) and the results for the system are shown below—

A's decision	*B's reaction*	*Comment*	*Result for the system*
1	1	No change	1
1	0	A considers a change unnecessary, but B overrules him	0
0	1	A wants a change, B agrees	0
0	0	A wants a change, B disagrees; result is no change	1

2. Suppose now that the initial conditions are—

A	B	System
0	0	1

After an event occurs, there are again the same four possibilities as enumerated in *1.*

3. Suppose the initial conditions are—

A	B	System
1	0	0

After an event, the four possibilities are—

A's decision	*B's reaction*	*Comment*	*Result for the system*
1	1	A believes the system should change to 1 (he held this view before the event) and B agrees	1
1	0	A wants a change, B disagrees; result is no change	0
0	1	A wants the system to remain in its present state and B agrees; result is no change	0
0	0	A wants no change, but B overrules him	1

4. Finally, if the initial conditions are—

A	B	System
0	1	0

the four possibilities would be similar to *3*, with the same results for the system.

Thus, irrespective of the initial state of the system, the correspondence between the controllers and the system complies with that stated in Tables 2 and 9. This conclusion can be easily verified for the AND* and OR* links.

Hierarchical aspects of control

THE introduction of sequential links provides an interesting new angle of looking at some facets of hierarchical control. The link AND* implies hierarchy in the sense that in A and* B the controller A is required to act in the first instance, but for his action to take effect he needs B's approval; if A does not wish to act, however, the matter is not even referred to B. In this case, B can overrule A only *when the latter decides to act*, but not otherwise, as shown in Table 9.

The link OR* has a similar hierarchical connotation, except that in A or* B the controller B does not need to confirm A's decision to act; but, if A decides not to act, then (and only then) B is approached and he must decide whether to uphold or to reverse A's decision. The link ANDOR*, however, requires that every decision of A is referred to B.

In addition to the way in which sequential linkages affect the participation of the individual controller in the decision-making process, they may well affect the type of decisions made by frontline controllers. For example, in the procedure A or* B, controller A knows that his decisions are subject to a close scrutiny by B only when A is OFF and, if he wishes to avoid such scrutiny, his judgment may well be impaired, resulting in his decisions being more biased towards the ON state than he would have been had he acted as a single controller. Such situations are not uncommon in administrative systems when the frequency of disagreement between A and B (when A is OFF and B is ON) is interpreted as a measure of A's incompetence or when A begins to anticipate B's decisions and ceases to switch to OFF if he expects B to overrule him.

It appears then that there are two distinct aspects of hierarchy. The first relates to the relative frequency with which a controller is called upon to act: if two controllers A and B are assigned to a system, B's position may be considered more privileged if he need not handle every problem that the system is presented with. Controller A is the one to act in the first instance and there are matters that he can handle satisfactorily without reference to B, whose involvement is therefore less frequent than A's.

The other aspect of hierarchy is that of authority or the ability of one controller to overrule decisions of others. Thus, B may be regarded as superior to A if B has the authority to reverse A's decisions, but not vice versa.

Consider the various linkages that were discussed earlier.

1. Time-independent linkages

AND Each controller has the power to stop an activity.

OR Each controller has the power to start an activity.

ANDOR Each controller can either start or stop an activity.

In addition, each controller can act as frequently as any other. There is therefore no special privilege (either in terms of frequency of action or in terms of authority) that any one controller enjoys and that is denied to others. Similarly, the procedure $Z(x)$ for control gives equal opportunity for all members.

2. Sequential linkages (for two controllers)

AND* In $\overline{\text{A and* B}}$, the latter acts less frequently than the former, also B can overrule A, but only for certain decisions of A.

OR* Here, the position is similar to the AND* linkage, but the type of decisions that are referred to B are different.

ANDOR* Both controllers are called upon to make decisions in every case, but, while B can overrule A (in $\overline{\text{A andor* B}}$), A has no authority to overrule B. Thus, B enjoys a privileged position of authority, though not that of lower frequency of action.

Clearly, then, time-dependent linkages introduce hierarchical control. Linkages AND* and OR* remove B from the front line and give him some authority over A; ANDOR* gives B full authority, but to exercise it he must scan *all* the decisions of A. It would seem that remoteness from the scene of action and complete authority are not compatible; to achieve the former, some of the latter must be sacrificed.

Sequential gates are, incidentally, relevant in designing training programmes. When a trainee A is connected to a trainer through an AND* or OR* linkage, only some of A's actions are examined by B, either those that are more important to the working of the system or those for which A's training is considered to be incomplete, whereas the ANDOR* linkage calls for all actions of the trainee to be checked by the trainer.

The multi-role controller

A CONTROLLER need not be confined to looking after the performance of a single system or a single task. He may be assigned to control several systems and his role may change from system to system or from task to task and so may his relationships to other controllers change, depending on the control procedures in force.

Take as an example two controllers A and B who jointly supervise three systems and suppose the controllers are linked as follows—

For system 1	A and B
For system 2	A and* B
For system 3	B or* A

The role of A in relation to B is different for each of the systems. In this example, even hierarchical relationships are not maintained. In system 1, no hierarchy is specified; in system 2, B has a hierarchical advantage; but, in system 3, A enjoys an advantage.

In establishing what hierarchical relationships exist between two managers in an organisation, it is therefore necessary to list their control linkages with respect to the many tasks that they are associated with; the wider the variety of such linkages, the more difficult it is to describe the relationship between the managers in a simple and concise statement. Such a state of affairs also raises the problem of the 'carryover effect'. Human controllers do not behave like automatons that are capable of switching from one mode of control to another without any after-effects. An ambivalent hierarchical relationship between two controllers that depends on how they are connected at any moment in time may well lead to an erosion of the intended procedures, whereas a more consistent hierarchical relationship reduces the frequency with which each controller needs to switch and to adapt to a new role and this is therefore more conducive to the maintenance of the designed procedures.

Multi-control procedures for the the same system

FOR any given set of circumstances, only one control procedure should apply, otherwise ambiguities may arise over the relationships between controllers and their system. This does not mean that all such relationships need to be determined by a single procedure. There could be certain circumstances that require one procedure to be used and others that call for another. So long as the two sets of circumstances are mutually exclusive, only one control procedure will apply at any one time.

Consider the following example. Two controllers are assigned to a system, but the linkages between the controllers depend on the circumstances, three classes of which are identified—

1. Quality characteristics that define the performance of the system need not be too stringent and the control procedure is then A or B.
2. Quality characteristics are not too stringent, but, because B has many other duties to attend to, the procedure becomes A or* B.
3. Quality characteristics are stringent and it is therefore necessary to institute the procedure A andor* B.

In this example, the two controllers are linked in three different ways, depending on the prevailing circumstances. It is as if a control procedure of a higher level is superimposed on the system. The supercontrol procedure consists of a supercontroller X whose task is to monitor certain parameters of the environment and of the system and to identify with the aid of this information the class of circumstances that prevail (see Fig. 4). If the class of circumstances changes, the supercontroller switches to another state, which corresponds to a new control procedure that the two controllers should adopt. Thus, as long as class *1* prevails, the controllers behave according to A or B; when the supercontroller perceives that the circumstances have changed to *2*, he switches the controllers to A or* B and so on. There is a one-to-one correspondence between the classes of circumstances and the control procedures of the system.

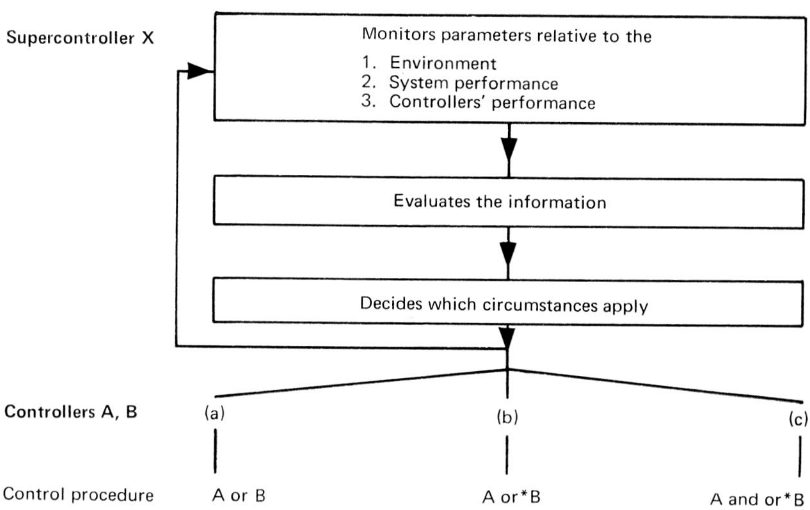

Fig. 4—An example of a supercontroller determining which control procedure to apply

Multi-controllers

As THE number of controllers of a system increases, the number of alternative procedures that can be devised increases very rapidly. In the case of a single controller in charge of a two-state system, the number of states of the controller is two, whereas the number of possible procedures is four as shown in Table 1. For two controllers, the number of possible combinations of the controllers' states is four, whereas the number of possible procedures is 16, listed in Table 5.

Consider now the case of three controllers A, B and C. There are eight possible combinations of their states—

A	B	C
1	1	1
1	1	0
1	0	1
0	1	1
1	0	0
0	1	0
0	0	1
0	0	0

In general, the number of these combinations for n controllers is 2^n and the number of procedures that can be enumerated for the system is $2^{(2^n)}$ and this number becomes formidable even for modest values of n as shown in the following examples.

Number of controllers n	*Combinations of controllers' states,* 2^n	*Possible procedures,* $2^{(2^n)}$
1	2	4
2	4	16
3	8	256
4	16	65 536

The number of possible procedures is compounded even further when the concept of sequential links between the controllers is introduced and naturally even further when a system with more than two states is examined.

In determining control procedures for administrative systems, not all the possible alternatives need of course be considered. For example, in the two-controller, two-state system described in Table 5, two alternatives (1 and 16) need no controllers and four alternatives (6, 7, 10 and 11) need only one controller, so that the choice of a control procedure narrows down from 16 to 10 alternatives and for administrative systems possible to six (AND, OR, ANDOR, NAND, NOR, NANDOR) or even to three (AND, OR, ANDOR). Nevertheless, the point should be made that, for a system with a larger number of states or controllers, there is a plethora of alternatives to choose from in the design of control procedures.

Conclusions

I HOPE that this paper has demonstrated some of the intricacies that may be implied in assigning several controllers to look after a single task or a single system. The need for meticulous specifications for the linkages between controllers becomes self-evident and this is particularly relevant to administrative systems for which we often find that rigorous definitions are lacking or inconsistent, with the resultant uncertainties both from the point of view of the individual controllers and from the point of view of the system.

References

1. Culbertson, J. T., *Mathematics and Logic for Digital Devices* (D. Van Nostrand, New York, 1958)
2. Hoernes, G. E. and Heilweil, M. G., *Introduction to Boolean Algebra and Logic Design* (McGraw-Hill, New York, 1964)
3. Eilon, S., 'Linkages between controllers': *J. Mgt. Studies*, 1969, **6** (1), 45–52
4. Eilon, S., 'Control systems with several controllers': *J. Mgt. Studies*, 1965, **2** (3), 259–268

GLOSSARY OF COMPUTER CONTROL TERMS

Abbreviations

CPU	Central processor unit
CRT	Cathode ray tube
CSMP	Continuous systems simulation language
DDA	Digitally directed analog—that is, set point supervisory control
DDC	Direct digital control
EDP	Electronic data processing
ID	Integral, derivative two-term controller
PAM	Pulse amplitude modulation
PDI	Proportional, derivative and integral three-term controller
PDM	Pulse duration modulation
PI	Proportional, integral two-term controller
PPM	Pulse position modulation

Access time The time it takes a computer to produce a bit of information from its memory section (also called *read time*) or the time it takes a computer to store information in its memory section (also called *write time*).

Accumulater A special storage register in the arithmetic unit that receives the results of arithmetic operations.

Accuracy The ability of a measuring device or system to give indicated values that closely approach true values of the quantities being measured. The measure of *control system* accuracy is often termed *system deviation*, the difference between the actual value of the controlled variable and its ideal value.

Address A numerical or other expression (cue) that designates the particular location of information in a computer's memory circuit for immediate location when needed. Also called *symbolic address* to denote the symbol or cue.

Algorithm A set of rules or a sequence of well-defined operations (steps) that leads to the solution of a problem (usually a mathematical problem).

Analog A continuously varying measurement signal (such as an electrical voltage, current or resistance, pneumatic pressure or rotational speed) is the analog of the numerical value of that physical variable.

Analog computer A computer that handles information that is analogous to some value of a controlled process. For example, the signal to an analog computer could be an electrical voltage that increases or decreases with the rise or fall of a process temperature. The voltage is analogous to the temperature.

Arithmetic unit The section of the computer in which arithmetic and logical operations take place.

Artificial language A computer language specifically designed for ease of communication in a particular area of endeavour, but one that is not yet natural to that area. This is contrasted with a natural language, which has evolved through long usage.

Assemble To gather, interpret and co-ordinate data required for a computer program, translate the data into computer language and project it into the final master routine or program for the computer to follow.

Assembler A method of instruction and subroutines that integrates relevant data into a computer program or converts from a low level language into a machine code.

Assembly language The operational language used in a specific machine-oriented computer program. There are various assembly language systems such as EASY, APRIL, EASYCODER, denoting different programming systems. Second level of computer language. See *Machine language* and *Programming language* for other levels.

Attenuation The decrease in the strength of a signal as it passes through a control system: the opposite of *gain*.

Automatic computer A computer that can process a specified volume of work, its assigned function, without requiring human intervention, except for program changes.

Automatic controller A device or instrument that is capable of measuring and regulating by receiving a signal from a sensing device, comparing this data with a desired value and issuing signals for corrective action.

Automatic error correction A technique for detecting and correcting errors that occur in data transmission or that occur within the system itself.

Automatic quality control Technique for evaluating the quality of a product being processed by checking it against a predetermined standard and automatically taking the proper corrective action if the quality falls below the standard.

Automation Automatically controlled operation of an apparatus, process or system by mechanical or electronic devices that take the place of human observation, effort and decision.

Auxiliary function In automatic machine tool control, a machine function other than the control of the motion of a work-piece or cutter. Control of machine lubricating and cooling equipment are typical auxiliary functions.

Backlash In a mechanical operation, the play between interacting parts (such as two gears) as a result of tolerance.

Band Range or scope of operation. In communications, a range of frequencies, as between two specified limits. In instrumentation, a range of values representing the full control authority of an instrument.

Binary Having only two possibilities, one of which is selected in either/or fashion. Computer information is handled as one of two possible digits (0 or 1) expressed as *binary digits* or *bits*. A system constructed on base-two arithmetic.

Binary digit Defined as either 0 or 1.

Binary number system A number system in which the only digits used are 0 and 1, instead of 0, 1, . . . 9 as in the decimal system.

Bit Binary digit.

Bit rate The speed at which bits travel over a communications channel, expressed in bits per second.

Boolean algebra Problem-solving system based on a series of either/or or yes/no (binary) conclusions. Named after George Boole (1815–64), English mathematician.

Buffer Intermediary storage area between two data processing storage or data handling systems with different access times or formats. An interim system to facilitate interface between two other systems.

Buffer storage An intermediary stage between two computer data storage systems; usually necessary to synchronise or co-ordinate non-compatible elements of the two systems. The buffer storage stage may receive information from external sources and translate it into language that the internal storage unit can accept.

Cascade control An automatic system in which the control units are linked as in a chain. The output from one controller is used to regulate the next controller (that is, adjust its set point).

Channel In communications, an electrical transmission path among two or more stations or channel terminations in telephone or telegraph company offices, furnished by wire, radio or a combination of both. In computer terminology, a route along which signals can travel or that portion of a computer memory to which a particular output station has access.

Closed loop An automatic control system in which the measured value of the variable is used as feedback to link a controlled process back (close the loop) to the original command signal and bring this value back to the set point. Contrast with *open loop*.

Code A system of symbols that represent data values, forming a special language that the computer can understand and handle.

Comparator A device for checking the accuracy of transcribed data by comparing it with a second transcription, noting any variation between the two.

Compiler A method or program that translates and evaluates the source information language (in which the computer programs are written) into a simpler language that the computer can use. The change is from a high level language into a low level language.

Compiler language A computer language system made up of various subroutines that have been evaluated and compiled into one routine that the computer can handle. FORTRAN, COBOL and ALGOL are compiler language systems. Compiler language is the third level of computer language: see *Machine language* for other levels.

Computer A machine designed to receive data, perform programmed operations on this data and supply the results. See *Analog computer* and *Digital computer*.

Computer program Procedure for solution of a problem on a computer.

Control constant Value of proportional, integral and derivative constants.

Control point The value such as the temperature of an oven or the position of a cutting tool that is the result of some control action. It is not necessarily the desired value.

Control register Element in a digital computer that controls the proper sequence of information being fed in for processing. The control register keeps the next word always ready for feed-in.

Control terms See *conventional control*, *derivative action*, *integral action*, *proportional action*, *one-term controller*, *two-term controller*, *three-term controller*, *control constant, feedback control, feedforward control, on-line, off-line, closed loop, open loop, isolated, direct digital control, cascade control.*

Controller An instrument that holds a process or condition at a desired value such as a thermostat that holds room temperature at, say, 70° by sensing the actual room temperature (actual value), comparing it with the thermostat setting (desired value) and causing corrective equipment to eliminate any variation between the two. See also *one-term, two-term* and *three-term controller.*

Conventional control Use of mechanical, pneumatic or electronic devices to operate control valves to minimise difference ('error') between the required value ('set point') and actual value of a measured variable on the plant.

Convert To translate data from one form of expression to another.

Cross-coupling An effect in which an output is a result of more than one input. For example, pressure applied to the gimbal of a spinning gyroscope will cause the gyro to tilt in a different direction from the applied pressure. The gimbal's reaction is a resultant of its spin plus the applied pressure.

Cross-talk Interference in a given transmitting or recording channel that has its origin in another channel.

Cybernetics The theory of control and communication in the machine and the animal.

Damping A characteristic built into an instrumentation system to reduce over-response, instability or oscillation by slowing up reaction time. A simple example of damping would be the shock absorbers on a motor-car that damp the reactions of the springs.

Data A general term used to denote any or all facts, numbers, letters and symbols or facts that refer to or describe an object, idea, condition, situation or other factors. It connotes basic elements of information that can be processed or produced by a computer. Sometimes data is considered to be expressible only in numerical form, but information is not so limited.

Data reduction The process of converting large quantities of information into more manageable form. This usually results in a considerable reduction in data volume, as well as simplification of format.

Data reduction system Automatically operated equipment engineered to simplify the use and interpretation of a large quantity of data gathered by instrument installations.

Dead band A neutral zone in which a proportioning controller is 'satisfied' and calls for no corrective action. A temperature controller set for 72°, for example, might call for more heat at a room temperature of 71°; less heat at room temperature of 73°; thus operating with a central dead band of 2°. The term *dead band* is used in connection with proportioning control and is equivalent to *differential gap* in on/off control. See also *neutral zone.*

Dead time The time taken for a change at the input of a system to become apparent at the output. Applied also to any definite delay deliberately placed between two related actions to avoid overlap that might cause confusion or to permit a particular different event such as a control decision, switching event or similar action to take place.

Decimal digit Defined as one of the numbers 0, 1, 2 . . . 9 that are used in the counting system with the base 10.

Derivative action Control operation in which the rate of change of output from one controller is proportional to the rate of change of error between set point and measured value, so the speed of a correction depends on how fast a condition is deviating from the set point. See also *rate action.*

Differential analyser A type of analog computer (either electronic or mechanical) that solves differential equations by assigning analogies to mathematical functions.

Differential gap In on/off control, the differential gap is the span between the *on* and *off* switching points. A room thermostat, for example, that is set to control for 70° might switch the furnace on at 68° and off when the temperature has risen to 72°, giving it a 4° differential. See *dead band.*

Digital Refers to data expressed in numerical form. A system in which variables are represented by numbers of electrical pulses rather than by a continuous signal.

Digital computer A calculating machine that deals only with numerical information, rather than analogous relationships. An adding machine or abacus are simple digital computers; numbers are fed in and numbers are read out. Large digital computers process numerically coded business data at fantastic speeds.

Digital output An output consisting of a set of discrete quantities coded to represent digits in a numbering system. Most common example of digital output is an electric typewriter.

Digits See *binary digit*, *bit*, *decimal digit* and *octal digit.*

Direct digital control A control system in which a digital computer calculates the required control valve movement and outputs digital signals to obtain this movement.

Dump Programs for listing the contents of the memory; may also refer to the list itself.

End point control Quality control through continuous, automatic analysis. In highly automatic operations, final product is analysed; if there are any undesirable variations, the control system automatically brings about the necessary changes.

Error The general term refers to any deviation of a computed or a measured quantity from the theoretically correct or true value. The part of the total error from a particular identifiable source may be specifically referred to—say, a truncation error. In a restricted sense, it may refer to that deviation arising from unavoidable random disturbances or from the use of finite approximations to what is defined by an infinite series.

Executive routine A set of coded instructions that controls loading and relocation of other routines and, in some cases, makes use of instructions that are unknown to the general programmer. Effectively, an executive routine is a non-hardware part of the machine itself, except that it is superimposed on all lower level programs and instructional sets.

Feedback The signal (or signals) fed back from a controlled process to denote its response to the command signal. Feedback is derived by comparing actual response to desired response and any variation is fed as an *error signal* into the original control signal to help enforce proper system operation. Systems employing feedback are termed *closed loop* systems, with feedback closing the loop.

Feedback control Control action based on measured values of the variable to be controlled.

Feedforward control Control action based on values of the variable to be controlled predicted from current values of other process variables.

Frequency response analysis The use of alternating or pulsating signals to excite a control system that is being studied. The system's response to different frequencies permits analysis of its operating characteristics.

Gain The increase or amplification of a signal as it passes through a control system, given as the ratio between the input and output of a system.

Hardware Mechanical, electrical or electronic components that make up the computer—magnetic cores, printers, tape readers, etc. (See *software*, which denotes written information.)

Hunting A control system's continuous, cyclical search for a desired or ideal value. Very rapid hunting is usually termed 'oscillation'.

Hybrid (1) An electronic circuit having both valves and transistors. (2) A mixture of thin-film and discrete integrated circuits. (3) A computer having both analog and digital capability.

Identification The technique used for on-line calculation of the unknown constants in the transfer equations—done by making a deliberate change in one of the main variables and measuring the changes in others, the computer then calculating the constants from the measured changes.

Immediate access Ability of a computer to put data in (or remove it from) storage without delay.

Input The data being supplied to a computer for processing or the device employed to accomplish this transfer of data.

Instruction A coded directive for a certain operation and the values or locations of stored information being addressed by this directive.

Integral action Rate of change of the output from the controller is proportional to the error difference between measured value and set point. Thus, doubling the error will double the rate at which the control valve, connected to the output of the controller, will open.

Interface The common boundary provided by the equipment through which the computer is connected to the plant—includes input and output scanners, amplifiers, noise rejection devices and signal converters.

Iterative process A repeating cycle of operations, in computation, in which the successive solutions are closer and closer to the desired result.

Isolated Irrespective of inputs to the computer, it is taking no control action.

Laplace transform $F(p)\int_{\infty}^{o} \exp(-pt) f(t)dt$

This integral transform sets up a correspondence between the time domain and the general complex variable domain. It has, by virtue of the special properties of the exponential functions noted above, the useful property of transforming a constant coefficient differential equation relationship between functions of time into an algebraic relationship between their Laplace transforms. This enables such differential equations to be solved by simple algebraic manipulations.

Language See *artificial language*, *assembly language*, *machine language* and *programming language*.

Linear programming A technique for finding the best solution(s) to a problem involving many variables that can be expressed as mathematical equivalents, which must be linear to (that is, must change in direct relation to) any change in the variables for the solutions to be valid.

Linearisation The process of scaling plant signals for use in the computer.

Linearity The ability of a device or system to change its output in exact proportion to changes of input—for example, if an electrical meter were perfectly linear, 1 per cent of current would deflect the indicating needle 1 per cent of total travel at any point throughout the travel of the mechanism.

Load What the process calls for in fuel or energy input.

Location A storage position in the computer memory.

Logger An instrument that automatically scans certain values in a controlled process and logs or records the readings for future record.

Logic In computers and in information-processing networks, the systematic method by which information is processed, usually with each successive step influencing the next one.

Loop In electronics, a complete electrical circuit. In computer terminology, a series of instructions being performed repeatedly until a terminal condition prevails. In automatic control, loop is the path followed by command signals that direct the work to be done and feed back signals that flow back to the command point to indicate what is actually being done. See *closed loop* and *open loop*.

Magnetic core A computer element for switching or storing one bit (binary digit) of information. Thousands of cores may be used in a large computer system to store information that needs to be handled in a few millionths of a second.

Magnetic core storage The process of storing information on magnetic memory elements for later use by computer. A bit (binary digit) of information is stored in each element by magnetic polarity (plus or minus polarity designating *ones* or *zeros*).

Machine language A series of 'bits' written as such to instruct computers. The first level of computer language. See *assembly language* (second level) and *compiler language* (third level).

Mathematical model A mathematical representation that simulates the behaviour of a process, device or concept.

Manipulated variable The one variable (value, condition, quantity, etc.) of a process that is being controlled. By manipulating this variable, the process can be controlled.

Matrix In an automation system, an arrangement of circuit elements such as wires, relays and diodes that performs logic functions in data handling. A device to route, distribute or, in some cases, decode data signals.

Memory Usually used to denote data retention inside the computer in contrast to *storage* for data retained outside the computer. In either case, the term means collecting and holding pertinent information until needed by the computer.

Micro-code A system of computer coding that includes sub-operations such as multiplication and division that are not ordinarily accessible in programming. A list of very small program steps.

Micro-program Program of basic subcommands to be handled by a computer to provide data for a higher level program; also the program that has been constructed from these precomputed subcommands.

Micro-programming Setting up basic sub-operations for the computer to handle, after which they are combined by the programmer and again presented to the computer in a higher level program. If a computer has only basic instructions for adding, subtracting and multiplying, for example, the instruction for dividing would be defined by micro-programming.

Model See *mathematical model.*

Modifier A component or subsystem that alters an instruction without changing the form of energy—that is, changes electrical input energy to electrical output signals.

Monte Carlo method Any method for obtaining a statistical estimate of a desired quantity by random sampling.

Multiplex (1) To execute several functions simultaneously in an independent, but related manner. (2) To interleave or simultaneously to transmit two or more messages on a single channel.

Multi-processing Two or more processes running simultaneously in one computer system.

Multi-programming Two or more programs operating simultaneously in one computer system.

Neutral zone A range in the total control zone where the controller does not respond to changes in the controlled process. A dead zone, usually at the mid-point of the control range. See also *dead zone*.

Noise In electronic systems, the undesirable background signals that ciutter up the control signal. Originally called noise, because the spurious signals in audio systems created static that interfered with the amplified sound signal.

Normaliser A computer operation that shifts all digits (bits) of a word over to right or left to accommodate the maximum number of digits.

Numerical analysis To provide convenient methods for obtaining useful solutions to mathematical problems and for extracting useful information from available solutions that are not expressed in tractable forms.

Numerical control A method of controlling a process, machine tool or assembly line by means of digital instruction carried on punched tape, magnetic tape or other such transmission system.

Octal digit Defined to be one of the numbers, 0, 1, 2, . . . 7 that are used in the counting system with the base 8.

Off-line operation In control terminology, a system in which the computer is not immediately or directly responsive to the moment-to-moment events of the controlled process, because they are not directly connected. Some time may elapse between the occurrence of an event and the computer's reaction to it. In computer terminology, operation pertaining to equipment or devices not under direct control of the central processing unit.

Offset The difference between the value or condition desired and that actually attained.

On-line operation In control terminology, operation in which data is fed directly from the controlled process into the computer and the computer exercises direct control based upon this data. Since the computer's reaction is moment to moment without appreciable time lag, results are said to be in real time (q.v.).

One-term controller Controller having only proportional action.

Open-ended In computer terminology, referring to a system or process which is receptive to augmentation.

Open loop A control system in which the information of the measured variable is relayed to the operator and any control action must therefore be taken by the operator (that is, there is no automatic control or self-correcting action for error). Contrast with *closed loop*.

Operations research Solution of large-scale or complicated problems arising from any operation (production, distribution process, etc.) in which a major decision must be based on solving a large number of variables.

Optimisation The specified level of operation is chosen to achieve some particular objective, such as general operating profit being maximised.

Output In electronics, the power, energy or signal delivered by a device or system. In computer terminology, the processed information being delivered by a computer or the device employed to accomplish this data delivery.

Parallel operation Type of information transfer within a digital computer in which all digits of a word are handled simultaneously.

Parallel processing The operation of a computer so that programs for more than one run are stored simultaneously in its storage and executed concurrently; also operation of computer hardware in parallel.

Parameter An inherent characteristic of the plant or process.

Patch In electronics, a temporary electrical connection. In computer terminology, an expedient modification of a computer program.

Peripheral equipment Equipment that is external to and not a part of the central processing instrumentation. Includes such equipment as tape punches and readers, magnetic tape or disc storage units, digital printers, graphic recorders and typewriters.

Precision The degree of agreement among repeated measurements of a given quantity, taken by the same method and under the same conditions.

Probability theory A measure of likelihood of occurrence of a chance event, used to predict behaviour of a group, not of a single item in the group.

Program A set of instructions that can be accepted by a computer to perform a task. It is given in a language usable by the computer.

Program control A control system that automatically holds or changes the process variables on the basis of time to follow a prescribed program for the process.

Programming language Method of writing programs. The terms *machine level* and *low level* refer to languages that can be readily understood by the computer, because one language instruction generally generates one word of instruction in the machine. Higher level languages employ symbols more nearly related to human language and are therefore more easily understood by non-specialists.

Proportional control Control in which corrective action is always proportionate to any variation of the controlled process from its desired value. For example, instead of snapping directly open/closed in the manner of two-position control, the proportional valve will be always positioned at some point between open and closed, depending on the system's flow requirement at any given moment.

Pulse modulation Use of a series of pulses, modulated or characterised to convey information. Types of pulse modulation include amplitude (PAM), position (PPM) and duration (PDM) systems.

Punched card Paper card used to store data that can be read by a computer. The standard card is $3\frac{1}{4}$ in $\times$ $7\frac{3}{8}$ in and contains eighty columns and twelve rows in which numeric information, alphabetic information or special characters may be punched.

Punched tape A tape, usually paper, upon which data is stored in the form of punched holes, for reading by a computer. Same principle as the *punched card.*

Quantum The smallest unit of measure employed in a system.

Queuing theory Research technique concerned with the correct sequential orders of moving units. May include sequence assignments for bits of information, whole messages, assembly line products or motor cars in traffic.

Random access Technique of storing information so that its recovery time will not be influenced by how recently it was stored. A random access device permits information fecovery at will, with no appreciable time penalty for the order in which it was last handled.

Random access storage In random access storage, information can be recovered immediately, regardless of when it was stored. For example, magnetic core memory devices will usually yield any bit of information with almost no time penalty, regardless of where it is located, whereas magnetic tape must be run until the required information can be found.

Random numbers Array of independent digits having no logical interrelationships, so that the occurrence of any particular one is totally unpredictable.

Rate action Control action in which the rate of correction is proportional to the rate of deviation from a standard value; also called *derivative action.*

Read In computer terminology, the procedure of obtaining data from a particular source.

Real time The actual time for a physical process to take place. A system in which the time taken by the computer to process information from the plant and to send out signals to the plant is related to the duration of the events controlled by the calculation.

Record In computer terminology, a group of related facts or fields of information treated as a unit; thus, a listing of information, usually in printed or printable form. Also, the procedure of putting data into a storage device.

Register A device capable of storing a specified amount of data, such as one word.

Reset action Type of control in which correction is proportional to both the time and the amount that a controlled process has deviated from its desired value and for which provision is made to guarantee return of the process to its set point.

Reset rate Number of corrections a control system makes per unit of time (usually per minute).

Routine Set of coded instructions that direct a computer to do a certain task. A functionally related block of instructions.

Serial access Process of storing or recovering information in relation to a time sequence so that successive bits are located in accordance to the order of their storage.

Serial operation Computer operation in which all digits of a word are handled sequentially, rather than simultaneously.

Servo-mechanism A control system that provides the following—a command instrument to control or program the final process, amplification to strengthen and modify the command signal, work instrumentation to manipulate the controlled process and feedback provision to initiate corrective action when needed. Since feedback signals go from the controlled process back to the original command station, a servo system is said to operate *closed loop*.

Set point The required or ideal value of a variable to which it is desired to control it. This value is usually preset into the computer or system controller.

Signal conditioning Modifying signals into the form required for its proper use by measuring, recording or controlling instruments. Such conditioning may include amplification, linearisation, attenuation and isolation.

Simulation Practice of substituting instrumentation (often a computer) for actual operational conditions so that a real process can be simulated for valid data.

Simulator A computer or other model of an actual system that so closely imitates the system's operation that usable data can be extracted from it.

Software Jobs, systems, flow charts, programs, routines, codes and other written information used to operate a digital computer. Contrast with *hardware*, the equipment itself.

Solid state Term given to electronic components made of solid materials, as opposed to valve or electromechanical relays. Transistors and magnetic computer cores, for example, are solid-state components.

Standby A separate system of controllers that enables control to be maintained on failure of the principal control system or in emergencies or when the computer is not available. It may range in complexity from a simple manual facility to an auxiliary computer.

Stochastic An approach to analysis of a process employing probability theory.

Storage The means by which the computer retains information for subsequent use. Ferrite cores are usually used to provide *direct access* storage. Discs or magnetic tape are used to provide large volumes of storage for material not required to be instantly available. This is sometimes referred to as *back-up* storage.

Subroutine A set of packaged instructions for computing some particular function, therefore only part of a complete routine. It may be used many times during the computation of a program, but is written only once in the whole code.

System A collection of parts or devices making an organised whole, through some form of regular interaction or interdependence.

Three-term controller Controller having proportional, integral and derivative action.

Time lag The time during which a variable undergoes an exponential change.

Time sharing Using a device, such as a computer, to work on two or more tasks, alternating the work from one task to another. Thus, the total operating time available is divided among several tasks, using the full capability of the device.

Transfer function A mathematical relationship between a set of conditions at two different times, typically between the input and output of a system.

Transfer rate The rate at which data may be transferred between the computer registers and storage, input or output devices. Usually stated in number of characters per second.

Transient state Condition in which a system or process is in a state of change—for example, in a control system, a change of set point will induce a transient state that will persist until stable conditions are re-established.

Two-term controller Controller having proportional and integral action or integral and derivative action.

Value theory Process of assigning numerical significance to the worth of alternative choices.

Variable Any process quantity that can change with time and can be measured or altered by a control system. Examples are pressure, flow, liquid level, humidity and weight.

Vector A set of fixed or variable numbers—for example, the vector $X = (x_1, x_2, x_3 \ldots x_n)$.

Word A group of characters forming the contents of one storage location. In most digital computers, a word is composed of 24 bits (binary digits) or of a combination of these bits with letters.

z-Transform $\mathcal{Z}u_r = u^o(z) = \sum_{-\infty}^{\infty} u_r z^{-r}$

It is the transform of a sequence u_r obtained by multiplying it by a function z and summing it from $r = -\infty$ to $r = \infty$.

Biographical notes of speakers, chairman of sessions and committee personnel

Dr A. Al-Shaikh, Senior Research Engineer, Westvaco Corporation, P.O. Box 5207, N. Charleston, S.C. 29406, U.S.A.
B.Sc., Ph.D., University of London
Member of IEEE, ISA and Institute of Measurement and Control
Appointed in 1965 by Science Research Council to work with Dr J. I. Soliman on soaking pit control for steel industry. Joined Westvaco in 1966 (process control research laboratories) to work on computer control.
Fields of work—research, development and application in mathematical modelling, process identification, simulation, process computer control and application of control theory to industrial processes

O. Alsholm, Process Computer Manager, Billeruds AB., Gruvöns Bruk, 664 00 Grums, Sweden
M.Chem.Eng., Chalmers University of Technology (Sweden); M.S.(Chem.Eng.), Columbia University, N.Y.
Joined Billeruds in 1961, spent 1964 as systems analyst with IBM Corporation, San José, Calif.

R. H. Anderton, Management Scientist, IPC (Group Management) Ltd., Feltham, Middx.
B.Sc.(Eng.) in electronic engineering, University of London
Fields of work—servomechanism development and control theory lecturing at RAF Technical College; systems engineering in aerospace industry; operational research as consultant

G. E. Annergren, Head of Pulping Research Dept., Svenska Cellulosa AB., Sundsvall, Sweden
Studied at Royal Institute of Technology, Stockholm
Field of work—research, mostly in pulping

D. Attwood, Head Scientist, Engineering & Scientific Services Division, Bowaters U.K. Paper Co. Ltd., Northfleet, Kent
B.Sc.(Physics), University of London
Joined Bowaters in 1953; worked on all aspects of research and development, now head of Scientific Services Division
Awarded 'Silver Jubilee' medal by Technical Section in 1961
Member of Technical Section London Division, Fundamental Research and Systems Engineering Committees

A. K. Bjerring, Central Systems Engineer, Stapleton, Dowdeswell, Kelly & Hunt Ltd., Montreal, Canada
B.A.Sc. (Electrical Engineering), University of British Columbia
M.A.Sc. (Electrical Engineering, Central Systems), University of Toronto
Field of work—computer applications, development of real time executive, development of a dynamic model of froth flotation, preparation of pulp, paper and mining industry courses on process control and development of control systems in kraft pulping and bleaching

Sir Lawrence Bragg, Professor Emeritus, The Royal Institution, 21 Albemarle Street, London W1X 4B2
C.H., O.B.E., M.C., M.A., D.Sc., F.R.S.
Born in Adelaide, Australia and educated there at St Peter's College and Adelaide University; Allen Scholar at Trinity College Cambridge, in 1914, where he was a Fellow and lecturer in natural sciences. Nobel prizewinner for physics in 1915. Elected F.R.S. in 1921. Technical advisor to G.H.Q. France, on sound ranging, 1915–18. Awarded M.C. and O.B.E. in 1918. Langworthy professor of physics, Victoria University of Manchester, 1919–37. Director of National Physical Laboratory, 1937–38. Cavendish Professor of Experimental Physics, Cambridge, 1938–53. Fullerian Professor of Chemistry, Royal Institution, 1953–66. Director of the Royal Institution, 1964–66
Many honorary degrees, medal awards and honorary memberships of societies and clubs at home and overseas. Numerous scientific papers on crystal structure and other subjects—*X-rays and crystal structure*, 1915 (with Sir W. H. Bragg); *The crystalline state*, 1934; *Electricity*, 1936; *Atomic structure of minerals*, 1937; *Crystal structures of minerals*, 1965 (with W. F. Claringbull). Many educational Royal Institution lectures and publications for school children and adults

Dr D. B. Brewster, Vice-President, Stapleton, Dowdeswell, Kelly & Hunt Ltd., Montreal, Canada
B.Sc. (Chemical Engineering), University of Leeds, Ph.D., University of Leeds
Field of work—Production and process experience in petrochemical and chemical industries; pioneering of paper industry computer control; Research Director with Westvaco for papermaking, converting and control

Dr N. K. Bridge, Paper & Board Division, Manager, PIRA, Kenley, Surrey
B.Sc.(Physics), Ph.D.(Physical Chemistry), University of Manchester
Fields of work—spectroscopy and photochemistry; U.S. Atomic Energy Commission; sheet formation of paper; latex; on-line process control computer papermachine installation

Prof. J. F. Coales, Professor of Engineering (Control), University of Cambridge
B.A.(Maths & Physics), M.A., University of Cambridge; Reader (1958), Professor (1965); Fellow of Clare Hall, Cambridge
Participated on several educational and professional governing bodies; Chairman of Measurement Section (1954–55), Chairman of Control and Automation Division (1965–66), currently Vice-President, Institution of Electrical Engineers; President, Society of Instrument Technology (1958–59); Chairman, Automation Council (1964–66); President, International Federation of Automatic Control (1963)
Principal fields of work—radar for gunnery; computer development, microwave instruments; consultant on research and development, particularly in automation

D. L. Cooper, Head of Computer, Mathematics and Statistics Section, PIRA, Kenley, Surrey
B.Sc.(Mathematical Science), University of Edinburgh; Fellow of Royal Statistical Society; Fellow of British Computer Society; Research Officer, Wool Industries Research Association (1954–58); Principal Computer Projects Officer, British Railways Board (1965–67); Statistician (1959–64) and Head of Computer Section (from 1967), PIRA

H. D. Cyprus, Senior Scientist, Engineering & Scientific Services Division, Bowaters U.K. Paper Co. Ltd., Northfleet, Kent
B.Sc.(Chemical Engineering), University of Leeds
Previously with British Titan Products Co. Ltd. (4 years)

G. Donkin, Head of Control Systems Group, Automation Systems Division, Ferranti Ltd., Gatley, Ches.
B.Sc.(Electrical Engineering), University of Durham
Served in Royal Navy as Electrical Officer, then joined Ferranti Ltd. in 1946 to work on missile control systems. Transferred to Automation Systems Division on its formation in 1960 to work on industrial applications of Argus process control computers

Prof. S. Eilon, University Professor, Dept. of Mechanical Engineering, Management Engineering Section, Imperial College of Science & Technology, London SW7
D.Sc.(Eng.), Ph.D., D.I.C., F.I.Mech.E., F.I.Prod.E., member of Operation Research Society. Educated in Haifa and London; lectured and studied in engineering subjects in Israel, London and U.S.A. Has served and advised on many professional and educational bodies, committees; edited technological and engineering publications; currently external examiner for University of Birmingham and London Graduate School of Business Studies; for Brunel University (1961–64), Manchester Business School (1967–68) and Durham University Business School (1967–68); consultant to several industrial firms, a college of technology and a banking house

H. W. Emerton, Director, Paper & Board Research and Development, Reed Paper Group Ltd., Aylesford, Maidstone, Kent
M.A.(Natural Sciences), Lincoln College, University of Oxford; Senior Executive Course, Manchester Business School; Fellow of Institute of Physics; previously member of Council and Vice-President, Royal Microscopical Society; member of TAPPI, Technical Section CPPA and R & D Society (London)
Research on fibre structure and beating, electron microscopy, etc. with PIRA, Kenley (1949–58)
Publications on fibre structure and properties; the beating process and electron microscopy

P. H. Engelstad, Project Manager, Work Research Institutes, Oslo, Norway
Civil Engineer; degree in chemical engineering, Technical University, Trondheim, Norway; studies in sociology, University of Oslo
Employed previously as process technologist; technical adviser on production and quality control; Research Assistant, Institute of Industrial Social Research, Trondheim

Dr A. R. Farmer, Systems Engineer, AEI/Elliott Automation, Kidsgrove, Stoke-on-Trent
Mechanical Sciences Tripos, Ph.D.(Electrical Engineering), University of Cambridge
Field of work—Radio equipment and RAF instructor, control systems development, latterly with the development and application of analog modules in papermachine automation

G. Gavelin, Technical Director, AB. Lorentzen & Wettre, Box 49 006, Stockholm 49, Sweden
Graduate of Royal Institute of Technology, Stockholm
Worked for 15 years on process and quality control in papermills in Sweden, Canada and U.S.A., also for 10 years as head of technical services at the Central Laboratory of the Swedish Papermills

Dr J. Grant, Consultant, Hehner & Cox Ltd., London EC3
B.Sc.(Chemistry), M.Sc. and Ph.D., University of London;
Fellow of the Royal Institute of Chemistry
Awarded 'Silver Jubilee' medal of the Technical Section in 1958
Experience with analysts, as papermill chief chemist and managing director
Author of technical books on pulp and paper

J. S. Harris, Technical Manager, Wiggins Teape Ltd., Chartham Paper Mills, near Canterbury, Kent
B.Sc.(Chemical Engineering), University of London; A.C.G.I.
On project and development work at Chartham since 1964

S. Hem, Control Engineer, Reed Research and Development Centre, Aylesford, Maidstone, Kent
B.Sc.Tech. and M.Sc.Tech., University of Manchester; Ph.D. candidate, University of Cambridge
Two years at Norwegian papermill, followed by research work at two British mills since 1964

W. D. Hoath, Research Division Manager, Wiggins Teape Research and Development Ltd., Beaconsfield, Bucks.
Chartered Engineer, Member of the Institute of Electronic & Radio Engineers
Member of TAPPI; Member of London Division Committee and Executive Committee of Technical Section BP & BMA
Field of work—Advancement of improved on-line measurement and control of the papermaking and paper coating processes

Prof. R. A. Holm, Director, Division of Industrial and Environmental Systems, The Institute of Paper Chemistry, Appleton, U.S.A.
Ph.D.(Chemical Engineering), University of Delaware; Research Associate and Associate Professor at Institute of Paper Chemistry
Principal fields of work—refractory metals, textiles (polymer spinning), process control education and chemical engineering research

Dr F. L. Hudson, Senior Lecturer in Paper Science, University of Manchester Institute of Science & Technology, Manchester 1
B.Sc. & M.Sc., University of Manchester; Ph.D., University of California
Principal fields of work—research in paper and plastics for papermaking, lecturing in papermaking
Publications cover patents and papers on wet strength, dimensional stability, sorption of sulphur in relation to the permanence of paper

H. R. Hutley, Group Chief Chemist, Peter Dixon & Son Ltd., Grimsby, Lincs.
Fellow of Institution of Plant Engineering
Principal fields of interest—education in papermaking technology, standard methods of test for paper and board

R. E. Johnston, Assistant Lecturer in Paper Science, University of Manchester Institute of Science & Technology, Manchester 1
Fellowship Diploma of Chemical Engineering from the Royal Melbourne Institute of Technology

C. G. Jolliff, Computer Control Systems Supervisor, Southern Kraft Division, International Paper Co., P.O. Box 2328, Mobile, Alabama 36601, U.S.A.
B.A. (Vanderbilt University, Tennessee)
Studied at medical school; graduate work in chemistry
Employed in the paper industry successively as chemist on statistical quality control and computer process controller, with responsibility for training and development

M. A. Keyes, Director of Engineering, Control Systems Division, Beloit Corporation, Beloit, Wis., U.S.A.
B.Sc.(Chemical Engineering), Stanford University; graduate work in electrical engineering and systems science, University of Illinois
Previously worked in pulpmill

Dr I. D. McFarlane, Process Control Engineer, Wiggins Teape Research and Development Ltd., Butler's Court, Beaconsfield, Bucks.
B.A.(Physics), Lincoln College, Oxford; Ph.D.(Applied Physics), University of Lancaster
Previously worked in electronics as applied to communications systems and to control

M. I. MacLaurin, Technical Manager and Computer Research Manager, Wiggins Teape Ltd., Dartford Paper Mill, Dartford, Kent
B.Sc.(Physics), University of Sheffield; Fellow of the British Computer Society; Member of the Institute of Measurement and Control; Associate Member of the Institute of Mechanical Handling; Council member of the Scientific Instrument Research Association; Senior member of Technical Section CPPA

J. Mardon, Technical Director, Pulp & Paper Group, Macmillan Bloedel, Vancouver, B.C.
M.A., University of Cambridge; City & Guilds of London Insignia Award; TAPPI Fellow; Weldon Gold Medals (Technical Section CPPA) 1955 and 1966; Prince Philip medal (City & Guilds of London Institute) 1968; Beilby Memorial Medal and Prize (Royal Institute of Chemistry/Society of Chemical Industry) 1968
Principal fields of work—papermaking technology and research, particularly on the papermachine wet end
Author of 70 technical papers, 100 educational articles and six books

N. B. Morgan, Economic Advisor, Department of Economic Affairs, c/o Scientific Control Systems Ltd., Sanderson House, Berners Street, London W1
B.Sc.(Economics), London School of Economics and Political Science
Previously Statistician, Television Audience Measurement Ltd. and Statistician/Operations Research, CEIR Ltd. (now Scientific Control Systems Ltd.)

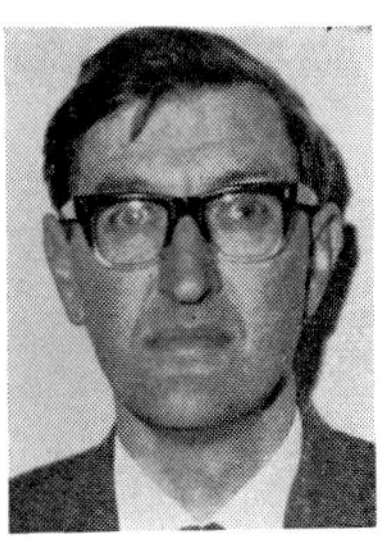

J. A. S. Newman, Technical Manager, Wiggins Teape Ltd., Fort William Mill, Corpach, Fort William, Scotland
B.Sc.(Physics), University of London
Principal fields of work—papermaking research and development technical management and production planning

P. H. Prior, Retired, Carshalton, Surrey
B.Sc.(Chemistry), University of London; Fellow of the Royal Insitute of Chemistry; Chairman of British Technical Section 1963–1964; Director of British Paper & Board Industry Research Association 1964–67; member of TAPPI and Technical Section CPPA
Principal fields of work—papermill chemist, manager of papermill group research and development department

A. L. Ramaz, Centre Technique de l'Industrie des Papiers, Cartons et Celluloses, CEDEX 17A, 38-Grenoble (Gare), France
Ingenieur EFP: Institut de Statistique de l'Université de Paris
Principal fields of work—laboratory instruments, documentation, statistical methods, instrumention and automation

Dr H. F. Rance, Technical Director, Wiggins Teape Ltd., 1 Watling Street, London EC4
B.Sc., Ph.D., Imperial College of Science & Technology, University of London
Ex-Chairman of Fundamental Research Committee, Technical Section BP & BMA
Principal fields of work—rheology, group research and development department

Dr S. A. Rydholm, Research Director and Departmental Technical Manager, Billeruds AB., 66 100 Säffle, Sweden
B.Sc.(Chemical Engineering) and Tekn.lic., Royal Institute of Technology, Stockholm
Previously carried out research at Swedish Wood Research Centre, Stockholm and Swedish Pulp Co., Sundsvall

Dr I. B. Sanborn, Manager of Process Development and Control, Research & Development Division, Consolidated Papers Inc., Wisconsin Rapids, Wis., U.S.A.
B.Ch.E., Renssalaer Poly Institute; M.S. and Ph.D.(Paper Chemistry and Engineering), Institute of Paper Chemistry
After graduating in 1961, has worked in a wide variety of posts in the technical and technical management fields, also closely associated with automation and computerisation projects over the past 5 years

C. R. Silversides, Chief, Logging Development Programme, Forest Management Institute, Dept. of Fisheries & Forestry, Ottawa, Canada
B.Sc.(Forestry) and Forest Engineer, University of Toronto
Previously Research Associate, PPRIC, Montreal and Director of Woodland Development, Abitibi Paper Co. Ltd., Toronto

Dr B. W. Smith, Systems Superintendent, Australian Paper Manufacturers Ltd., Melbourne
B.E.(Electrical), University of Melbourne; Ph.D., University of Cambridge
Previously in research work for electrical industry, then for paper industry

W. Stopper, Manager, IBM Information Records Division, Quality Engineering Laboratory (Europe), IBM Germany, Böblingen
Technician of Chemistry, National Technical Academy, Isny/Allgäu, Germany

U. J. G. Ström, Process and Control Engineer, A.Ahlström Oy., Varkaus, Finland
M.Sc.(Technical Physics), Finland Institute of Technology, Helsinki

G. F. Underhay, Consultant, Weybridge, Surrey
B.Sc.(Chemistry), University College London; Fellow of the Royal Institute of Chemistry; life member of Society of Chemical Industry and Royal Institutions; member of British and Canadian Technical Sections and APPITA
Principal fields of work—paper industry scientist and company director, with special responsibilities for central research and development

N. C. Underwood, Group Technical Director, Inveresk Paper Group, 19 Tudor Street, London, EC4
B.Sc., M.Sc., University College, Nottingham
Previously in paper industry research, part of the time as a director of research

Prof. J. A. Van den Akker, Professor of Physics, The Institute of Paper Chemistry, Appleton, Wis., U.S.A.
B.Sc., Ph.D.(Physics), California Institute of Technology; TAPPI Research and Development Award 1967; TAPPI Gold Medal 1968
Previously Instructor in Physics, Washington University, St. Louis, Professor of Physics, Chairman of Dept. of Physics and Mathematics, Institute of Paper Chemistry and Invitational Fulbright Scholar, University of Manchester Institute of Science & Technology 1961–62

A. J. Ward, Management Consultant (in dynamic systems), I.C.I. Central Management Services, Fulshaw Hall, Wilmslow, Cheshire
B.Sc.(Eng.) in Mechanical Engineering, University College, London
Previously worked in systems engineering, advanced control development and automation

P. E. Wrist, Vice-President, Research & Engineering, Mead Corporation, Dayton, Ohio 45402, U.S.A.
B.A., M.A., University of Cambridge; M.Sc.(Physics), University of London; Advanced Management Programme, Harvard Business School
Fields of work—research in British, Canadian and American paper industries

INDEX

A

B

C

D

G

H

I

J

K

L

M

N

O

P

Q

R

S

T

U

V

W

Type set by Gloucester Typesetting Co. Ltd.
Printed in Great Britain at
the Pitman Press, Bath